Computer-Aided Cancer Detection and Diagnosis

RECENT ADVANCES

Computer-Aided Cancer Detection and Diagnosis

RECENT ADVANCES

Jinshan Tang
Sos S. Agaian
Editors

SPIE PRESS
Bellingham, Washington USA

Library of Congress Cataloging-in-Publication Data

Computer-aided cancer detection and diagnosis : recent advances / Jinshan Tang and Sos S. Agaian, editors.
p. ; cm.
Includes bibliographical references and index.
ISBN 978-0-8194-9739-0
I. Tang, Jinshan, editor of compilation. II. Agaian, S. S., editor of compilation. [DNLM: 1. Neoplasms–diagnosis. 2. Early Detection of Cancer–methods. 3. Image Interpretation, Computer-Assisted–methods. QZ 241]
RC270
616.99'4075–dc23

2013032266

Published by

SPIE—The International Society for Optical Engineering
P.O. Box 10
Bellingham, Washington 98227-0010 USA
Phone: +1 360 676 3290
Fax: +1 360 647 1445
Email: spie@spie.org
Web: http://spie.org

Printed in the United States of America.
First printing

Table of Contents

Preface

Cancer is a significant threat to human life. Currently, the United States spends over 100 billion dollars annually on cancers, such as breast, lung, and prostate cancer. Based on statistics from the World Health Organization (WHO), deaths caused by cancer will reach approximately 12 million people in 2030. Thus, it has become a challenge to fight cancers in both medical practice and in the scientific research field. Imaging of cancer is an increasingly important component of understanding and treating cancer. Today, it is necessary to not only assess tumors morphologically but also provide information about the pathophysiological and metabolic aspects of tumor behavior with functional imaging techniques. Over the last two to three decades, the field of diagnostic cancer imaging has witnessed remarkable evolution that has affected virtually every aspect of research and clinical management of cancer. This evolution has been the result of innovations in three main aspects: innovative instrumentation (including a new class of scanners); development of new contrast agents and radiolabeled tracers; and imaging tools (including computer-aided detection or diagnosis technologies) for the detection, evolution, staging, and prognosis of many types of cancer. Current standard imaging techniques cannot accurately detect early diseases, and they provide limited information for disease staging.

The major goals of current cancer imaging are as follows:

- Provide more reliable disease characterization through the synthesis of anatomic, functional, and molecular imaging information;
- Refine and optimize imaging capabilities in oncology;
- Establish new imaging modalities and findings, and discover the potential use of these techniques;
- Find more individualized assessment of tumor biology, personalized treatments, and response to treatment;
- Develop image-processing-based cancer control systems; and
- Explore imaging capabilities and strategies to streamline cancer drug development.

Six levels of assessment to determine the efficacy of diagnostic imaging should be considered:*

- Technical performance – the ability to obtain a high image quality.
- Diagnostic performance – the ability to identify a disease correctly.
- Diagnostic impact – a measure of sensitivity, specificity, positive predictive value, negative predictive value, and accuracy.
- Therapeutic impact – the influence an imaging result has on clinical diagnostic confidence.
- Alteration in management based on the results of imaging.
- Impact on health – the influence of imaging on the disease outcome.

Computer-aided detection or diagnosis (CAD) technologies play a key role in the detection of cancers and help reduce the death rate; as such, they have greatly advanced over the past decades. The aim of this book is to publish and promote high-quality research in key technologies used in computer-aided cancer detection and diagnosis systems. The following 11 chapters cover different types of cancers, including skin cancer, breast cancer, prostate cancer, colon cancer, etc.; they also span different scientific fields, such as biomedicine, imaging, image processing, pattern recognition, system analysis, etc.

Colonic polyps are fleshy growths that appear on the inside of the large intestine, and certain types of polyps grow large enough and can become cancerous. Screening for colon polyps and removing them before they become cancerous can reduce the risk of colon cancer. Chapter 1 reviews computer-aided systems and technologies for colonic polyp detection using CT colonography. It introduces the history, preparation, imaging protocol, and clinical value of CTC and related image processing technologies, including colon segmentation, supine–prone registration, colon unfolding, polyp segmentation and characterization, classification, and content-based image retrieval. It also summarizes the performances and limitations of various CAD systems.

Digital image processing technologies have important applications in computer-aided cancer imaging systems, and they play a key role in cancer detection. These technologies include image enhancement, image segmentation, image compression, image encryption, etc., all of which are needed to provide information about the extent of disease and help plan treatment of the cancer. However, these technologies are not fully developed, and further investigation is needed to improve the accuracy of computer-aided cancer detection systems. Chapter 2 examines three image processing technologies

*[R. MacKenzie and A. K. Dixon, "Review: measuring the effects of imaging: an evaluation framework," *Clin. Radiol.* **50**, 513–518 (1995); H. V. Fineberg, J. Wittenberg, and J. T. Ferruci, "The clinical value of body computed tomography over time and technologic change," *Am. J. Roentgenol.* 141, 1067–1072 (1983)]

(image enhancement, image compression, and image encryption) that are often adopted by computer cancer detection systems. Based on nontraditional representation of images in the form of 1D independent signals, new approaches for enhancement, compression, and encryption are presented as a preprocessing tool for computer-aided imaging systems.

Chapter 3 presents an overview of recent advances in multimodality imaging technologies for diagnostic radiology and image-guided radiotherapy. In particular, it discusses the expanding role of multimodality imaging in cancer detection and segmentation for radiation oncology. Using complementary information from multimodality images significantly improves the robustness and accuracy of tumor volume definitions in radiotherapeutic treatments of cancer. The chapter also provides working examples for developing algorithms for multimodality target volume definitions in different cancers and highlights the potential opportunities in this field for computer-aided detection and image-guided treatment.

Mammography plays a key role in fighting breast cancer, and research has found that screening has reduced breast cancer mortality by up to 44%. However, low-dose x rays will generally reduce the contrast of the mammograms. In order to resolve this issue, Chapter 4 introduces a new nonlinear unsharp masking (NLUM) scheme for enhancing suspicious regions in mammograms. The NLUM method offers users the flexibility to embed different types of filters in the nonlinear filtering operator, to choose different linear or nonlinear operations for the fusion processes, and to optimize the NLUM parameters manually or by using a quantitative enhancement measure. The chapter also introduces the new second-derivative-like measure of enhancement (SDME). The comparison and evaluation of enhancement performance demonstrates that NLUM can improve the disease diagnosis by enhancing suspicious regions in mammograms with no *a priori* knowledge of the image contents.

Skin cancer is the most common of all cancers, accounting for nearly half of all cancers in the United States. Automatic detection of skin cancer is a key technology in computer-aided skin cancer diagnosis. Chapter 5 studies skin lesion detection based on color information. Several color spaces are studied, and the detection results are compared. Experimental results show that the YUV color space can achieve the best performance. Furthermore, the chapter develops a distance-histogram-based threshold selection method that is proven to be better than other adaptive threshold selection methods for color detection. Based on the aforementioned methods, a hybrid skin-lesion detection algorithm is presented. The book chapter also investigates GPU techniques for skin lesion extraction, and the results show that GPUs have potential applications in skin lesion extraction.

Chapter 6 presents an incremental learning method for lesion detection using endoscopy videos. With advances in data acquisition technology, data has become large and dynamic. A large number of examples often reduces the generalization error of the trained model. In the deployment of new image-based diagnosis tools such as capsule endoscopy, new examples continue to be acquired, which enriches the understanding of the imaging modality and could potentially alter previous beliefs. Therefore, efficient and scalable learning approaches are needed that can modify the model structure without having to revisit all of the previously processed examples. The incremental learning method presented in this chapter is developed based on geometric support vector machines (SVMs). The chapter describes the concept of the skin of convex hulls and a method to identify it (only the examples within the skin are retained in the incremental training, which is approximated with the extreme points). The set of extreme points are found via a recursive process by searching along the direction defined by a pair of extreme points. When additional examples become available, they are used along with the retained ones within the skin of the convex hull constructed from the previous data set. This process results in a small number of instances used in incremental training steps and, hence, improved memory efficiency to handle a large amount of data, as well as robustness that exhibits competitive performance.

Chapter 7 provides a comprehensive review of a melanoma screening system, including various imaging technologies, publicly available skin lesion data sets, and image analysis methods such as lesion segmentation, feature extraction and selection, and classification. This chapter also describes in detail a method to bridge the gap between the domain knowledge of physicians (i.e., dermatologists) and computer-generated features representing size, shape, spatial relationship, and texture. Comprehensive comparison using publicly available skin lesion data sets demonstrates the advantage of incorporating domain knowledge.

Microcalcifications are tiny deposits of calcium that appear as small, bright spots on mammograms, and the detection of microcalcifications is an extremely challenging task. In Chapter 8, a novel, hybrid 2D complex-wavelet-transform-based (2D-CWT-based) multifractal feature extraction system is proposed for the detection of microcalcification clusters (MCCs) in digital mammograms. A hybrid feature set, including a set of texture-based features and a set of 2D-CWT-based multifractal features, is presented as the input to a SVM classifier for the detection of the MCCs. The 2D-CWT algorithm and its 2D-CWT-based novel multifractal feature extraction scheme are proposed in the book chapter. Experiments demonstrated a good MCC detection rate and a satisfactory ratio of the true positive fraction to the false positive fraction. The proposed MCC detection system with hybrid features provides an adequate framework for MCC detection.

Chapter 9 focuses on challenges in accurately and automatically detecting and validating suspected prostate cancer lesions in biopsy images. Despite recent improvements in detection and treatment, prostate cancer continues to be the most-common malignancy and the third-leading cause of cancer-related mortality in American men. Evaluation of prostate cancer can be divided into detection, localization, classification, grading, and staging; accurate assessment is a prerequisite for optimal clinical management and therapy selection. Current diagnosis of prostatic adenocarcinoma is conducted by experienced pathologists using visual analysis of biopsy tissue samples: pathologists assess glass slides under a microscope in order to detect the presence of tumors and to assign a grade according to the architecture of prostatic glands. However, the grading process is time-consuming and error-prone, as well as highly influenced by pathologist experience, pathologist fatigue, and variability in the image interpretation. Therefore, CAD prostate cancer diagnosis has been developed to assist pathologists in the analysis of histopathology images. Prostate biopsy imaging has been accepted as a primary imaging modality for evaluating prostate cancer grades. In the coming decade, the main aim for prostate cancer imaging is more-accurate disease description, characterization, and interpretation through the synthesis of functional, anatomic, and molecular imaging information; therefore, in order to make accurate diagnoses, it is important to thoroughly understand their advantages and limitations, histological background related with image findings, and their clinical relevance for evaluating prostate cancer. Chapter 9 provides an overview of the current clinical approach for detecting and grading prostate cancer and describes the current status and future potential of CAD technology applied to prostate cancer, which is intended to be a support tool in cancer diagnosis and management. The chapter also presents some future perspectives and new strategies in pursuit of better prostate cancer CAD systems, and so on.

Chapter 10 investigates mass analysis using fractal dimension and shape factors in order to differentiate benign masses and malignant tumors. Fractal dimension (FD) and several shape factors—including compactness, convex deficiency, a measure based on Fourier descriptors, fractional concavity, and spiculation index—were calculated from the contour of a mass and used to estimate whether the mass is benign and malignant. The results indicate that shape analysis can lead to efficient discrimination between benign breast masses and malignant tumors. The results also show that fractional concavity gave the highest individual AUC (the area under the receiver operating characteristic curve).

Chapter 11 deals with tomographic imaging, and a new approach for reconstructing images from a finite number of projections is presented. In the new approach, the ray integrals of the image are transformed uniquely into the ray sums of the discrete image on the Cartesian lattice. The experimental results of image reconstruction from a finite number of projections are illustrated in the book chapter and demonstrate the effectiveness of the proposed approach.

Jinshan Tang
Sos Agaian
November 2013

List of Contributors

Mohamed Abouelenien
University of North Texas, USA

Sos Agaian
University of Texas–San Antonio, USA

Kwabena Agyepong
Alcorn State University

Antonietta Ancona
San Paolo Hospital of Bari, Italy

Shantanu Banik
University of Calgary, Canada

C. L. Philip Chen
University of Macau, China

Loredana Chieppa
University of Bari, Italy

Balathasan Giritharan
City Grid Media, USA

Artyom M. Grigoryan
Unversity of Texas–San Antonio, USA

Yanliang Gu
Michigan Technological University, USA

Samuela L'Abbate
University of Bari, Italy

Jianguo Liu
University of North Texas, USA

Fabio Mangieri
San Paolo Hospital of Bari, Italy

Clara Mosquera-Lopez
University of Texas–San Antonio, USA

Issam El Naqa
McGill University, Canada

Karen Panetta
Tufts University, USA

Maria Luisa Pepe
University Hospital – Policlinico of Bari, Italy

Grazia Raguso
University of Bari, Italy

Rangaraj M. Rangayyan
University of Calgary, Canada

Ning Situ
Microsoft, USA

Jinshan Tang
Michigan Technological University, USA

Ian Thompson
University of Texas Health Science Center, USA

Alejandro Velez-Hoyos
Pablo Tobon Uribe Hospital, Colombia

Jianhua Yao
National Institutes of Health, USA

Xiaohui Yuan
University of North Texas, USA

Xiaojing Yuan
University of Houston, USA

Ping Zhang
Alcorn State University, USA

Yicong Zhou
University of Macau, China

George Zouridakis
University of Houston, USA

Chapter 1
Computer-Aided Detection of Colonic Polyps in CT Colonography

Jianhua Yao
Radiology and Imaging Sciences Department, Clinical Center, National Institutes of Health, Bethesda, Maryland

1.1 Colonic Polyps and Colon Cancer

Colonic polyps are fleshy growths that occur on the inside of the large intestine, also known as the colon. Polyps in the colon are extremely common, and their occurrence increases as individuals grow older.[1] It is estimated that 50% of the people over the age of 60 will have at least one polyp. The significance of polyps is that when certain types of polyps grow large enough, they can become cancerous. Colon cancer is a cancer from uncontrolled cell growth in the colon; it is the third most commonly diagnosed cancer in the world, but it is more common in developed countries: approximately 60% of cases were diagnosed in the developed world. Colon cancer is the second leading cause of death from cancer in the United States. It is estimated that there were 143,460 new cases of colorectal cancer in the United States in 2012, with 51,690 deaths.[2]

The risk of a colon polyp becoming cancerous increases as the size of the polyp increases.[3] Therefore, screening for colon polyps and removing them before they become cancerous should markedly reduce the incidence of colon cancer.[4] Many doctors in the U.S. are recommending colonoscopies to screen healthy subjects with an average risk of developing colon cancer.[5] Colonoscopies are recommended beginning at the age of 50 and every 7–10 years thereafter if no colon polyps or cancers are found. Unfortunately, many patients do not undergo screening due to the perceived inconvenience and

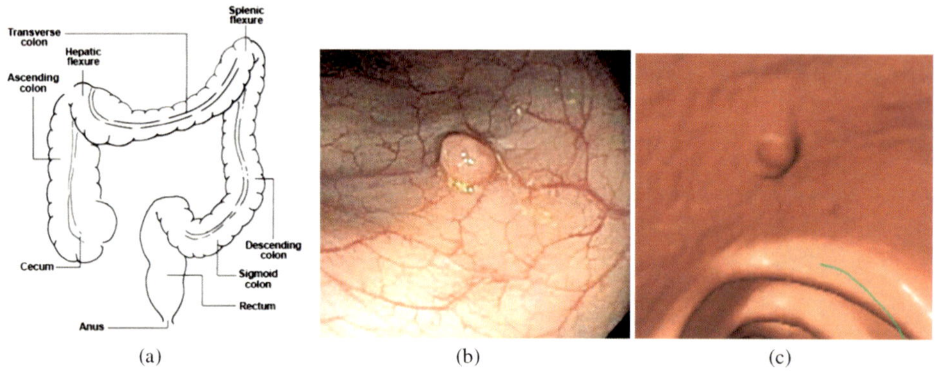

Figure 1.1 Colon and colonic polyps: (a) Illustration of colon anatomy. (b) Optical and (c) 3D virtual colonoscopy images of a 0.8-cm polyp in the sigmoid colon of a 60-year-old man.

discomfort of existing screening tests. CT colonography (CTC), a CT-scan-based imaging method, has been under study for the past ten years and shows promise as a method of colorectal cancer screening that may be acceptable to many patients.[6,7] Figure 1.1 shows the colon anatomy and a colonic polyp shown in the optical colonoscopy and CTC.

1.2 CT Colonography

The use of CT imaging for the detection and staging of colon cancer was first proposed as early as 1980.[8,9] In 1983, Coin et al.[10] proposed that CT had potential as a mass screening method for colorectal polyps. The terms "virtual colonoscopy" and "CT colonography" were formally introduced in 1994.[11] Since then, clinical studies began including more patients, and technologies were advanced based on new hardware and software. Colon phantoms were employed to determine the optimal CTC scanning parameters. Optical colonoscopy was used as the reference standard to determine the feasibility of using CTC in a clinical setting.[12] After that, multiple investigations have been conducted to improve the scanning parameters, the stool and fluid tagging techniques, and the colon preparation techniques.[13,14]

The standard preparation for a CTC may begin several days before the procedure. One to two days prior, patients undergo a standard 24-hour clear-liquid-diet colonic preparation. Patients are orally administered laxatives: 90 mL of sodium phosphate [Fleet 1 preparation (Fleet Pharmaceuticals, Lynchburg, VA)] and 10 mg of bisacodyl. Patients also consume oral contrast agents in divided doses; a total of 500 mL of barium sulfate [Scan C (Lafayette Pharmaceuticals, Lafayette, IN), 2.1% by weight] is administered for solid fecal tagging; and a 120-mL solution of diatrizoate meglumine and diatrizoate sodium [Gastrografin® (Bracco Diagnostics, Princeton, NJ)] is administered to opacify any luminal fluid.[15]

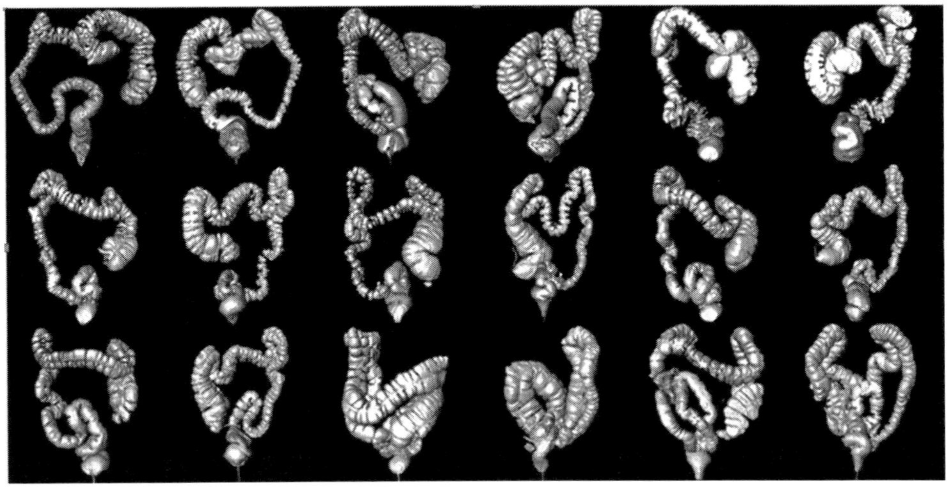

Figure 1.2 Anteroposterior projection images of the 3D reconstructions of colons using CTC from nine patients. There are two consecutive images from each patient, one supine and the other prone. Observe the diversity of colonic shapes and distensions, and the completeness of the colonic segmentation.

The insertion of a small, flexible rectal catheter allows the patient to self-insufflate the colon immediately before scanning. Insufflating the colon with air or carbon dioxide (CO_2) allows for polyps to be seen on a CT scan because of the large contrast difference between air and soft tissue. The use of CO_2 has been shown to decrease immediate and delayed pain compared the use of air.[16]

Patients are currently scanned in two positions: supine and prone. The protocol required only one 20–25-second breathhold per scanning position. Conducting a prone scan in addition to a supine scan has been shown to increase the sensitivity of detecting polyps by 13–15%, but it has little effect on or decrease of specificity.[17] The optimal settings for CTC aim to decrease scanning time, decrease radiation exposure, and increase image quality. A high pitch value decreases the scanning time, a lower tube current reduces the radiation exposure, and a smaller slice thickness improves image quality. Typical CT scanning parameters include a 1.25–2.5-mm collimation, a table speed of 15 mm per second, a reconstruction interval of 1 mm, and scanner settings of 100 mAs and 120 kV.[18] Figure 1.2 shows several examples of 3D colon reconstruction using CTC, and Fig 1.3 shows examples of polyps on CTC.

1.3 Computer-Aided Detection Using CTC

CTC is a promising diagnostic tool for detecting colorectal polyps and preventing colon cancer; however, CTC requires a trained radiologist to perform a lengthy interpretation of the CT images, which is both costly and

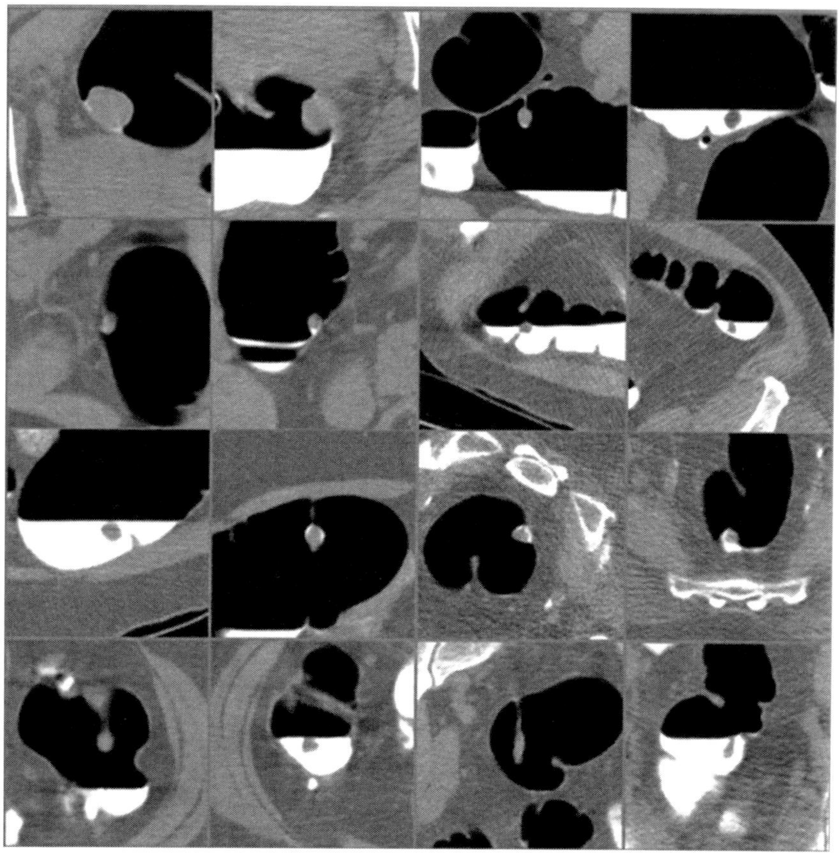

Figure 1.3 Examples of 16 2D CTC images showing eight adenomatous polyps 1 cm or larger. Each polyp is shown twice: first in the prone view and then in the supine view.

prone to human error.[19] Challenges associated with CTC include a lack of consistency in results between radiologists and difficulty detecting smaller polyps (6–9 mm). Proposed solutions to these obstacles include double readings or a computer-aided detection (CAD) system. CAD has the potential to decrease the time needed to complete an interpretation and increase the accuracy of the diagnosis.

The objective of the CAD system is to identify and mark suspicious lesions on the CTC scan. Radiologists can use the results from the CAD system along with 3D or 2D CT images to make a final diagnosis. Computer vision and computer graphics techniques, such as colon and polyp segmentation, supine–prone registration, novel unfolded view, and virtual fly-through, can also help radiologists dictate the cases.

1.3.1 CAD pipeline

The pipeline of a typical CTC CAD system is shown in Fig. 1.4. First, a CTC data set is read into the program. The colon surface is extracted using fuzzy

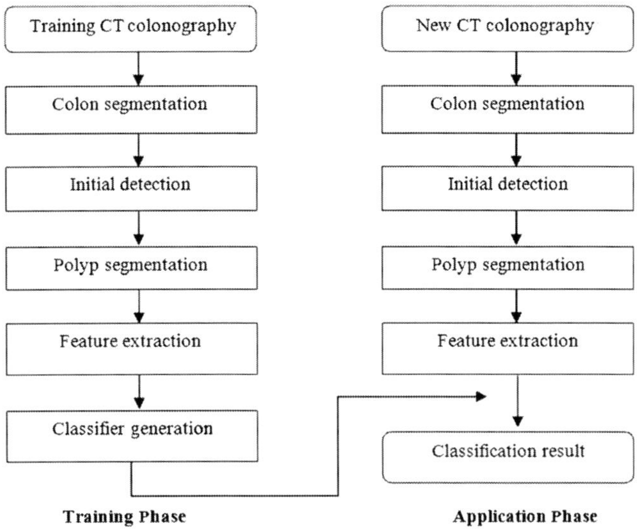

Figure 1.4 CTC CAD pipeline.

connectness and isosurface techniques.[20] A curvature filter is applied to every vertex on the surface to get the initial detections. The passed vertices are then clustered according to their connectivity, and the centroid of each cluster is treated as one detection. The detections are then sent to a polyp segmentor for computation of boundaries, 3D shape, and attenuation features.

A CAD system has two phases: training and application. The system needs to be trained (using training data) before being applied to new data. Training data is usually manually or semi-automatically preprocessed to identify lesions and structures of interest; a computer program is then developed and optimized to detect as many true lesions and as few false detections as possible. The trained CAD system generally has a segmentation module, a feature-extraction module, and a classifier. In the application phase, previously unseen data first undergoes the segmentation and feature-extraction process. The features of potential detections are then fed to the classifier to determine whether they are true lesions or false ones. The following subsections describe a few key components in a CAD system.

1.3.2 Colon segmentation

Colon segmentation is an essential component of the CTC CAD software that extracts a 3D representation of a colonic surface from the CTC data. Colon segmentation is very important because incorrect segmentation may impair interpretation using 3D visualization programs.

In clinical practice, an oral contrast agent may be given to patients for CTC. The oral contrast agents contain iodine or barium, which eventually tag residual colonic fluid and stool remnants. This appears advantageous when CT scans are visually inspected by radiologists because it helps reveal hidden areas of the colon surface. However, the contrast agent also creates problems for segmentation because it converts a two-class (air and tissue) segmentation problem into a three-class (air, contrast, and tissue) problem.

Most colon segmentation algorithms rely on thresholding and connectivity to extract the colon.[21] Different schemes include region growing, fuzzy connectivity, level set, active contour models, etc.[22] To account for the oral contrast, Franaszek et al.[23] proposed a hybrid technique consisting of eight steps:

1. Modified threshold-based region-growing segmentation,
2. Extraction of individual air/fluid pockets,
3. Identification of air–fluid boundaries,
4. Construction of the pocket tree,
5. Pruning of the tree,
6. Fuzzy connectedness segmentation,
7. Filling gaps and holes in fuzzy space, and
8. Level-set segmentation applied to fuzzy space.

Information collected in the first three steps is used in step 4 to organize pockets of air and fluid into the tree. In step 5, the tree is pruned to mark leaked regions of segmented colon. The two last steps provide final corrections and smoothing. As input of the first step, original CT images are read, and the location of one or more starting seeds is provided. As output of the last step, the colonic wall approximated by an isosurface in level-set space is generated. The results of the hybrid colon-segmentation algorithm are shown in Fig. 1.2.

1.3.3 Supine–prone registration

In CTC, a patient will be scanned twice—once supine and once prone—to improve the sensitivity for polyp detection. This improves CTC sensitivity by reducing the extent of uninterpretable collapsed or fluid-filled segments. Figure 1.5(a) shows pairs of supine and prone CTC data. Because the colon moves between the prone and supine scans, the two CT scans need to be registered to match detections.[24]

One way to reduce the complexity of the registration problem is to register the centerlines of the colon on prone and supine scans. Van Uitert et al.[25] proposed a subvoxel, precise centerline-extraction method that utilizes information of the colon outer wall to determine the colon centerline. Initially, segmentation of the colon is performed to obtain a subvoxel representation of the colon. The discrete segmentation is used as an initial surface for a narrow-band level-set segmentation to more-accurately determine the location of the colon inner wall and smooth the boundary between the air- and fluid-filled regions of the colon. From the level-set segmentation, a subvoxel distance field is computed using the fast marching method. The centerline of a colon is then computed based on the distance field.

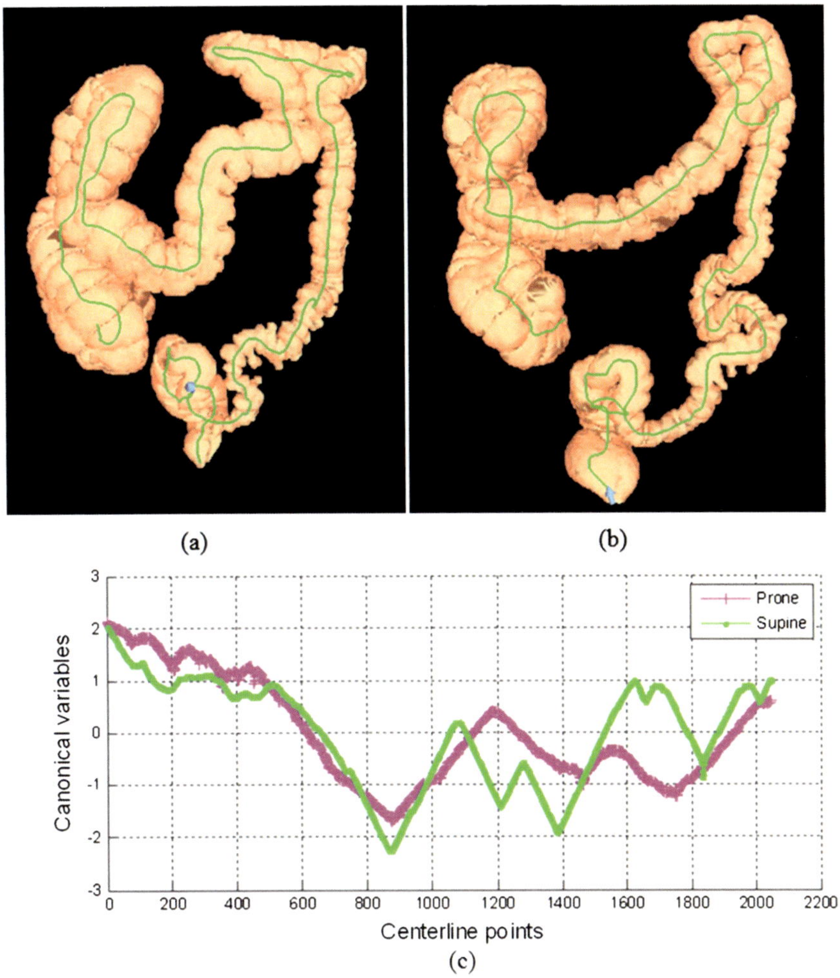

Figure 1.5 Supine and prone registration: (a) supine scan, (b) prone scan (lines inside the segmented colon indicate centerlines), and (c) matched centerline.

To make full use of shape information of the colon for registration, Nain et al.[26] proposed a centerline registration algorithm based on dynamic time warping (DTW) and colon distension along the centerline. Nappi et al.[27] proposed a region-based supine–prone correspondence method to reduce false-positive CAD polyp candidates in CTC. Li et al.[28] proposed a heuristic algorithm for the colon centerline registration by employing the coordinate information of the centerline.

Wang et al.[24] proposed an automated method for colon registration based on correlation optimized warping (COW)[29] and canonical correlation analysis (CCA).[30] Four anatomical, salient points on the colon are distinguished first, and then a COW method is applied to the segments defined by the anatomical landmarks to find better global registration based on local correlation of segments. To utilize more features along the colon centerline, researchers extended the COW method by embedding canonical correlation analysis into it for correlation calculation of colon segments. Figure 1.5(b) shows the matching of centerlines from supine and prone studies.

1.3.4 Colon unfolding

The colon is a tubular structure. The traditional way to navigate the colon along the centerline is not efficient due to the limited view angle. A more-efficient way is to open and flatten the colon, and view the surface from above. Although this is not possible in optical colonoscopy, it can be implemented in CTC via computer graphics techniques.

Many techniques have been proposed to unfold the colon; one main category is based on raycasting. Vilanova et al.[31] first unfolded the colon locally by using a local projection and then globally unfolding the colon using a suitable parameterization. Wang et al.[32] proposed a technique using the electrical field of a charged centerline to transform the colon into a straight cylinder-like shape and uniformly sample the planar cross-sections orthogonal to the centerline. Sudarsky et al.[33] presented an efficient method based on skeletal subspace deformation and uniform raycasting along the central path. It is difficult to choose the right sampling rate in raycasting-based techniques, and structures behind the first object encountered by rays may be obstructed. Another category of approaches is based on conformal mapping. Haker et al.[34] used angle-preserving conformal mapping to map the entire colonic surface onto a flat plane. Hong et al.[35] proposed a conformal mapping method based on minimizing the harmonic energy to achieve angle preservation and minimal distortion. These techniques have to deal with texture distortion and surface parameterization; they often require high-quality surfaces and are computationally expensive. Another type of approach is based on local projection. Paik et al.[36] proposed various map-projection techniques, including cylindrical and planar projection for the fly-through of virtual colonoscopy. Vos et al.[37] projected six orthogonal images onto an unfolded cube to render the complete FOV.

A reversible colon unfolding technique has been proposed by Yao et al.[38] Given a CTC data set, the 3D colonic surface is first segmented using thresholding, region growing, and level sets.[23] The centerline of the colon is then extracted based on a fast-marching level set and topographical thinning.[25] Rotation-minimizing frames (RMF) are then established along the centerline. After that, a recursive ring-set technique is applied to map vertices on the colon's surface to their corresponding centerline points. Next, mesh skinning is employed to straighten the colon, followed by the application of cylindrical projection to flatten the colon. Finally, reverse transformation is computed for each vertex. Mesh skinning is a skeleton-driven deformation technique widely used in computer animation; this technique can be applied to straighten the colon using its centerline as the skeleton of the colon. The result of the colon unfolding is shown in Fig. 1.6.

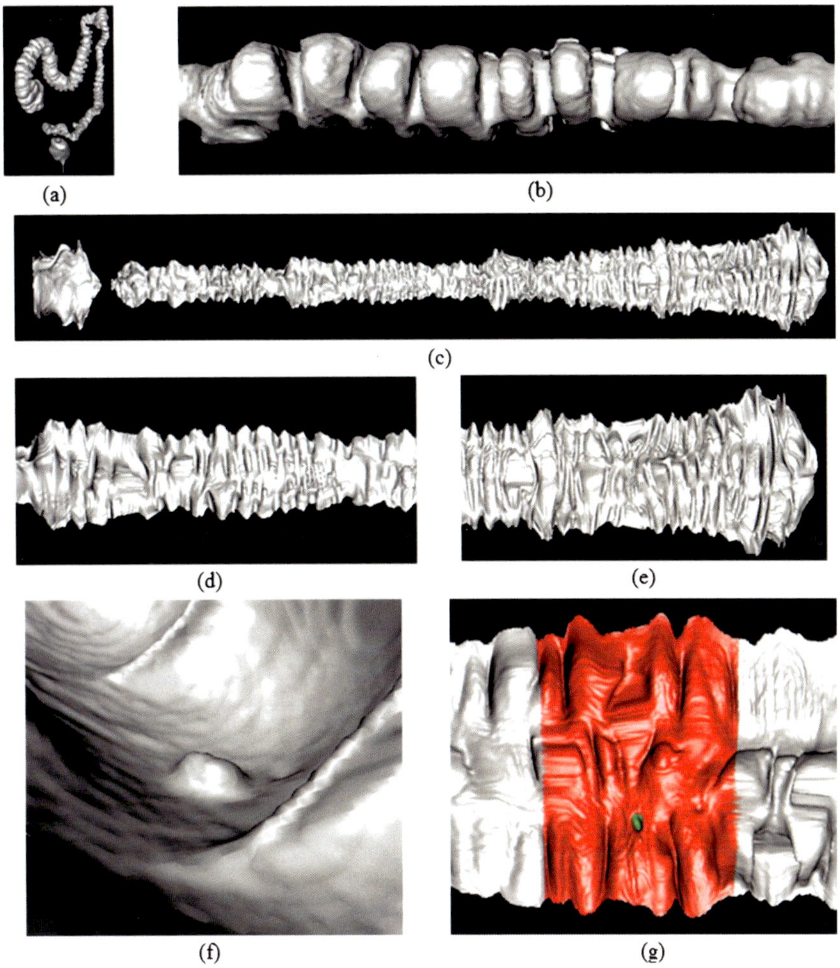

Figure 1.6 Colon unfolding: (a) 3D colon surface, (b) straightened colon, (c) unfolded colon, (d, e) close-up views of the unfolded colon, (f) endoluminal view of polyps, and (g) unfolded view of the same polyp in (f).

1.3.5 Polyp segmentation

Polyp segmentation is essential in the CAD pipeline because it provides the entire voxel set of a polyp, which can be used to quantify the polyp's characteristics. Several comprehensive volumetric features and statistical analysis can be obtained, such as the density distribution within the polyp, the volume and dimension of the polyp, and its relationship with surrounding tissues. Once the segmentation is obtained, additional analysis (such as texture analysis) can be performed.[39]

Colonic polyp segmentation is a challenging task due to several reasons. First, polyp shapes are irregular. Second, polyp sizes vary greatly. Third, the surrounding regions are complex. Methods utilizing both shape and densitometry information are necessary for a successful segmentation.

Jerebko et al.[40] used the Radon transformation and Canny edge operators to detect polyp boundaries. They first applied Canny edge operators to locate potential polyp–lumen boundaries, and then they used Radon transformation to connect the boundaries and identify round-shape structures. Their method was primarily 2D and only worked well on round polyps. Dijkers et al.[41] proposed a method to segment polyps using the colon's surface. After a seed patch was placed, the patch grew over the surface based on criteria of the surface normal and several stopping conditions according to the polyp shape. The method relies heavily on the surface normal and is sensitive to image noise. The assumption of polyp shape also makes it difficult to apply to a wide range of polyps. The segmentation cannot be automated because several key parameters need to be manually set according to the character of a polyp.

Deformable models were applied in polyp segmentation.[42] An initial model of the polyp is first placed at the seed location and the initial parameters are set, and then an iterative process is started. During each iteration, deformation forces are computed, and the model is updated according to the forces. The force weights and other control parameters are then adaptively updated. The model resolution and topology is also adaptively maintained. The iterative process is repeated until all forces reach a balance (the model remains unchanged between iterations) or until a maximum number of iterations are executed. The deformable model is represented as triangular meshes. The model is driven by the combination of internal force, image force, external force, and a counter force in the haustral fold regions. Internal spline forces tend to maintain the smoothness and continuity of the model.[43] Polyp boundaries tend to have larger gradients than other regions; therefore, the gradient of the edge map can be used as the image force. Furthermore, an expansion force is exerted to push the model from its initial state and prevent it from collapsing. Polyps on haustral folds are more difficult to segment because the deformable model often leaks into the fold region. Figure 1.7 shows the polyp segmentation results.

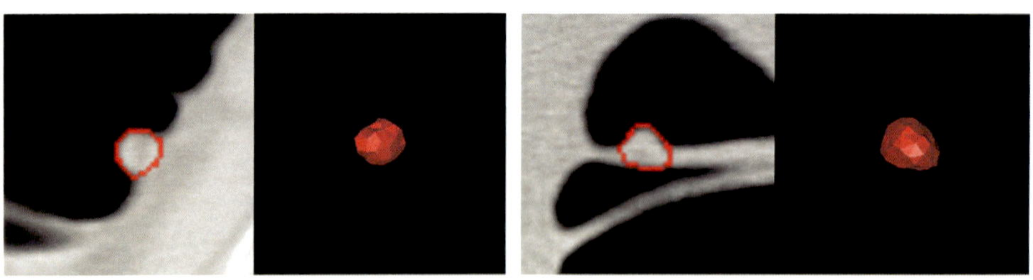

(a) (b) (c) (d)

Figure 1.7 Polyp segmentation: (a) and (c) are 2D axial images superimposed with segmentation result, and (b) and (d) are 3D surface reconstructions of the segmentation. (a) and (b) show segmentation of a 7-mm sessile polyp, and (c) and (d) show segmentation of a 10-mm polyp on-fold.

1.3.6 Polyp characterization and features

Colonic polyps appear as elliptical protrusions on the inner surface of the colon. Curvature-based features for colonic polyp detection have proved to be successful in the CTC CAD systems. Curvatures are important shape attributes;[44] the two principal curvatures and the derived Gaussian and mean curvatures can be computed from the surface using partial-differential-equation- or kernel-based methods directly from the image.[45] Curvedness and shape index[44] are more meaningful as the shape indicator compared to the curvatures. The curvedness is a positive number that specifies the amount of curvature. Shape index captures the intuitive notion of shape and classifies the shape as one of the following: cup, rut, saddle, ridge, cap, trough, saddle rut, saddle ridge, or dome. The global shape index or the distribution of local shape index can be used as shape features.

Other useful features include shape features such as circularity, spherecity, compactness, irregularity, elongation, or texture features such as contrast, roughness, and texture attributes. The compactness of a shape measures how close a structure is to a circle or a sphere. Compactness in 2D can be written as $C = P^2 / (4\pi A)$, where P is the perimeter, and A is the area of the shape. In 3D space, compactness can be written as $C = A^3 / (36\pi V^2)$, where A is the surface area, and V is the volume of the shape. In either case, the compactness of the circle and sphere is 1, and that of other shapes is larger than 1. The distribution and statistics of pixel intensity within a polyp reveals the smoothness, contrast, regularity, or homogeneity of tissues. Texture analysis, such as statistical moments and the co-occurrence matrix, provides ways to describe the tissue appearance.[46] The statistical moments are computed based on the intensity histogram. The second moment of the histogram measures the intensity variance within the region, which correlates with the roughness perception. The third and fourth moments—skewness and kurtosis—reflect the asymmetry and uniformity of the intensity distribution. The co-occurrence matrix is also known as a spatial gray-level dependence matrix in the sense that it combines spatial information and intensity statistics. The inertia of the co-occurrence matrix characterizes the texture contrast of a region. The entropy of the matrix quantifies the level of randomness in the region. The angular second moment of the co-occurrence matrix can be used to describe the homogeneity of a region.

Colonic polyps are small growths that protrude outward from the colon wall and are characteristically round in contour. In contrast, other structures inside the colon such as haustral folds and other normal colonic structures tend to be circumferential and ridge-shaped. The unique topographic features of bumps can be used in characterize polyps.[47] Height maps are commonly used in geographic information systems (GIS), where they are also called digital elevation models.[48] The height map generation is illustrated in Fig. 1.8. The method is based on a ray casting technique. For every point p_i on the projection plane, a ray r_i is cast through p_i; the point at which ray r_i encounters

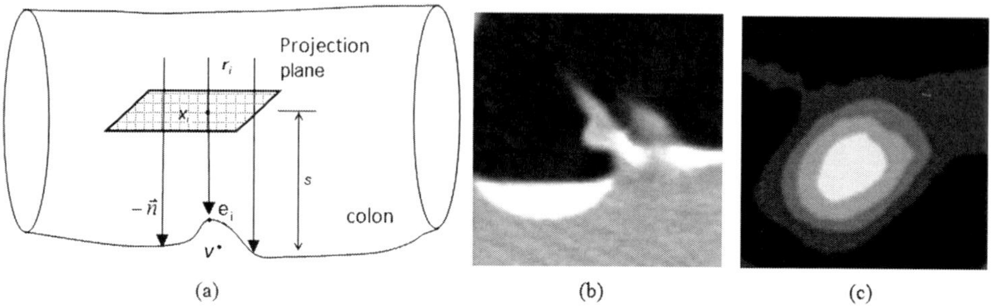

Figure 1.8 Polyp height map: (a) height-map generation based on ray tracing, (b) a 10-mm adenoma polyp, and (c) height map of (b).

the colon surface is e_i; and the distance d_i between p_i and e_i is recorded at p_i. The distance map can be converted to a digitized height map. Figure 1.8 shows an example of the height map. For a typical polyp, a concentric pattern can be observed from the map.

1.3.7 Machine learning and classification

Machine learning techniques translate an expert's knowledge into computer algorithms. In a CAD system, a classifier is a mathematical model that determines whether a detection is a true or false lesion. The classifier in a CAD system is usually a supervised learning system, i.e., the classifier is trained using annotated data by experts. Well-known classifiers, such as neural networks (NN)[49] and support vector machines (SVM),[50] have been widely used in CAD systems. In order to achieve an effective classifier, a subset of features needs to be selected from the entire feature space based on their individual or joint performances. Feature selection methods such as forward stepwise search (FFS) and genetic algorithm (GA) have been proposed.

SVM is a relatively new technique for data classification that uses hyperplanes in a high-dimensional feature space to separate data into different classes. SVM is trained with a learning system derived from statistical learning theory and is generalizable to unknown data. In the training phase, detections are given a class label (polyp, nonpolyp) to form feature-class pairs (x, y). Given a training set of S detections (x_1, y_1), (x_2, y_2), ..., (x_s, y_s) for p-dimensional feature space $x_i \in \Re^p$ and $y_i \in \{+1, -1\}$, a hyperplane can be optimized to separate the two groups of data (true and false). The decision function for a classification rule is then based on which side of the hyperplane the detection lies. An SVM in higher-dimensional space (more features) can lead to more-accurate classification. However, SVM in a very high-dimensional space may increase the complexity of the model, overtrain the data, and decrease the generality of the model. One solution uses an ensemble of classifiers, and each classifier only includes a small number of features. Figure 1.9 illustrates an ensemble of SVMs.

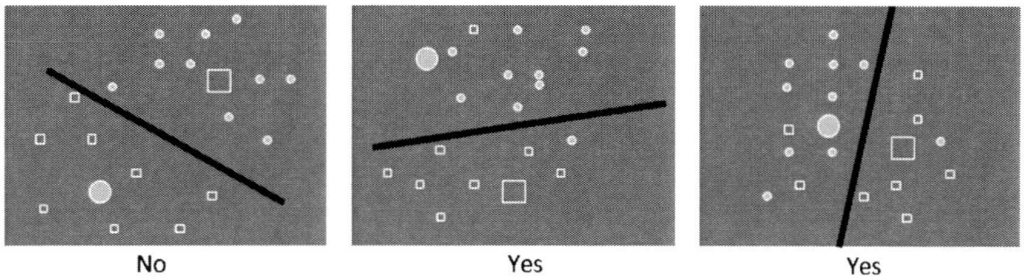

Figure 1.9 Ensemble of SVMs. Each plot illustrates one SVM; the classifier decision is reached by combining all classifiers.

Ensemble learning combines multiple trained classifiers under the assumption that multiple models are better than one if they are diverse. Popular ensemble approaches include boosting and bagging.[51] Combination strategies for the multiple decisions can be divided into two types: those that adaptively adjust to the training set based on the performance of previous models, as in boosting methods, and those that do not, as in bagging.[52] The bootstrap is widely used to estimate the standard error or confidence intervals of an estimate. Bagging is based on the bootstrap technique whereby the predictions on bootstrapped samples are aggregated to form an ensemble hypothesis. Boosting combines the predictions from re-sampled data based on the previous model's performance such that harder data samples for the system are more likely to be sampled. Bagging has been shown to reduce the variance of classifiers, whereas boosting can reduce both variance and bias. However, it is rarely shown that bagging and boosting combined with feature selection can significantly reduce ensemble training time in practice.

1.3.8 Content-based image retrieval

Content-based image retrieval (CBIR) is a computer vision technique for searching for similar images within an image database. It has been used in applications such as image searching[53] and artwork retrieval.[54] Recently, CBIR has shown potential as a diagnostic tool in medical applications.

CBIR systems describe images as a set of features directly computed from the images and then categorize the images into several categories. The scale-invariant feature transform (SIFT) was first proposed by Lowe[55] in the applications of natural-scene and facial recognition. It has the advantage of describing the local image feature with a scale- and rotation-invariant representation. The bag-of-words (BoW) model[56] was first introduced in natural language processing then in computer vision, especially for object categorization. The BoW model usually includes three steps: feature detection, feature description, and codebook generation. A representative database is built to store example images and their associated BoW

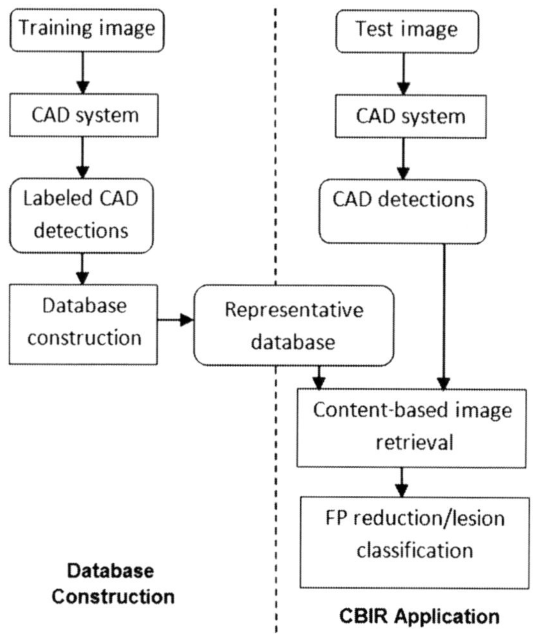

Figure 1.10 Flowchart of the application of CBIR on a CTC CAD system.

histograms. Given a new detection image, its BoW histogram is computed and compared against the representative database to retrieve the most-similar example images.

CBIR has been tested in a CTC CAD system to reduce the number of false positives[57] and to characterize the histology type (adenoma or hyperplastic)[58] and shape (pedunculated or sessile)[59] of a colonic polyp. Figure 1.10 shows the flowchart of the application of CBIR on a CTC CAD system. It consists of two stages: database construction and CBIR application. In the database-construction stage, CAD results from the training images are labeled by knowledgeable users and stored in a representative database. In the CBIR-application stage, CAD detections from the test images are sent to the CBIR for query. Based on the query results, certain false positives can be eliminated and polyp lesions can be classified based on their histology type and shape. Figure 1.11 shows the retrieval results of a true positive (TP) detection and a false positive (FP) detection. Based on the retrieval set, FPs could be eliminated.

1.3.9 CAD performance

The quality of a CAD system can be characterized by the sensitivity and specificity of the diagnosis. Sensitivity refers to the fraction of diseased cases correctly identified as positive in the system (true positive fraction, TPF). Specificity refers to the fraction of disease-free cases correctly identified as

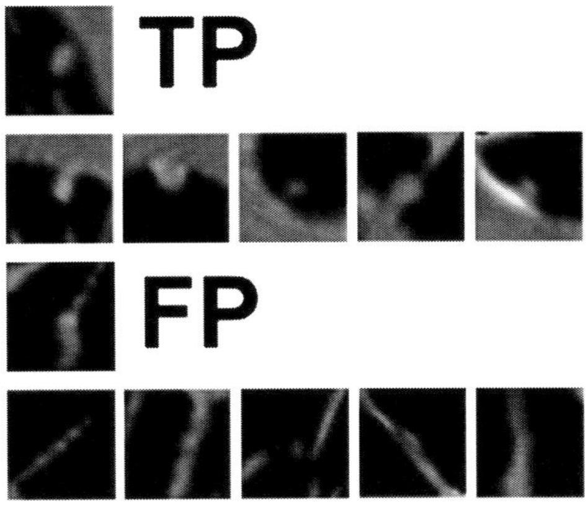

Figure 1.11 Retrieval sets of TP and FP detections in a CTC CAD system.

negative. "Receiver operating characteristic" (ROC) curves are used to describe the relationship between sensitivity and specificity. The ROC curves show the true positive fraction (TPF = sensitivity) versus the false positive fraction (FPF = 1 − specificity). In addition to ROC curves, free-response ROC (FROC) curves (TPF versus FP per images) were proposed to more-accurately represent the number of FP detections.[60] The area under the ROC and FROC curve is a measure of the quality of a CAD system. Note that there is a tradeoff between specificity and sensitivity. A successful CAD system should detect as many true lesions as possible and also retain as little FP detection as possible.

Because CAD is still under intensive development, most studies of CAD to date have reported its performance in the laboratory setting rather than in the radiology reading room. The largest such study to date was in 1186 patients for which the performance of CAD alone was compared to optical colonoscopy.[19] The patient data was randomly divided into a training set of 394 patients and test set of 792 patients for CAD to analyze. For polyps 10 mm or greater in size, the sensitivity of CAD on the test set was 89.3%, whereas the sensitivity of optical colonoscopy was only 85.7%. Similarly, there were two carcinomas found in the study; the detection rate of carcinomas was 100% for CAD and 50% for optical colonoscopy. Optical colonoscopy performed significantly better than CAD for polyps 6 mm or greater in size (87.2% compared to 66.1%) but comparably for adenomas 8 mm or greater in size (89.6% compared to 85.4%). CAD also had a FP rate of 7.9, 6.7, and 2.1 for polyps that were at least 6, 8, and 10 mm, respectively. There was a FP rate of 0.7 carcinomas per patient. The authors of this study concluded that CTC with CAD had similar results compared to optical colonoscopy for polyps that are 8 mm or larger in size. The FROC curve of the CAD system is shown in Figure 1.12.

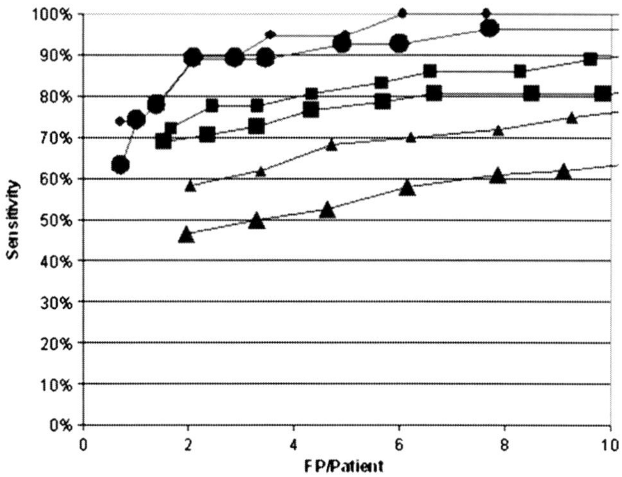

Figure 1.12 FROC curves for the training (small markers) and test (large markers) sets for adenomatous polyps 10 mm or larger (●), 8 mm or larger (■), and 6 mm or larger (♦).

1.4 Discussion

Currently, CTC is expensive for patients and not practiced regularly in most American imaging centers. Patients prefer CTC over optical colonoscopy, but for Americans to embrace CTC as their preferred colon cancer screening method, insurance companies must begin reimbursing for it. If the cathartic part of the bowel preparation can be eliminated, it is likely that patient compliance rates to CTC will increase, leading to reduced cost per life year saved and greater economic benefit to society. CTC is a low-risk method: the radiation dosage and risk of perforation are minimal when compared to the potential benefits.

Some studies have shown that CTC has the potential to perform well when compared to other screening methods. Still, more studies must be conducted for CTC to be an endorsed CRC screening method in the United States. The results of several large clinical trials currently underway are eagerly anticipated. Previous studies have shown that combined 2D and 3D displays and adequately trained readers are essential to the success of CTC as a reliable screening method. Further studies must be conducted on varying risk populations to establish the proper screening recommendations for the population as a whole. In addition, by maintaining uniformity in technology and reading technique in future trials, and ultimately in the community, radiology settings may allow for more accurate and consistent results.

With sensitivities approaching 90%, CAD shows promise in improving the sensitivity of polyp detection beyond the capability of a radiologist alone. CAD may enable CTC to overcome large barriers that CTC alone would potentially face in the clinical setting. A good CAD system may reduce the dependency that CTC has on an experienced, well-trained radiologist. In addition, CAD may greatly reduce the amount of time a radiologist needs to spend on each CTC case, thereby making CTC more cost-effective and reducing the number of significant polyps missed during a CTC. The results of future CTC and CAD studies may provide the necessary foundation for the acceptance of CTC as a consistent and reliable screening method for CRC.

Several studies have been conducted to determine the efficacy of CTC as a screening method for colon cancer. Unfortunately, the results of various studies have a wide range in sensitivity values for CTC. There are several factors that may contribute to the inconsistent results between CTC studies: different stool tagging and fluid opacification protocols were used, or none were used at all; different display methods were used for interpretation; image processing techniques such as electronic fluid cleansing were used in some studies and not in others; double readings were performed in some studies and not in others; multiple risk populations were used; and, depending on the study, radiologists had varied experience and training with CTC.

References

1. D. Cunningham et al., "Colorectal cancer," *Lancet* **375**(9719), 1030–1047 (2010).
2. R. Siegel, D. Naishadham, and A. Jemal, "Cancer Statistics," *CA: A Cancer Journal for Clinicians* **62**(1), 10–29 (2012).
3. S. Winawer et al., "Colorectal cancer screening and surveillance: clinical guidelines and rationale-Update based on new evidence," *Gastroenterology* **124**(2), 544–560 (2003).
4. A. Qaseem, T. Denberg, and R. J. Hopkins, "Screening for colorectal cancer: A guidance statement from the American College of Physicians," *Annals of Internal Medicine* **156**(5), 378–386 (2012).
5. A. Stein, D. Atanackovic, and C. Bokemeyer, "Current standards and new trends in the primary treatment of colorectal cancer," *Eur. J. Cancer* **47**(Suppl 3), 312–314 (2011).
6. T. Gluecker et al., "Colorectal cancer screening with CT colonography, colonoscopy, and double-contrast barium enema examination: prospective assessment of patient perceptions and preferences," *Radiology* **227**(2), 378–384 (2003).
7. R. van Gelder et al., "CT colonography and colonoscopy: assessment of patient preference in a 5-week follow-up study," *Radiology* **233**(2), 328–337 (2004).
8. J. Ellert and L. Kreel, "The value of CT in malignant colonic tumors," *J. Comput. Tomogr.* **4**(3), 225–240 (1980).
9. J. Husband, N. Hodsun, and C. Parsons, "The use of computed tomography in recurrent rectal tumors," *Radiology* **134**(3), 677–682 (1980).

10. C. Coin, F. Wollett, and J. Coin, "Computerized radiology of the colon: a potential screening technique," *Comput. Radiol.* **7**(4), 215–221 (1983).
11. D. Vining et al., "Virtual colonoscopy," *Radiology* **193**, Suppl P (1994).
12. A. Hara et al., "Detection of colorectal polyps by computed tomographic colography: feasibility of a novel technique," *Gastroenterology* **110**(1), 284–290 (1996).
13. S. M. Frentz and R. M. Summers, "Current status of CT colonography," *Acad. Radiol.* **13**(12), 1517–1531 (2006).
14. P. Pickhardt et al., "Computed tomographic virtual colonoscopy to screen for colorectal neoplasia in asymptomatic adults," *N. Engl. J. Med.* **349**(23), 2191–2200 (2003).
15. M. Macari, M. Lavelle, and I. Pedrosa, "Effect of different bowel preparations on residual fluid at CT colonography," *Radiology* **218**(1), 274–277 (2001).
16. J. Holemans et al., "A comparison of air, CO2 and air/CO2 mixture as insufflation agents for double contrast barium enema," *Eur. Radiol.* **8**(2), 274–276 (1998).
17. J. Fletcher, C. Johnson, and T. Welch, "Optimization of CT colonography technique: Prospective trial in 180 patients," *Radiology* **216**(3), 704–711 (2000).
18. S. Taylor, S. Halligan, and C. Bartram, "Multi-detector row CT colonography: Effect of collimation, pitch, and orientation on polyp detection in a human colectomy specimen," *Radiology* **229**(1), 109–118 (2003).
19. R. M. Summers et al., "Computed tomographic virtual colonoscopy computer-aided polyp detection in a screening population," *Gastroenterology* **129**(6), 1832–1844 (2005).
20. W. Lorensen and H. Cline, "Marching cubes: a high resolution 3D surface construction algorithm," *ACM SIGGRAPH Comput. Graphics* **21**(4), 163–169 (1987).
21. P. Sahoo, S. Soltani, and A. Wong, "A survey of thresholding techniques," *Comput. Vis. Graphics Image Process.* **41**(2), 233–260 (1988).
22. J. K. Udupa and S. Samarasekera, "Fuzzy connectedness and object definition: Theory, algorithms, and applications in image segmentation," *Graphical Models Image Process.* **58**, 246–261 (1996).
23. M. Franaszek et al., "Hybrid segmentation of colon filled with air and opacified fluid for CT colonography," *IEEE Trans. Med. Imag.* **25**(3), 358–368 (2006).
24. S. Wang et al., "Registration of prone and supine CT colonography scans using correlation optimized warping and canonical correlation analysis," *Med. Phys.* **36**(12), 5595–5603 (2009).

25. R. van Uitert and R. M. Summers, "Automatic correction of level set based subvoxel precise centerlines for virtual colonoscopy using the colon outer wall," *IEEE Trans. Med. Imaging* **26**, 1069–1078 (2007).
26. D. Nain et al., "Intra-patient Prone to Supine Colon Registration for Synchronized Virtual Colonoscopy," in *MICCAI*, Springer-Verlag, Berlin (2002).
27. Nappi, J. et al., "Region-based supine-prone correspondence for the reduction of false-positive CAD polyp candidates in CT colonography," *Acad. Radiol.* **12**(6), 695–707 (2005).
28. P. Li et al., "Registration of central paths and colonic polyps between supine and prone scans in computed tomography colonography: Pilot study," *Med. Phys.* **31**(10), 2912–2923 (2004).
29. N. P. V. Nielsen, J. M. Carstensen, and J. Smedsgaard, "Aligning of single and multiple wavelength chromatographic profiles for chemometric data analysis using correlation optimised warping," *J. Chromatography* **805**, 17–35 (1998).
30. D. R. Hardoon, S. Szedmak, and J. Shawe-Taylor, "Canonical correlation analysis: An overview with application to learning methods," *Neural Computation* **16**(12), 2639–2664 (2004).
31. A. Vilanova and E. Groller, "Geometric Modeling for Virtual Colon Unfolding," in *Geometric Modeling for Scientific Visualization*, G. Brunnett et al., eds., Birkhäuser, Berlin (2004).
32. G. Wang et al., "GI tract unraveling with curved cross sections," *IEEE Trans. Med. Imag.* **17**(2), 318–322 (1998).
33. S. Sudarsky et al., "Colon Unfolding Via Skeletal Subspace Deformation," in *MICCAI*, Springer-Verlag, Berlin (2008).
34. S. Haker et al., "Nondistorting flattening maps and the 3D visualization of colon CT images," *IEEE Trans. Med. Imag.* **19**(7), 665–670 (2000).
35. W. Hong et al., "Conformal virtual colon flattening," *Proc. ACM SPM 2006*, 85–93 (2006).
36. D. S. Paik et al., "Visualization modes for CT colonography using Cylindrical and Planar Map Projections," *J. Comput. Assisted Tomogr.* **24**(2), 179–188 (2000).
37. F. M. Vos et al. "A New Visualization Method for Virtual Colonoscopy," in *MICCAI*, Springer-Verlag, Berlin (2001).
38. J. Yao et al., "Reversible projection technique for colon unfolding," *IEEE Trans. Biomed. Eng.* **57**(12), 2861–2869 (2010).
39. J. Yao et al., "Colonic polyp segmentation in CT colonography based on fuzzy clustering and deformable models," *IEEE Trans. Med. Imag.* **23**(11), 1344–1352 (2004).

40. A. Jerebko, M. Franaszek, and R. Summers. "Radon transform based polyp segmentation method for CT colonography computer aided diagnosis," *Radiology* **225**(P), 257 (2002).
41. J. J. Dijkers et al. "Segmentation and Size Measurement of Polyps in CT Colonography," in *MICCAI*, Springer-Verlag, Berlin (2005).
42. J. Yao and R. M. Summers, "Adaptive deformable model for colonic polyp segmentation and measurement on CT colonography," *Med. Phys.* **34**(5), 1655–1664 (2007).
43. M. Kass, A. Witkin, and D. Terzopoulos, "Snakes: Active contour models," *Int. J. Comput. Vision* **1**(4), 321–331 (1988).
44. J. J. Koenderink and A. J. van Doorn, "Surface shape and curvature scales," *Image and Vision Computing* **10**(8), 557–565 (1992).
45. A. Huang, D. Roy, and R. M. Summers. "Teniae coli guided navigation and registration for virtual colonoscopy," *IEEE Visualization Conf.*, Baltimore, MD (2005).
46. I. N. Bankman, T. S. Spisz, and S. Ravlopoulos, "Two-Dimensional Shape and Texture Quantification," in *Handbook of Medical Imaging, Processing and Analysis*, I. N. Bankman, ed., Academic Press, San Diego, CA (2000).
47. J. Yao, J. Li, and R. M. Summers, "Employing topographical height map in colonic polyp measurement and false positive reduction," *Pattern Recognition* **42**(6), 1029–1040 (2009).
48. J. P. Wilson and J. C. Gallant, "Digital Terrain Analysis," in *Terrain Analysis: Principles and Applications*, J. P. Wilson and J. C. Gallant, eds., John Wiley & Sons, New York (2000).
49. W. G. Baxt, "Application of artificial neural networks to clinical medicine," *Lancet* **346**(8983), 1135–1138 (1995).
50. N. Cristianini and J. S. Taylor, *An Introduction to Support Vector Machines*, Cambridge University Press, Cambridge, UK (2000).
51. E. Bauer and R. Kohavi, "An empirical comparison of voting classification algorithms: bagging, boosting, and variants," *Machine Learning* **36**(1–2), 105–139 (1999).
52. L. Breiman, "Bagging predictors," *Machine Learning* **24**(2), 123–140 (1996).
53. H. Müller et al., "A review of content-based image retrieval systems in medical applications—clinical benefits and future directions," *Int. J. Med. Info.* **73**(1), 1–23 (2004).
54. G. Ciocca and R. Schettini, "A relevance feedback mechanism for content-based image retrieval," *Info. Proc. Manag.* **35**(5), 605–632 (1999).
55. D. G. Lowe, "Distinctive image features from scale-invariant keypoints," *Int. J. Comput. Vision* **60**(2), 91–110 (2004).

56. J. M. Aman, J. Yao, and R. M. Summers, "Content-based image retrieval on CT colonography using rotation and scale-invariant features and bag-of-words model," *Int. Symp. Biomed. Imag.*, 1357–1360 (2010).
57. J. M. Aman, J. Yao, and R. M. Summers, "Reducing the false positive rate of computer-aided detection for CT colonography using content based image retrieval," *Proc. IEEE Int. Symp. Biomed. Imaging*, 915–918 (August 2009).
58. J. M. Aman, J. Yao, and R. M. Summers, "Prediction of polyp histology on CT colonography using content-based image retrieval," *Proc. SPIE* **7624**, 76240D (2010) [doi: 10.1117/12.844571].
59. J. M. Aman, J. Yao, and R. M. Summers, "Automatic colonic polyp shape determination using content-based image retrieval," *Proc. SPIE* **7963**, 79632G (2011) [doi: 10.1117/12.878196].
60. K. W. Bowyer, "Validation of Medical Image Analysis Techniques," in *Handbook of Medical Imaging, Volume 2. Medical Image Processing and Analysis*, M. Sonka and J. M. Fitzpatrick, eds., SPIE Press (2000) [doi: 10.1117/3.831079.ch10].

Jianhua Yao, PhD, is an associate scientist and manager in the Radiology and Imaging Sciences Department at the National Institutes of Health. Jianhua Yao received B.S., M.S., and Ph.D. in Computer Science from Tianjin University (China), Tsinghua University (China), and Johns Hopkins University, respectively. His research interests include clinical image processing, deformable models, nonrigid registration, computer-aided diagnosis, and CT colonography.

Chapter 2
Preprocessing Tools for Computer-Aided Cancer Imaging Systems

Artyom M. Grigoryan and Sos S. Agaian
Department of Electrical and Computer Engineering, University of Texas at San Antonio, TX, USA

Digital image processing methods are widely used in medical imaging cases where different types of cancer require complex processing of images (see Fig. 2.1). For many medical images, it is desirable to enhance the quality of images; detect, extract, and analyze the areas with tumors; accurately classify the type of the tumor; compress the images (when necessary); and encrypt images to protect patient data. All of these methods serve one purpose: provide information about the extent of disease and help plan treatment.

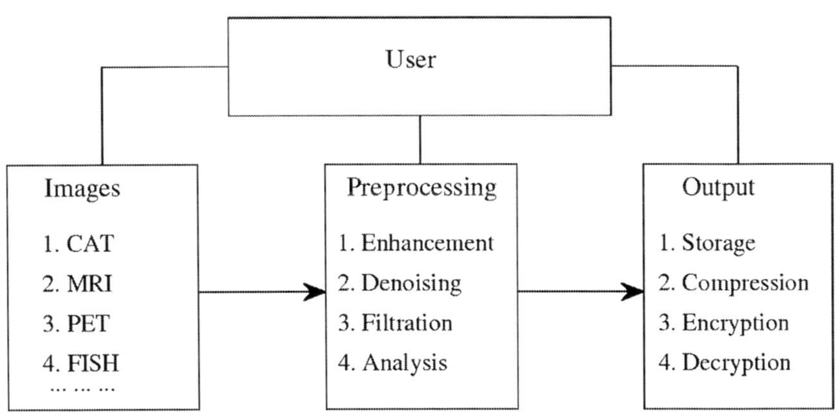

Figure 2.1 Schematic representation of a medical imaging system.

2.1 Introduction

Digital image enhancement is a powerful tool for many image processing applications when the critical details cannot be seen clearly enough. The purpose of image enhancement is to improve digital image quality and to support human perception;[1–7] for example, in medical imaging, computed tomography (CT), magnetic resonance (MRI), and fluorescence *in situ* hybridization (FISH) produce 3D images (or a stack of 2D images) of different organs and tissues.[8–13] There are many sources of interference in the production of medical images, such as the movement of a patient, insufficient performance, and noise produced by imaging devices. When the quality of an image is poor in contrast, enhancement methods can be used to improve the image quality, enhance edges, and reduce noise for diagnostic purposes. Examples of such low-quality medical images are shown in Fig. 2.2.

Unitary transforms play an important role in image processing, and they are used in different stages of processing such as filtering, coding, recognition, and restoration analysis.[14–21] This chapter focuses on the transform-based image enhancement methods, although different spatial domain methods operate directly on pixels, and both frequency and spatial domain methods are also used for image enhancement.[3,22–24,34–38] This chapter considers the tensor transform and Fourier transform methods of enhancement that are widely used and have been generalized for the Hartley, Hadamard, cosine, and other transforms.[21,26–32,37] These methods are based on manipulating all or part of the spectral components of the transform. The well-known methods of α-rooting[25–27] and log-α-rooting,[58,59,82,83] modified unsharp masking[32–34] and contrast entropy,[36,37] as well as methods based on wavelet transforms, are also mentioned.[60–64]

Image enhancement is considered the first important step in processing medical images, and the following steps in image processing are image compression[85] and image cryptography.[89–93] These three steps in medical image processing are considered from the same point of view, i.e., the images will be considered in new forms of representation when the information of the 2D images can be transferred uniquely into the set of 1D signals, and then

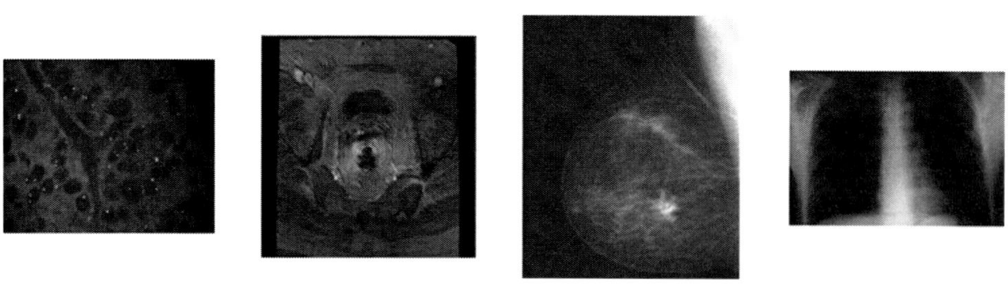

(a) (b) (c) (d)

Figure 2.2 (a) FISH image, (b) MRI prostate image, and (c) breast and (d) chest x-ray images.

these signals will be processed and transferred back to the spatial domain to form the processed image.

2.1.1 Novel view on image processing

Processing images in the frequency domain is very effective for many imaging applications, including image enhancement. The advantages of transform-based image enhancement techniques include low complexity of computations; high quality of enhancement; the presence of bright, visible zones in the frequency domain; the ease in using parameterized linear filters; and the critical role of fast unitary transforms in digital image processing. The main idea behind enhancement methods in the frequency domain consists of calculating a discrete unitary transform (DUT) of the image, which is denoted by Φ manipulating the transform coefficients by a parameterized operator $\mathbf{M}$; and then performing the inverse transform, as shown in Fig. 2.3.

As an example, Fig. 2.4. illustrates an image that is processed by the 2D discrete Fourier transform. The discrete breast x-ray image of 512×512 pixels is shown in (a), along with the amplitude of the transform $\{\Phi_{p,s};\ p,s = 0:511\}$ in (b), the amplification of the transform in (c), and the result of the inverse 2D DFT in (d). In (c), the 2D DFT was amplified by $M[\Phi_{p,s}] = [1 + k(p_2 + s_2)]\Phi_{p,s}$ with a factor of $k > 0$.

Because the calculation and analysis of the 2D discrete unitary transform of the image, including the 2D DFT, are the main steps of image enhancement, this section considers forms of splitting the 2D DFT mathematical structure that can be used effectively for image enhancement. The tensor and paired forms of image representation[65–74] can reduce the problem of image enhancement (as well as image filtration, compression, and encryption) to the processing of 1D signals. The image can be uniquely

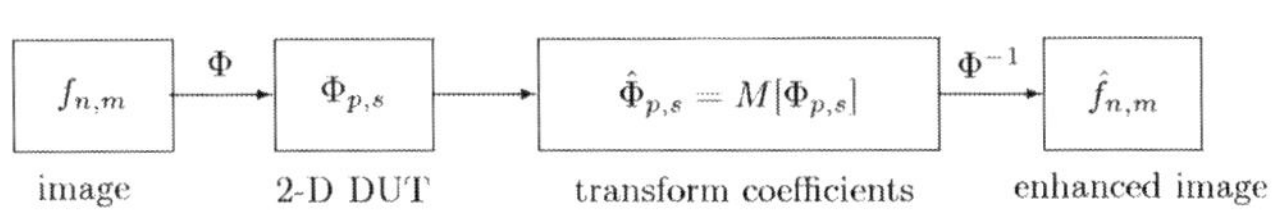

Figure 2.3 Block diagram of the Φ-transform-based image enhancement.

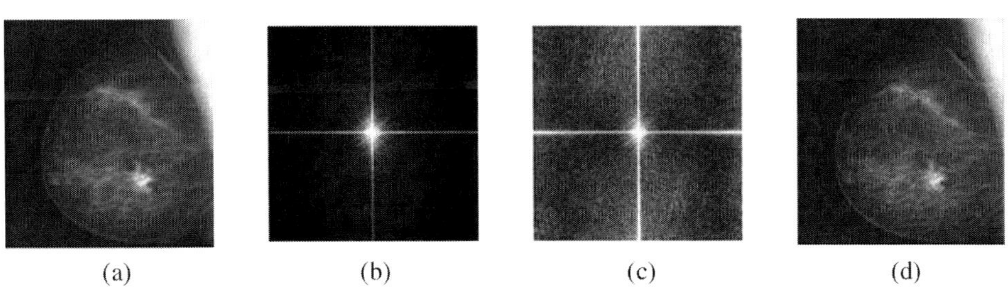

Figure 2.4 (a) Breast x-ray image, (b) 2D DFT of the image (in absolute scale and shifted to the center), (c) amplified transform, and (d) image of the inverse 2D DFT.

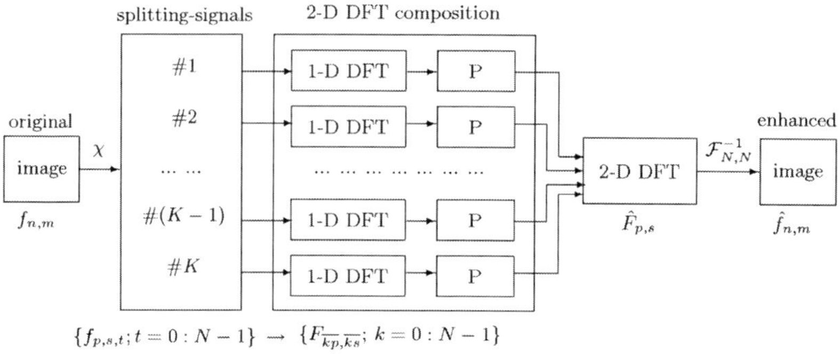

Figure 2.5 Enhancement of an image of size $N \times N$ by processing K splitting-signals.

described as a certain totality of 1D "independent" signals that carry the spectral information of the 2D DFT at frequencies of different or disjoint subsets in the frequency plane. These signals are called *splitting-signals* or *image-signals*, and they can be processed separately to achieve effective image enhancement. The enhancement is with respect to the quantitative measure of image enhancement based on the Weber definition of the contrast, which is described in detail in the literature.[21,75,83]

Figure 2.5 shows a diagram of the proposed method of processing an $N \times N$ image with the Fourier transform. The image is transferred to a totality of K splitting-signals that represent the image f in tensor representation $\chi[f]$. This number equals $N + 1$ and $3N/2$ when N is a prime and a power of two, respectively. For other N, the number of splitting-signals can also be calculated.[21] Rather than enhance the image by methods of the Fourier transform, one can process (P) them all separately, or only a few splitting-signals, and then calculate the 2D DFT of the enhanced image from 1D DFTs of the processed splitting-signals. The enhanced image can then be calculated by the inverse 2D DFT. This diagram can also be used when processing the image with other discrete unitary transforms, such as the 2D Hartley, Hadamard, and cosine transforms. For that, the corresponding 1D DUTs are used to process the splitting-signals, and the enhanced image is calculated by the inverse 2D DUT.

A similar diagram is used for image enhancement when using a more-advanced paired representation of the image, which is considered next for the case of most interest when N is a power of two. There is a redundancy in the tensor representation, which means that the different splitting-signals carry the spectral information of the image at subsets of frequency points, which may have intersections. The tensor transform is invertible but not orthogonal. The paired representation of the image removes the redundancy of the tensor transform, and it is orthogonal. The paired transform-based method of enhancement requires many fewer operations of multiplication when compared with the traditional transform-based methods. For instance, approximately $N/2(N - k)$ operations of multiplication can be saved when

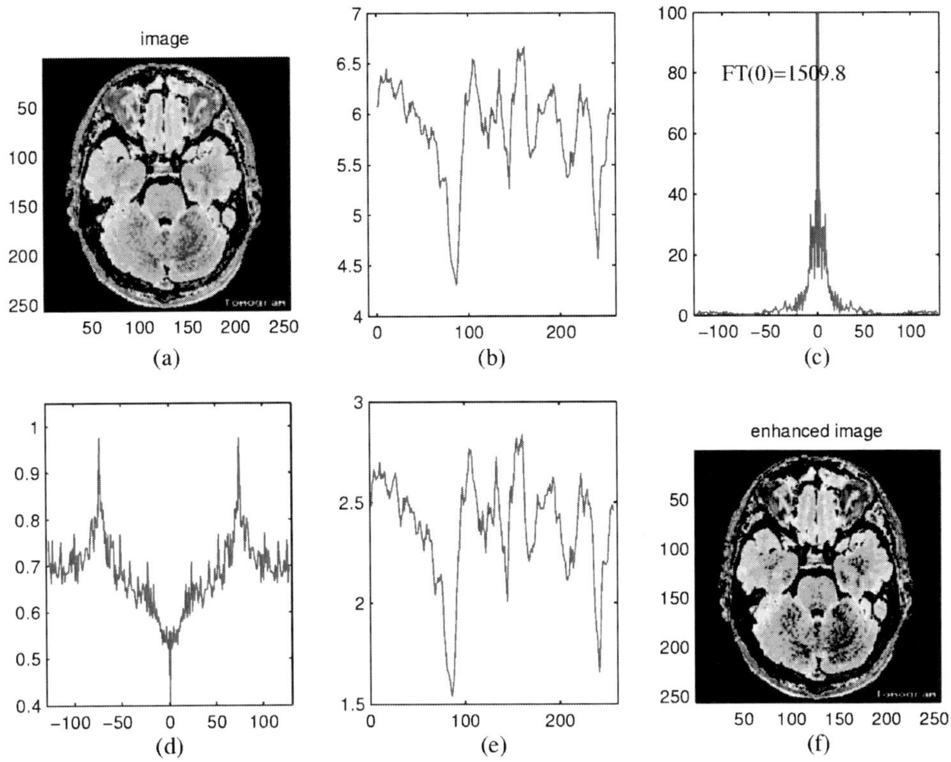

Figure 2.6 Frequency-time method of image enhancement: (a) the original image, (b) the splitting-signal, (c) the amplitude spectrum of the signal, (d) coefficients of amplification and the processed signal I, and (f) the enhanced image.

enhanced by k splitting-signals. By processing only one or a few splitting-signals, enhancement of the image can be achieved, which in many cases is greater than the traditional methods. For instance, the α-rooting method of image enhancement can be fulfilled by processing one splitting-signal. As an example, Fig. 2.6(a) shows a CT image of size 256×256, (b) shows one splitting-signal, and (c) shows its 1D DFT, which equals the 2D DFT of the image at frequency points located in the main diagonal. The 1D DFT of the signal is amplified by coefficients shown in Fig. 2.6(d) and the processed signal in I; the image changed by this one splitting-signal is shown in (f). As described in Section 2.5, this signal allows for processing the image along the diagonal direction. In other words, the enhancement of the image in this example is achieved by processing the image in the diagonal direction.

The block diagram in Fig. 2.5 illustrates the effective realization of the Fourier transform methods of image enhancement. Fast algorithms exist for calculating the tensor and paired transforms, and image enhancement can be performed without calculations of the DFT, i.e., by processing the splitting-signals and transforming them back to the image. Fourier transform methods of image enhancement modify the image in the frequency domain. The tensor and paired transforms are referred to as the 2D frequency and 1D time representations, respectively. In such a time-in-frequency domain, image processing is performed. This chapter presents effective formulas for inverse 2D $N \times N$-point tensor and paired transforms, where N is a prime and a power of 2, respectively. New concepts of direction and series images will be introduced that define the resolution and periodic structures of the image components, which can be packed in the form of the "resolution map" of the size of the image. Such a resolution map can be effectively used not only in image compression but also in image enhancement. Other applications of the 2D tensor and paired transforms can be found in medical imaging, including the image encryption that is described in Section 2.11.

2.2 Transform-Based Image Enhancement

The basic idea behind the frequency domain methods consists of computing a discrete unitary transform of the image (for instance, the 2D discrete Fourier transform), manipulating the transform coefficients by operator $\mathbf{M}$, and then performing the inverse transform, as shown in Fig. 2.7.

Transform-based image enhancement methods include techniques such as α-rooting, weighted α-rooting, modified unsharp masking, and filtering, which are all motivated by the human visual response.[27] Consider the 2D DFT of the image to be the transform Φ for image processing, which is defined as

$$F_{p,s} = \sum_{n=0}^{N-1}\sum_{m=0}^{N-1} f_{n,m} W_{np+ms}, p, s = 0 : (N-1),\ W = W_N = \exp(-j2\pi/N).$$

In the Fourier-transform-based method of image enhancement, $\mathbf{M}$ is an operator of magnitude, and the enhancement is described by

$$\{f_{n,m}\} \to \left\{F_{p,s} = |F_{p,s}|e^{-j\vartheta_{p,s}}\right\} \to \mathbf{M} \circ F \to \left\{\hat{F}_{p,s} = M[|F_{p,s}|]e^{-j\vartheta_{p,s}}\right\} \to \left\{\hat{f}_{n,m}\right\}.$$

In the α-rooting method of image enhancement, the magnitude of the Fourier transform is modified as

$$\mathbf{M}\big[|F_{p,s}|\big] = |F_{p,s}|^{\alpha}, \quad p, s = 0 : (N-1), \quad \alpha > 0,$$

where the parameter α is taken from the interval (0, 1). Thus, in the α-rooting enhancement, the coefficients of the Fourier transform are multiplied by coefficients $C_1(p, s) = |F_{p,s}|^{\alpha-1}$:

$$F_{p,s} \to \hat{F}_{p,s} = \mathbf{M}[F_{p,s}] = |F_{p,s}|^{\alpha} e^{-j\vartheta_{p,s}} = |F_{p,s}|^{\alpha-1} F_{p,s}.$$

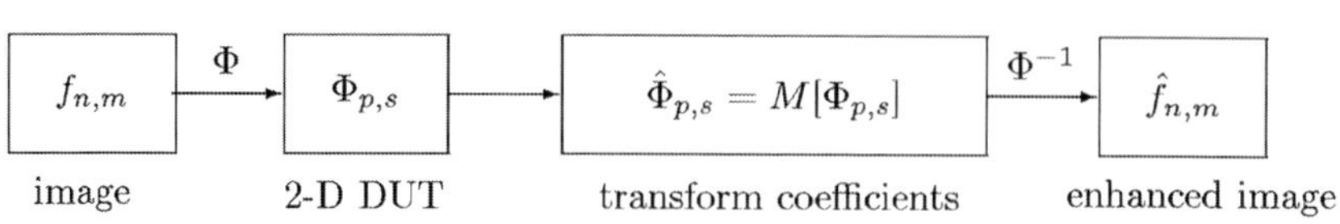

Figure 2.7 Block diagram of transform-based image enhancement.

This is one of the simplest parameterized methods used to effectively enhance images. The "best" value of α may be dynamically adjusted by the user, or it can be found automatically if the quantitative measure of image enhancement is used.[49–55,76] Operators $\mathbf{M}$ with two and three parameters can also be used for image enhancement in the log-magnitude and the combined α-rooting methods when the coefficients have the forms $C_2(p,s) = \log^{\beta}|X(p,s)|^{\gamma}$ and $C_3(p,s) = C_1(p,s) \cdot C_2(p,s)$, respectively (see Grigoryan and Agaian[21] and Agaian et al.[75] for more details).

2.2.1 Quantitative measure of image enhancement

To measure the quality of images and select the optimal processing parameters, consider the quantitative measure of image enhancement (described by Agaian et al.[59,75]) that relates to Weber's law of human visual system.[28,29,35,97] This measure can be used to select the best parameters for image enhancement by the Fourier transform, as well as other unitary transforms Φ. The measure is defined as follows: A discrete image $\{f_{n,m}\}$ of size $N_1 \times N_2$ is divided by $k_1 k_2$ blocks of size $L_1 \times L_2$, where $k_i = \lfloor N_i/L_i \rfloor, i = 1,2$, and where $\lfloor . \rfloor$ denotes the floor function. The quantitative measure of enhancement of the image processed by the Φ transform $M_{\alpha;\Phi}$: $\{f_{n,m}\} \to \{\hat{f}_{n,m}\}$ is defined by

$$EME_{\Phi}(\hat{f}) = EME_{\boldsymbol{\alpha},k_1,k_2;\Phi}(\hat{f}) = \frac{1}{k_1 k_2}\sum_{k=1}^{k_2}\sum_{l=1}^{k_1} 20\log\left[\frac{\max_{k,l} M_{\boldsymbol{\alpha};\Phi}(\hat{f})}{\min_{k,l} M_{\boldsymbol{\alpha};\Phi}(\hat{f})}\right], \tag{2.1}$$

where $\max_{k,l} M_{\boldsymbol{\alpha};\Phi}(\hat{f})$ and $\min_{k,l} M_{\boldsymbol{\alpha};\Phi}(\hat{f})$ are the maximum and minimum, respectively, of the image $\hat{f}_{n,m}$ inside the (k,l)th block, and $\boldsymbol{\alpha}$ is a vector parameter of the enhancement algorithm. $EME_{\Phi}(\hat{f})$ is called a *measure of enhancement* or *measure of improvement* of the image f with respect to the transform Φ. A parameter $\boldsymbol{\alpha}_0$ is defined such that

$$EME_{\boldsymbol{\alpha}_0,\Phi}(f) = \max_{\boldsymbol{\alpha}}\{EME_{\boldsymbol{\alpha},\Phi}(f)\}$$

is the best (or optimal) Φ-transform-based image enhancement vector parameter. When Φ is the identical operator $\Phi(f) = f$, the value of $EME(f) = EME_I(f)$ is called the enhancement measure of the image. Experimental results show that the discrete Fourier transform can be considered optimal when compared with the cosine, Hartley, Hadamard, and other transforms. Therefore, the enhancement measure $EME(f)$ will be considered with respect to the Fourier transform, i.e., when $\Phi = F$.

As an example, Fig. 2.8(a) shows an image, and 2.8(b) shows the measure of enhancement that was calculated for block sizes 5×5 and 7×7. Image enhancement by the α-rooting method has been parameterized by α varying in the interval $(0.35, 1]$. The curves have one pike with maximums at point $\boldsymbol{\alpha}_0 = 0.93$. The experimental results show that the parameter $\boldsymbol{\alpha}_0$ corresponds

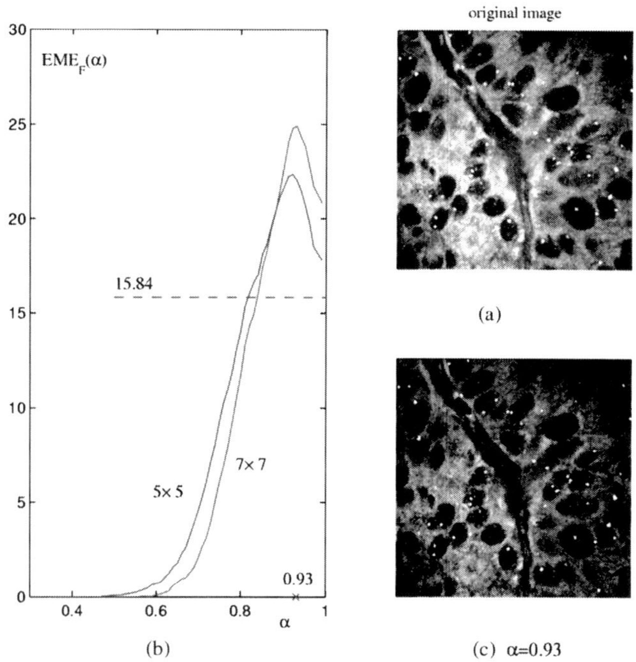

Figure 2.8 Parameterized image enhancement based on the 2D discrete Fourier transform by α-rooting.

to the best visual estimation of enhancement. Figure 2.8(c) illustrates the α-rooting enhancement of the original image f when $\boldsymbol{\alpha} = 0.93$, which yields the enhancement $EME(f_{0.93}) - EME(f) = 24.43 - 18.55 = 6.88$.

2.3 Tensor Representation of the Image

This section describes the main concepts of the theory of splitting the 2D discrete unitary transforms by 1D transforms of the signals that uniquely represent the image, which were developed by Grigoryan[65–70] and described in detail by Grigoryan and Agaian.[21,57] In many recent publications, these concepts (with various applications in digital image processing) were published under various names, such as the discrete Radon transform [Gertner (1988)], fast multidimensional Radon transform [Labunets (1999)], the finite Radon transform [Matúš and J. Flusser (1993)], a new discrete transform based on the exact discrete Radon transform [Guèdon, Barba, and Burger (1995)] or the mojette transform [Gueèdon and Normand (2005)], the discrete periodic Radon transform [Hsung, Lun, and Siu (1996)], the orthogonal discrete periodic Radon transform [Lun, Hsung, and Shen (2003); Kingston (2006)], and the generalized finite Radon transform [Kingston and Svalbe (2007)].

The tensor representation of the image $f_{n,m}$ of size $N \times N$ is defined as a set of splitting-signals of length N each:

$$\chi : \{f_{n,m}\} \to \{f_{T_{p,s}} = \{f_{p,s,t}; t = 0 : (N-1)\}\}_{(p,s)\in J}.$$

By definition, the components of these signals are calculated as the sums of the image along the parallel lines:

$$f_{p,s,t} = \sum_{(n,m)} \{f_{n,m}; np + ms = t \bmod N\}, t = 0 : (N-1).$$

In the notation of the splitting-signal $f_{Tp,s}$, the cyclic group

$$T_{p,s} = \{(kp \text{ mod } N, ks \text{ mod } N); k = 0 : (N-1)\}$$

is used because this signal defines the 2D DFT of the image at frequency points of $T_{p,s}$. This means that the 1D DFT of the splitting-signal coincides with values of the 2D DFT of the image at frequency points of $T_{p,s}$:

$$F_{kp \text{ mod } N,\, ks \text{ mod } N} = \sum_{t=0}^{N-1} f_{p,s,t} W^{kt}, k = 0 : (N-1), \tag{2.2}$$

where the kernel of the discrete Fourier transform is $W = W_N = \exp(-2\pi j/N)$. As an example, Fig. 2.9(a) shows the image of size 512 × 512, and (b) shows the splitting-signal of length 512, which is generated by the frequency-point $(p, s) = (3, 1)$. The 1D DFT of the splitting-signal in absolute scale is shown in Fig. 2.9(c), and the 2D DFT of the image and the location of all frequency points of the cyclic group $T_{3,1}$ is shown in (d).

X denotes the Cartesian lattice $X_{N,N} = \{(n, m); n, m = 0; (N-1)\}$. Given a triplet (p, s, t), where $(p, s) \in X$ and $t \in \{0, 1, 2, ..., N-1\}$, the following set of points (n,m) of the lattice are defined: $V_{p,s,t} = \{(n, m); n, m = 0 : (N-1), \overline{np + ms} = t\}$, where $l = l \bmod N$; consider its characteristic function:

$$X_{p,s,t}(n, m) = \begin{cases} 1, & \text{if}\,(n, m) \in V_{p,s,t}, \\ 0, & \text{otherwise.} \end{cases} \tag{2.3}$$

The set $V_{p,s,t}$, if it is not empty, is the set of points (n,m) along a maximum of $p + s$ parallel straight lines at the angle $\varphi = \arctan(s/p)$ to the horizontal axis. In the square domain $[0, N] \times [0, N]$, the equations for the set $L_{p,s,t}$ of parallel lines are $xp + ys = t + kN$, where $k = 0 : (p + s - 1)$. It is interesting to note that the direction of parallel lines of $L_{p,s,t}$ is perpendicular to the direction of frequency points of the group $T_{p,s}$.

Example 1

In the lattice $X_{8,8}$, consider the generator $(p, s) = (2, 1)$ and two sets of parallel lines L_1 and L_2. Each family contains three parallel lines. For the family L_1, the parallel lines are

$$y_1 : 2x + y = 1, \quad y_9 : 2x + y = 9, \quad y_{17} : 2x + y = 17.$$

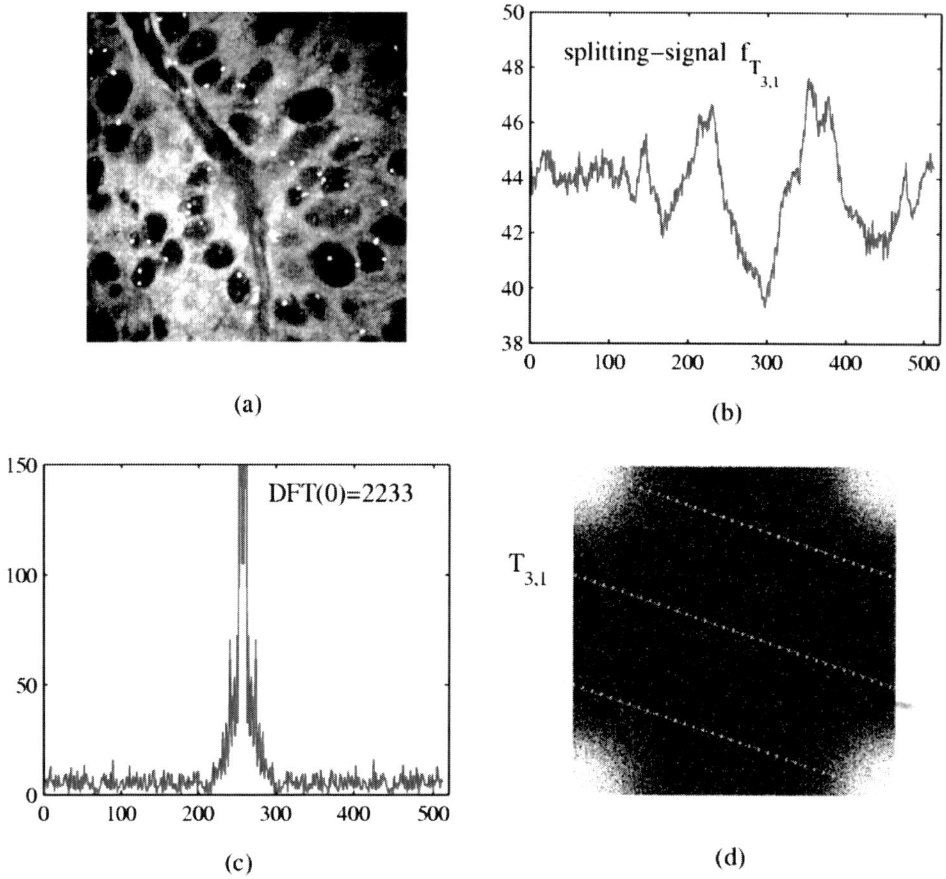

Figure 2.9 (a) The original image, (b) splitting-signal $f_{T3,1}$, (c) 1D DFT of the splitting-signal, and (d) arrangement of values of the 1D DFT in the 2D DFT of the image at points of the set $T_{3,1}$.

One point (0,1) of the set $V_{2,1,1}$ lies on the first line of L_1; four points (1, 7), (2, 5), (3, 3), (4, 1) lie on the second line; and three points (5, 7), (6, 5), (7, 3) lie on the third one. Therefore, $f_{2,1,1} = (x_{0,1}) + (x_{1,7} + x_{2,5} + x_{3,3} + x_{4,1}) + (x_{5,7} + x_{6,5} + x_{7,3})$. The parallel lines of the family L_2 are defined by

$$y_2 : 2x + y = 2, \quad y_{10} : 2x + y = 10, \quad y_{18} : 2x + y = 18,$$

and the component $f_{2,1,2}$ is calculated by $f_{2,1,2} = (x_{0,2} + x_{1,0}) + (x_{2,6} + x_{3,4} + x_{4,2} + x_{5,0}) + (x_{6,6} + x_{7,4})$. The disposition of the points lying on the parallel lines of these sets is given in Fig. 2.10. The location of the frequency points of the group $T_{2,1}$ is also shown. Two parallel lines pass through these frequency points, which are defined in the frequency plane (ω_1, ω_2) as $l_1 : 2\omega_2 - \omega_1 = 0$ and $l_2 : 2\omega_2 - \omega_1 = 8$. The

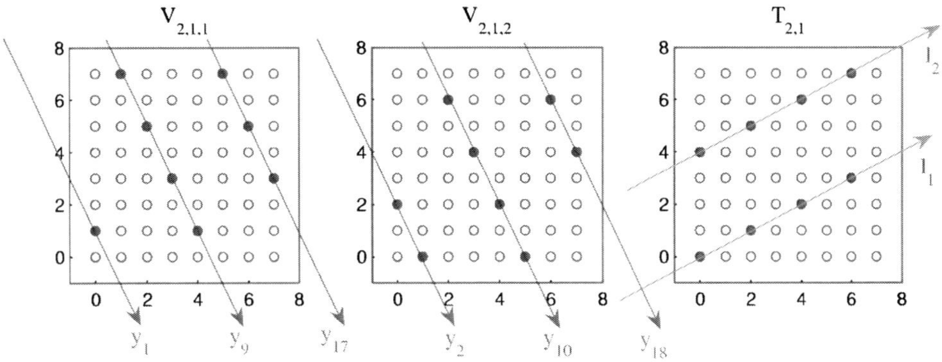

Figure 2.10 The locations of the points of sets $V_{2,1,1}$ and $V_{2,1,2}$ and the frequency points of the group $T_{2,1}$.

parallel lines l_1 and l_2 are perpendicular to the parallel lines of both sets L_1 and L_2.

From Eq. (2.4), the following can be stated:

1. The image can be presented as a set of splitting-signals $f^{(k)}$, $k = 1 : l$, each of length N:

$$\{f_{n,m}\} \leftrightarrow \{f^{(1)}, f^{(2)}, \ldots, f^{(l)}\}.$$

2. The 2D DFT $F_{N,N}$ of the image can be split by 1D transforms F_N of the splitting-signals:

$$F_{N,N}[f] \leftrightarrow \{F_N[f^{(1)}], F_N[f^{(2)}], \ldots, F_N[f^{(l)}]\}.$$

The number l of the splitting-signals depends on N and can be determined by the set of generators $J = J_{N,N}$ for the splitting-signals. This set is defined as a set for which the totality of cyclic groups $\{T_{p,s}; (p,s) \in J_{N,N}\}$ is an irreducible covering of the discrete lattice of frequency points $X_{N,N} = \{(p,s); p,s = 0 : (N-1)\}$. Given N, one can construct different such sets $J_{N,N}$; however, their cardinalities are equal. For cases where $N = 4$, 5, and 8, we obtain $l = 6$, 6, and 12, respectively, because the following sets of generators can be considered:

$$J_{4,4} = \{(1,0), (1,1), (1,2), (1,3), (0,1), (2,1)\},$$
$$J_{5,5} = \{(1,0), (1,1), (1,2), (1,3), (1,4), (0,1)\},$$
$$J_{8,8} = \{(1,0), (1,1), (1,2), (1,3), (1,4), (1,5), (1,6), (1,7), (0,1), (2,1), (4,1), (6,1)\}.$$

2.4 Decomposition by Direction Images

Consider the case where the size of the image is $N \times N$, and N is a prime. The tensor representation in this case has a minimum redundancy, and it can be used effectively to process images with other sizes. For instance, the problem of image enhancement of an image of size 256 × 256 can be reduced to processing the extended image of size 257 × 257. When N is a prime, $(N + 1)$ splitting-signals represent the image $f_{n,m}$. Indeed, the grid $N \times N$ can be covered by the family $(T_{p,s})$ of $(N + 1)$ cyclic groups. The set J of generators (p,s) for these groups can be defined as

$$J = J_{N,N} = \{(0,1),(1,1),(2,1),(3,1),\ldots,(N-1,1),(1,0)\}.$$

Figure 2.11(a) shows the FISH image of size 521 × 521, and (b) shows the set of all splitting-signals. The splitting-signals are written in rows of the image of size 522 × 521. The magnitude of the 2D DFT of the image is shown in Fig. 2.11(c), and the magnitudes of all 1D DFTs of the splitting-signals written in rows of the image are shown in (d). There is no redundancy in the tensor transform because all sums of splitting-signals equal the sum of the image:

$$\sum_{t=0}^{N-1} f_{p,s,t} = \sum_{n=0}^{N-1}\sum_{m=0}^{N-1} f_{n,m}, \forall (p,s) \in J_{521,521}.$$

Therefore, one complete splitting-signal can be written, e.g., for $(p, s) = (1,0)$, and only the first $(N - 1)$ components of the remaining splitting-signals need be written. As a result, the tensor transform will be written in the table of $N + N(N - 1) = N^2$ elements.

Given a generator (p,s), the following complex data $D = D(p, s)$ of size $N \times N$ is thus defined:

$$D_{p_1,p_2} = D(p,s)_{p_1,p_2} = \begin{cases} F_{\overline{kp},\overline{ks}}, & \text{if } (p_1,p_2) = (\overline{kp},\overline{ks}), k = 0:(N-1), \\ 0, & \text{otherwise,} \end{cases}$$

where $p_1, p_2 = 0 : (N - 1)$. The data D represents an incomplete 2D DFT of the image that is zero at all frequency points except the group $T_{p,s}$. The direction image d_{n_1,n_2} is defined as the inverse 2D DFT of D, which can be calculated as follows:

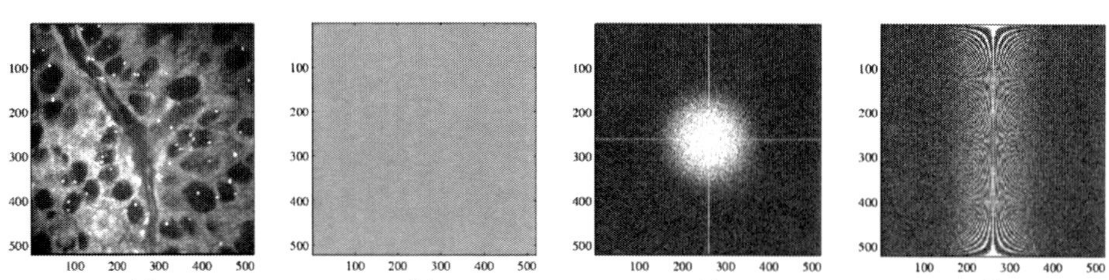

Figure 2.11 (a) FISH image of size 521 × 521, (b) image of all 522 splitting-signals, (c) the 521 × 521-point 2D DFT of the image, and (d) the image of 522 × 521-point 1D DFTs of the splitting-signals. (All DTFs are shifted to the center.)

$$d_{n_1,n_2} = d^{(p,s)}_{n_1,n_2} = \frac{1}{N^2}\sum_{p_1=0}^{N-1}\sum_{p_2=0}^{N-1} D_{p_1,p_2} W^{n_1p_1+n_2p_2} = \frac{1}{N^2}\sum_{k=0}^{N-1} F_{\overline{kp},\overline{ks}} W^{n_1(kp)+n_2(ks)}$$

$$= \frac{1}{N}\frac{1}{N}\sum_{k=0}^{N-1} F_{\overline{kp},\overline{ks}} W^{k(n_1p+n_2s)} = \frac{1}{N} f_{p,s,(n_1p+n_2s) \bmod N}.$$

Thus, N values of the splitting-signal are placed along the parallel lines in N^2 points of the image $N \times N$. As an example, consider an FISH image of size 521×521, shown in Fig. 2.12(a). The direction images for the generators $(p, s) = (p, 1)$, where $p = 0 : 10$, are shown in Figs. 2.12(b–l).

All $(N + 1)$ cyclic groups intersect only at frequency point (0,0). Therefore, taking the inverse 2D DFT of the sum of all incomplete 2D DFTs, the following image is obtained:

$$\sum_{(p,s)\in J} d^{(p,s)}_{n_1,n_2} = \sum_{(p,s)\in J} (F^{-1}_{N,N} \circ D(p,s))_{n_1,n_2} = \left[F^{-1}_{N,N} \circ (F_{p_1,p_2} + NF_{\delta(p_1,0),\,\delta(p_2,0)})\right]_{n_1,n_2}$$

$$= f_{n_1,n_2} + \frac{1}{N}F_{0,0} = F_{n_1,n_2} + NE[f],$$

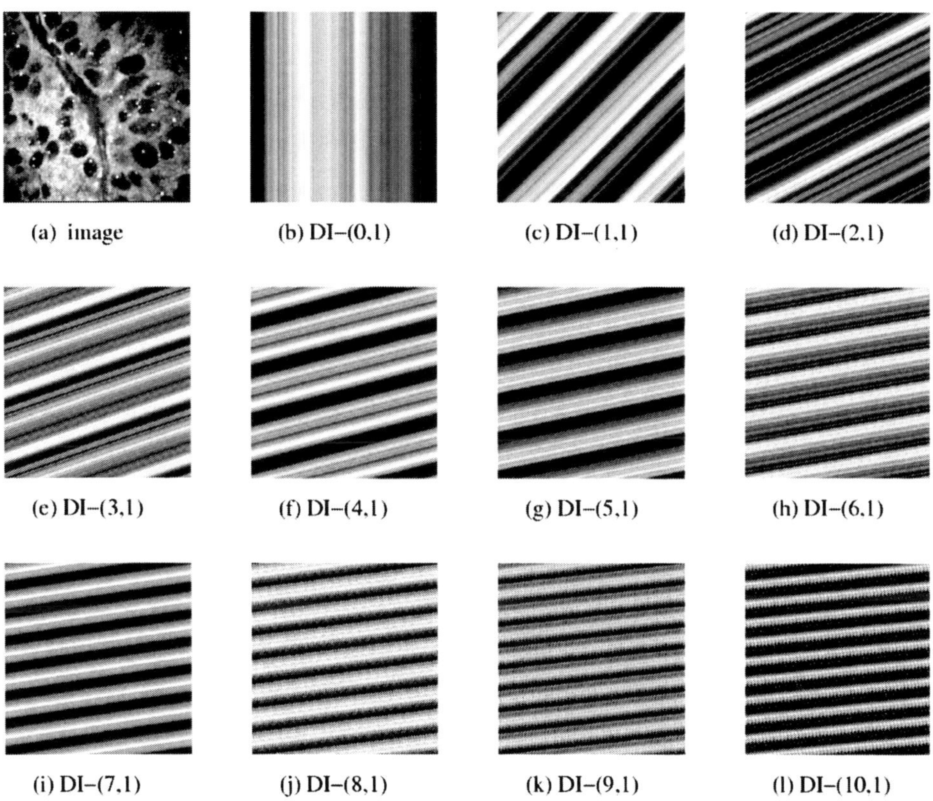

(a) image (b) DI–(0,1) (c) DI–(1,1) (d) DI–(2,1) (e) DI–(3,1) (f) DI–(4,1) (g) DI–(5,1) (h) DI–(6,1) (i) DI–(7,1) (j) DI–(8,1) (k) DI–(9,1) (l) DI–(10,1)

Figure 2.12 (a) The image of size 521 × 521 and (b)–(l) its eleven direction images in tensor representation. (All images are scaled.)

where $\delta(n,m)$ is the delta symbol of Kronecker, which is 1 when $m = n$, and 0 otherwise. Here, $E[f]$ denotes the mean of the image. If it is assumed that the image is centered, i.e., $f_{n_1,n_2} \to f_{n_1,n_2} - E[f]$, then

$$f_{n_1,n_2} = \sum_{(p,s)\in J} d^{(p,s)}_{n_1,n_2} = \frac{1}{N} \sum_{(p,s)\in J} f_{p,s(n_1p+n_2s) \bmod N}, \quad n_1, n_2 = 0:(N-1). \quad (2.4)$$

Statement 1 (Principle of Superposition): The image f_{n_1,n_2} of size $N \times N$, where N is odd and prime, can be composed from $(N+1)$ directional images or splitting-signals as follows:

$$f_{n,m} = \frac{1}{N}\left[\sum_{p=0}^{N-1} f_{p,1,(np+m) \bmod N} + f_{1,0,n}\right] - NE[f], \quad n,m = 0:(N-1).$$

Thus, the simple formula for reconstructing an image from its projections is obtained by using $(N+1)$ splitting-signals or direction images in tensor representation. Each direction image or splitting-signal is defined by the direction at the corresponding angle $\varphi(p,s) = \arctan(s/p)$ to the horizontal axis. $(N+1)$ is the required number of directions for the exact composition of the image. The interesting property of the tensor transform is derived. The splitting-signal is defined as the sum of the image along the parallel lines. The direction image is composed of N values of the splitting signal, which are placed at all points of the image $N \times N$ along another set of parallel lines (these two sets of lines are perpendicular to each other). As an example, Fig. 2.13(a) shows the image of size 521×521, (b) shows the

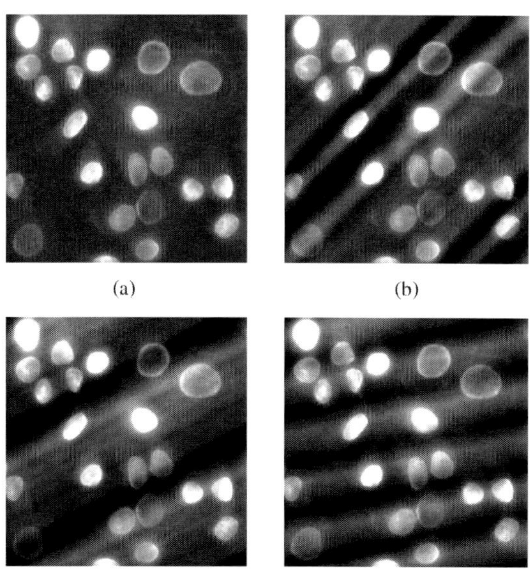

Figure 2.13 (a) The original cell image, and images with one amplified direction image for (p,s) equal to (b) (1, 1), (c) (1, 2), and (d) (1, 5).

image after amplifying the projection data that corresponds to the generator $(p, s) = (1, 1)$, (c) shows $(p, s) = (1, 2)$, and (d) shows $(p, s) = (1, 5)$. The same factor of 4 is used to amplify the corresponding direction images in Figs. 2.13(b)–(d). The direction of these images can be seen, defined by the angles $\operatorname{arctg}(s/p) = 45$ deg, 26.56 deg, and 11.30 deg, respectively.

2.5 Tensor Transform Method of α-Rooting

The Fourier transform method of image enhancement can be performed by processing the splitting-signals $f_{T_{p,s}}, (p, s) \in J_{N,N}$. In other words, the problem of enhancing 2D images can be reduced to processing 1D splitting-signals. Consider the following implementation of the α-rooting method of image enhancement by splitting-signals.

Tensor algorithm of image enhancement

Step 1: Calculate the splitting-signals $f_{T_{p,s}}$.

Step 2: Perform the 1D DFTs of the splitting-signals:

$$f_{T_{p,s}} \to F_k = \sum_{t=0}^{N-1} f_{p,s,t} W^{kt}, \quad k = 0 : (N-1).$$

Step 3: Multiply the coefficients of the splitting-signal transforms by coefficients $C_k = A|F_k|^{\alpha-1}$. The new transform is $\hat{F}_k = C_k F_k$, $k = 0 : (N-1)$. Consider $A = 1$, or find the value of A from the condition

$$F_{0,0} = F_0 = \sum_{t=0}^{N-1} f_{p,s,t} = \sum_{t=0}^{N-1} \hat{f}_{p,s,t} = \hat{F}_0 = C_0 F_0,$$

and therefore $C_0 = 1$ and $A = |F_k|^{1-\alpha}$.

Step 4: Fill the 2D DFT with new 1D DFTs at frequency points of sets $T_{p,s}$. Only sets with generators $(p, s) \in J_{N,N}$ are considered.

Step 5: Perform the inverse 2D DFT.

Rather than process all splitting-signals by the 1D α-rooting method with a fixed parameter α, the splitting-signals can be processed separately by different values of α to achieve an optimal enhancement. The optimality is with respect to the enhancement measure $EME_F(f)$. Thus, step 3 may be changed in the algorithm by using different or optimal $\alpha = \alpha(p, s) \in (0, 1]$ for each splitting-signal. The preliminary results show that there is not much need to process all splitting-signals; only a few can be processed for image enhancement. For instance, consider splitting-signals with high energy. The energy carried by splitting-signal $f_{T_{p,s}}$ is defined as

$$E(p, s) = \sum_{t=0}^{N-1} f_{p,s,t}^2 = \frac{1}{N} \sum_{k=0}^{N-1} |F_{\overline{kp},\overline{ks}}|^2. \tag{2.5}$$

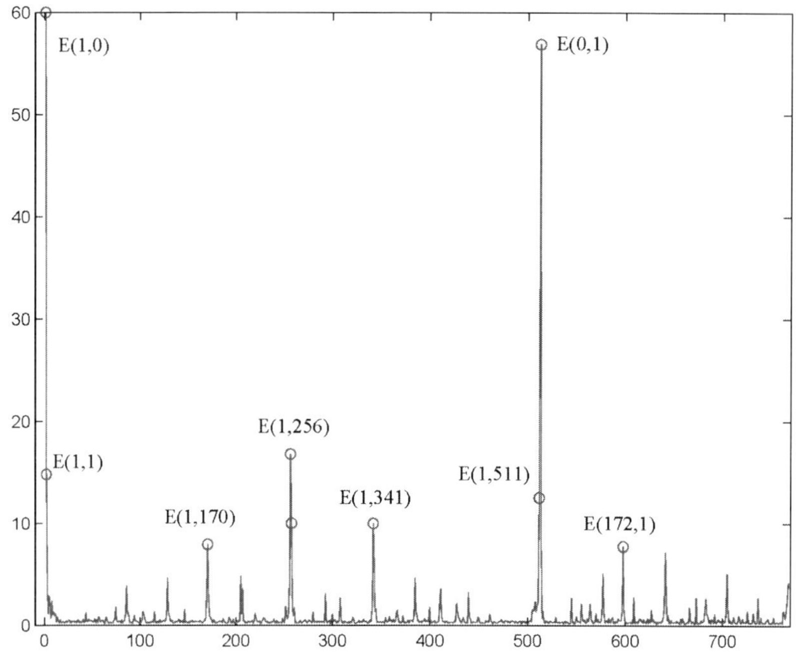

Figure 2.14 The energy curve of the splitting-signals of the image of Fig. 2.3(a). (The image was centered.)

As an example, Fig. 2.14 shows the graph of the energy $E(p,s)$ for all generators (p,s) of the groups $T_{p,s}$ in the order given in the construction of the covering $\sigma = \{\{T_{1,s};\ s = 0:511\}, \{T_{2p,1};\ p = 0:255\}\}$. The splitting-signal with the maximum energy 76.09 is $f_{T_{1,0}}$. The next two signals of high energy are $f_{T_{0,1}}$ and $f_{T_{1,256}}$.

Figure 2.15(a) shows the graph of the enhancement measure $EME(n;\alpha_0)$, $n = 0:767$, calculated after processing only one—the nth splitting-signal—for $\alpha_0 = 0.97$. The maximum of $EME(n;\alpha_0)$ equals 19.78 and is achieved for $n = 361$, which corresponds to the splitting-signal $f_{T_{1,361}}$. This splitting-signal is shown in Fig. 2.15(b); coefficients $C_k, k = 0:511$ appear in (c), and the image enhanced by this splitting-signal is shown in (d).

One can see that by processing only one splitting-signal, significant enhancement of the image can be achieved. The function $EME(361;\alpha)$ has a spike at the point $\alpha = 0.97$. Value 0.97 is considered to be optimal for this splitting-signal. Note that the traditional α-rooting method yields the optimal value of α to be 0.92, and image enhancement equals 22.3. In general, one may process $m \geq 1$ selected splitting-signals and achieve enhancement close to or even greater than this estimate. Moreover, for each of the splitting-signals, the best value of α can be used. Thus, for an image of size $2^r \times 2^r$, approximately $2^{r-1}(2^r - m)$ operations of multiplication can be saved when compared with the traditional Fourier-transform method of image

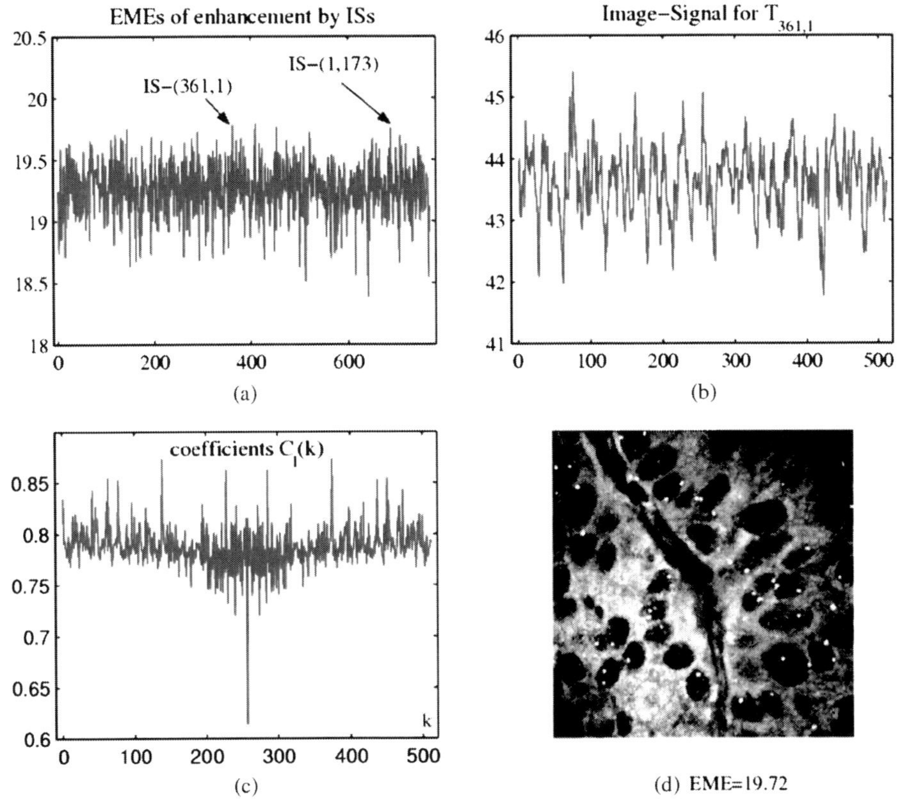

Figure 2.15 (a) The enhancement measure function $EME(n; \alpha_0)$ calculated for $\alpha_0 = 0.97$, when $n = 0 : 767$. (b) Splitting-signal $f_{T_{1,361}}$. (c) Coefficients $C_1(k), k = 0 : 511$ of the 1D α_0-rooting enhancement. (d) Image enhanced by the splitting-signal.

enhancement. As shown in Arslan and Grigoryan,[77] the tensor method of image enhancement via splitting-signals is more effective than the wavelet transform methods of image enhancement, including the methods of wavelet thresholding[62] and wavelet scaling.[63]

2.5.1 Effective formula for image enhancement

When processing only one or a few splitting-signals, the image enhancement can be performed in the following way. If a splitting-signal, for instance, the signal with generator $(p, s) = (1, 1)$:

$$\{f_{1,1,t}; t = 0 : (N-1)\} \rightarrow \{\hat{f}_{1,1,t}; t = 0 : (N-1)\},$$

is processed with the condition

$$\sum_{t=0}^{N-1} \hat{f}_{1,1,t} = \sum_{t=0}^{N-1} f_{1,1,t} = N^2 E[f],$$

then the new image, denoted by $\hat{f}_{n,m}$, can be calculated as follows:

$$\hat{f}_{n,m} = \frac{1}{N}\left[\sum_{p=0(p\neq 1)}^{N-1} f_{p,1,(np+m) \bmod N} + \hat{f}_{1,1,(n+m) \bmod N} + f_{1,0,n}\right] - NE[f]$$

$$= \frac{1}{N}\left[\sum_{p=0}^{N-1} f_{p,1,(np+m) \bmod N} + f_{1,0,n}\right] - NE[f] + \frac{1}{N}\left[\hat{f}_{1,1,(n+m) \bmod N} - f_{1,1,(n+m) \bmod N}\right]$$

$$= f_{n,m} + \frac{1}{N}\left[\hat{f}_{1,1,(n+m) \bmod N} - f_{1,1,(n+m) \bmod N}\right], \quad n,m = 0:(N-1).$$

Thus, the direction image that is calculated from the difference of the splitting-signals $\Delta f_{1,1,t} = \hat{f}_{1,1,t} - f_{1,1,t}, t = 0:(N-1)$ can be used to calculate the processed image:

$$\hat{f}_{n,m} = f_{n,m} + \frac{1}{N}\Delta f_{1,1,(n+m) \bmod N}, \quad n,m = 0:(N-1).$$

A similar inverse formula can be used for image enhancement via one splitting-signal generated by another frequency point $(p,s) \in J_{N,N}$. For instance, the enhancement of the image in Fig. 2.15 is due to the α-rooting of the splitting-signal that is generated by frequency point (1,361). The image in Fig. 2.15(d) can be calculated by

$$\hat{f}_{n,m} = f_{n,m} + \frac{1}{512}\Delta f_{1,361,(n+m) \bmod 512}, \quad n,m = 0:511.$$

Thus, the enhancement is achieved by changing one image component directed by the angle arctan(361).

Now analyze the general equation for the α-rooting image enhancement when processing the splitting-signal generated by (p,s):

$$\hat{f}_{n,m} = f_{n,m} + \frac{1}{N}\Delta f_{p,s,np+ms \bmod N},$$

where $\Delta f_{p,s,t} = \hat{f}_{p,s,t} - f_{p,s,t}$, $t = 0:(N-1)$. The directional image $\Delta_{n,m} = \Delta f_{p,s,np+ms \bmod N}$ has N different values Δ_t, $t = 0:(N-1)$, which can be calculated in the spatial domain as the circular convolution of the splitting-signal. Indeed, the following calculations hold:

$$\Delta_t = \frac{1}{N}\sum_{k=0}^{N-1}(\hat{F}_k - F_k)W^{-tk} = \frac{1}{N}\sum_{k=0}^{N-1}(C_k F_k - F_k)W^{-tk} \quad \left(C_k = A|F_k|^{\alpha-1}\right)$$

$$= \frac{1}{N}\sum_{k=0}^{N-1} F_k(C_k - 1)W^{-tk} = (f_{T_{p,s}} \otimes c)(t), \quad t = 0:(N-1).$$

Here, the sequence $c(t)$ is calculated by

$$c(t) = c_{p,s,\alpha}(t) = \frac{1}{N}\sum_{k=0}^{N-1}(C_k - 1)W^{-tk} = \frac{1}{N}\sum_{k=0}^{N-1} C_k W^{-tk} - \delta_{t,0}, \quad t = 0:(N-1),$$

where $\delta_{t,0}$ is the delta symbol that equals 1 when $t = 0$, and 0 otherwise. Thus, the set of coefficients $c(t) + \delta_{t,0}$ is the inverse 1D DFT of the coefficients C_k. The enhancement of the image by one splitting-signal represents the circular convolution of the signal with the sequence or impulse response $c(t)$. This impulse response is defined by the splitting-signal itself and is parameterized by the frequency point (p,s) and parameter α of enhancement. In image enhancement, the processing of one splitting-signal can therefore be described as a linear discrete system with impulse response $c(t)$. These calculations hold not only for the α-rooting but also for other Fourier-transform-based methods of processing splitting-signals.

As an example, Fig. 2.16(a) shows the enhanced FISH image with $EME = 20.05$, which is due to the processing of the splitting-signal $\{f_{0,1,t};\ t = 0 : 511\}$. The EME curve of α-rooting for this signal is shown in Fig. 2.16(b); the maximum of EME is at point $\alpha = 0.975$. The coefficients $C_k, k = 0 : 511$ are shown in Fig. 2.16(c), and the impulse response $c_{0,1,0.975}(t),\ t = 0 : 511$ appears in (d).

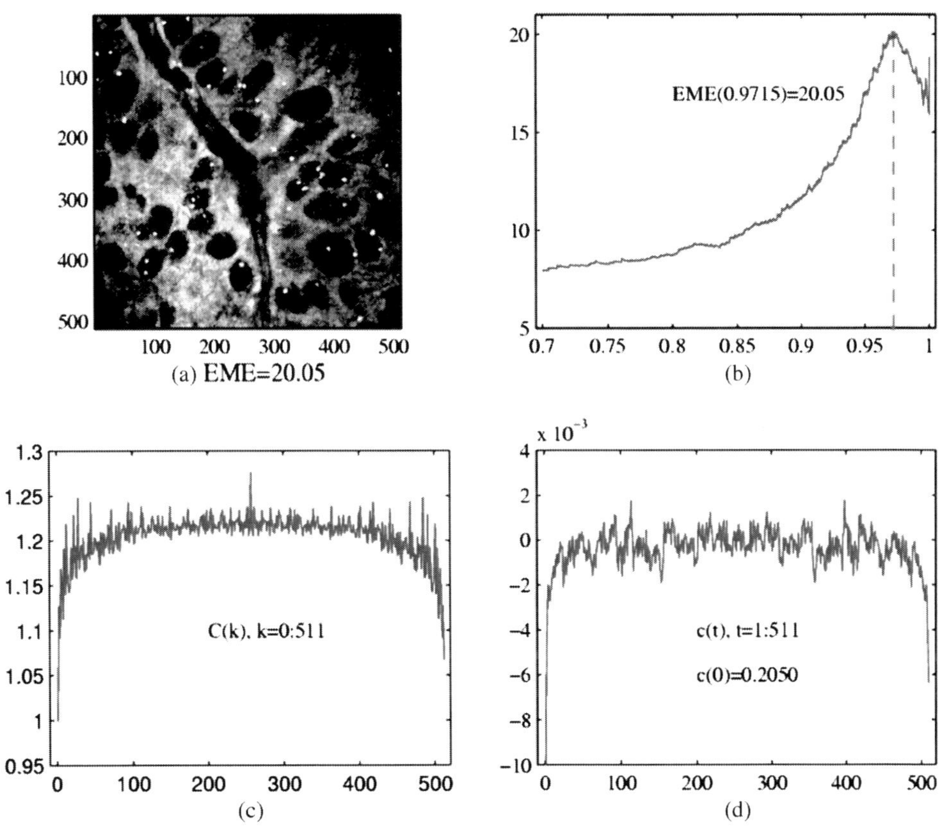

Figure 2.16 (a) Image enhanced by the 0.975-rooting of one splitting-signal generated by $(p, s) = (0, 1)$, (b) EME curve of the 1D α-rooting, (c) coefficients C_k, and (d) impulse response $c_{0,1,0.975}(t)$.

2.5.2 Algorithm of image enhancement by 1D α-rooting

The fast-splitting α-rooting enhancement via one splitting-signal with number (p_0, s_0) presented earlier is described by the following algorithm:

Algorithm I: Image enhancement by one splitting-signal.

Step 1: Calculate the splitting-signal $f_{T_{p_0,s_0}}$.

Step 2: Calculate the N-point DFT of the splitting-signal:

$$f_{T_{p0,s0}} \to F_k = \sum_{t=0}^{N-1} f_{p_0,s_0,t} W^{kt}, \quad k = 0:(N-1).$$

Step 3: Calculate the inverse N-point DFT by

$$\Delta_t = \frac{1}{N}\sum_{k=0}^{N-1} F_k(C_k - 1)W^{-tk}, \quad t = 0:(N-1).$$

Step 4: Calculate the new image by

$$\hat{f}_{n,m} = f_{n,m} + \frac{1}{N}\Delta_{np_0+ms_0 \bmod N}, \quad n,m = 0:(N-1).$$

For the α-rooting method of enhancement, the coefficient $C_k = A|F_k|^{\alpha-1}$, $k = 0:(N-1)$. This algorithm can also be used for other transform-based methods of image enhancement, including enhancement by the log-magnitude and the combined α-rooting methods, and by modified unsharp masking and filtering, all of which are motivated by the human visual response.[17,27,75–77]

2.6 Decomposition by the 2D Paired Transform

Consider the case of most interest, where the size of the image $f = \{f_{n,m}\}$ equals $N \times N(N = 2^r$, and $r > 1$ is integer). The tensor transform $\{f_{p,s,t}; t = 0:(N-1),\ (p,s) \in J_{N,N}\}$ is redundant; the groups $T_{p,s}$ cover the Cartesian lattice and have intersections at many frequency points. Therefore, consider the modification of the tensor transform, which is called the paired transform and is orthogonal.

The unitary 2D discrete paired transformation represents the image as a family of $(3N-2)$ 1D splitting-signals:[21–66–67]

$$\chi'_{N,N}: \{f_{n,m}\} \to \left\{\{f'_{p,s,2^k t}; t = 0:(N/2^{k+1}-1)\}, 2^k = \text{g.c.d.}(p,s), (p,s) \in J'_{N,N}\right\}.$$

Components of splitting-signals are numbered by the triplets $(p, s, 2^k t)$, where (p,s) are the frequency points that generate the signals, and $2^k t$ is the time parameter that runs the interval $[0, N/2-1]$ with the step 2^k, which depends on the frequency (p,s). The discrete paired transformation is thus a unitary transformation from the 2D image domain into the 2D frequency and 1D time domain.

The set of frequency points or generators (p, s) of the paired splitting-signals can be defined as

$$J'_{N,N} = \bigcup_{k=0}^{r-1} \{2^k J_{N/2^k, N/2^k}\} \cup \{(0,0)\},$$

where subsets

$$J_{N/2^k, N/2^k} = \{(p,1); p = 0 : (N/2^k - 1)\} \cup \{(1,2s); s = 0 : (N/2^{k+1} - 1)\}.$$

The elements $f'_{p,s,t}$ of the paired representation of the image $f_{n,m}$ are defined by

$$f'_{p,s,t} = f_{p,s,t} - f_{p,s,t+N/2}, \quad t = 0 : (N/2 - 1). \tag{2.6}$$

Thus, the pair of elements of the tensor representation $f_{p,s,t}$ and $f_{p,s,t+N/2}$ defines $f'_{p,s,t}$. Figure 2.17(a) shows an image, (b) shows the paired splitting-signal $\{f'_{1,2,0}, f'_{1,2,1}, f'_{1,2,2}, \ldots, f'_{1,2,511}\}$ of length 256, and (c) shows the tensor splitting-signal $\{f_{1,2,0}, f_{1,2,1}, f_{1,2,2}, f_{1,2,3}, \ldots, f_{1,2,511}\}$ of length 512. The first $3N/2$ splitting-signals of the paired transform with length $N/2$ each can be obtained directly from the tensor transform, i.e., the tensor splitting-signals $f_{T_{p,s}}, (p,s) \in J_{N,N}$. The next $3N/8$ paired splitting-signals $\{f'_{p,s,t}\}$, $(p,s) \in J_{N/2,N/2}$ are also calculated by (2.10), but $f_{2p,2s,2t}$, $t = 0 : N/2 - 2$ are not components of the tensor transform. In other words, the splitting-signals $f_{T_{2p,2s}}$ are not calculated in this transform, nor are splitting-signals $f_{T_{2^k p}, {}_{2^k s}}$ when $(p,s) \in J_{N/2^k, N/2^k}$, $k = 2, 3, \ldots, (r-1)$. However, it should be noted that the paired splitting-signal, or the set of components $f'_{2p,2s,2t}$ of the paired transform, can be calculated by the tensor transform as follows:

$$f'_{2p,2s,2t} = f_{p,s,t} - f_{p,s,t+N/4} + f_{p,s,t+N/2} - f_{p,s,t+3N/4}.$$

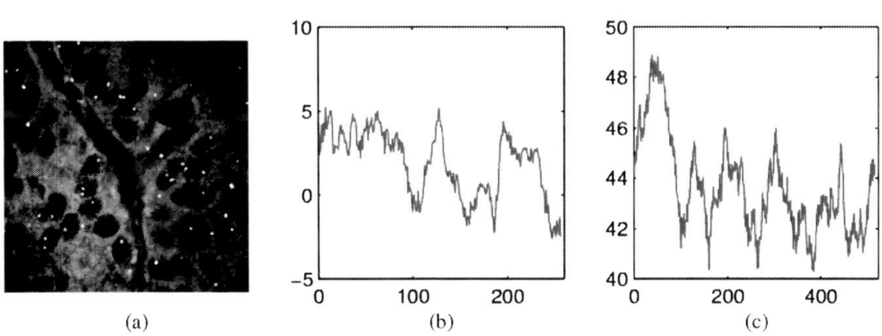

Figure 2.17 (a) FISH image of size 512 × 512, (b) splitting-signal $f_{T'_{1,2}}$, and (c) splitting-signal $f_{T_{1,2}}$.

Similarly, the components of the splitting-signals generated by frequencies $(\overline{4p}, \overline{4s}), (\overline{8p}, \overline{8s}), \ldots$ can be calculated from the same tensor splitting-signal $\{f_{p,s,t}; t = 0 : (N-1)\}$. The calculation of these $(r+1)$ paired splitting-signals, which together compose N components, is reduced to the fast 1D paired transform of the tensor splitting-signal.[71–72–83] Therefore, although the complete family of paired splitting-signals consists of $(3N-2)$ signals, all of these signals can be calculated from $3N/2$ splitting-signals in tensor representation. In other words, the 2D paired transform of the image can be calculated from the 2D tensor transform.

2.6.1 Fourier transform splitting theorem

The 2D paired representation of the image as a set of splitting-signals of different lengths, with a total of N^2 components, allows discovery of the structure of the 2D DFT of the image. The 2D DFT is calculated by $(3N-2)$ 1D DFTs without a redundancy. All subsets $T'_{p,s}$, where $(p,s) \in J'_{N,N}$, do not intersect.

Given a frequency point (p,s), the following holds for the $N \times N$-point 2D DFT:[66–67–72]

$$F_{\overline{(2m+1)p},\overline{(2m+1)s}} = \sum_{t=0}^{N/2^{k+1}-1} \left(f'_{p,s,2^k t} W^t_{N/2^k}\right) W^{mt}_{N/2^{k+1}}, \quad m = 0 : (N/2^{k+1} - 1), \tag{2.7}$$

where 2^k = g.c.d. (p,s). The 2D DFT at frequency points of the subset

$$T'_{p,s} = \left\{\overline{(2m+1p)}, \overline{(2m+1)s}; m = 0 : (N/2^{k+1} - 1)\right\}$$

is defined by the $N/2^{k+1}$-point DFT of the splitting-signal $f_{T'_{p,s}}$ modified by the vector of twiddle coefficients

$$\left\{f'_{p,s,2^k t}; t = 0 : (N/2^{k+1} - 1)\right\} \to \left\{f'_{p,s,2^k t} W^t_{N/2^k}; t = 0 : (N/2^{k+1} - 1)\right\}.$$

In paired representation, the 2D DFT splits into a set of $(3N-2)$ 1D DFTs of different orders: $N/2, N/4, N/8, \ldots, 4, 2$, and 1. As an example, Fig. 2.18(a) shows a discrete image of size 512×512, and Fig. 2.18(b) shows the magnitude of the 2D DFT of the image. In the paired representation, this image is described by $3(512)/2 = 768$ splitting-signals, as shown in Fig. 2.19(a), and the 2D DFT of the image is calculated by 1D DFTs of these splitting-signals modified by the twiddle coefficients, as shown in Fig. 2.19(b). The splitting-signals of the same length are united and separated from others according to the order of the paired transform components, which is given in the definition of the set $J'_{N,N}$. In Fig. 2.19(a), the first part of size 768×256 represents the set of paired-splitting signals of size 256 each. This set of splitting-signals are generated by $(p,s) \in J_{512,512}$. The second part of size 384×128 represents the set of paired-splitting signals of size 128 each, which are

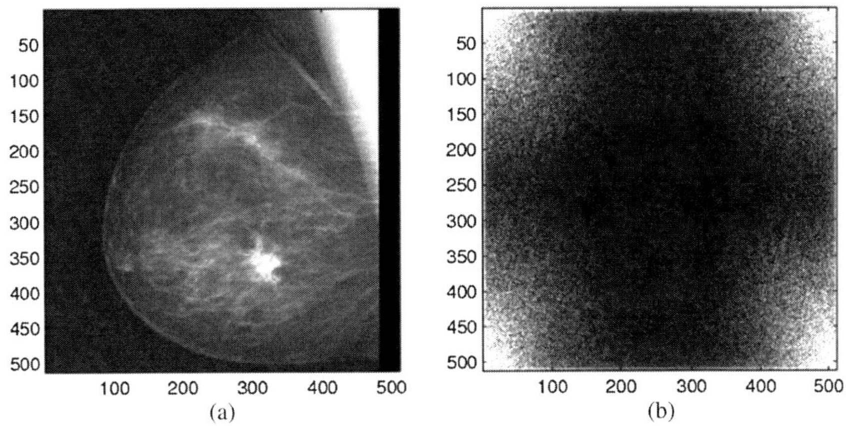

Figure 2.18 (a) Image of size 512 $\times$ 512 and (b) the 2D DFT of the image in absolute scale.

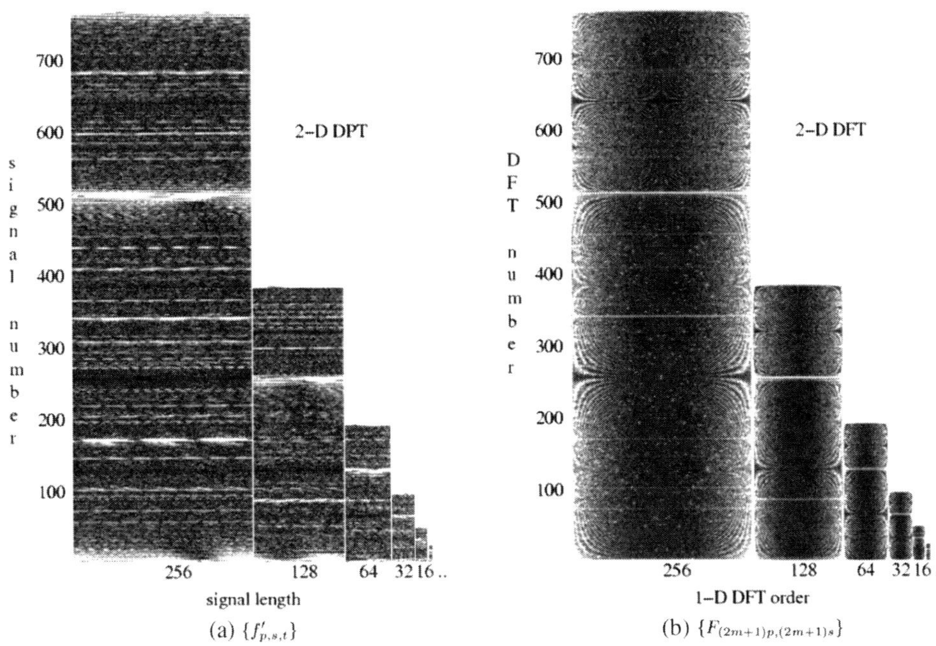

Figure 2.19 (a) The set of paired splitting-signals of lengths 256, 128, 64, 32, 16, 8, 4, 2, 1, and (b) the set of 1D DFTs of the modified splitting-signals (the transforms are shown in the absolute scale).

generated by $(p, s) \in 2J_{256,256}$, and so on. The set of 1D DFTs of modified splitting-signals is shown in a similar way.

Both sets of paired splitting-signals and the corresponding 1D DFTs can be written in form of $N \times N$ matrices. Figure 2.20 shows the same sets of the splitting-signals and 1D DFTs, which are written into the matrices of size 512 $\times$ 512 each. All DFTs were shifted to the centers.

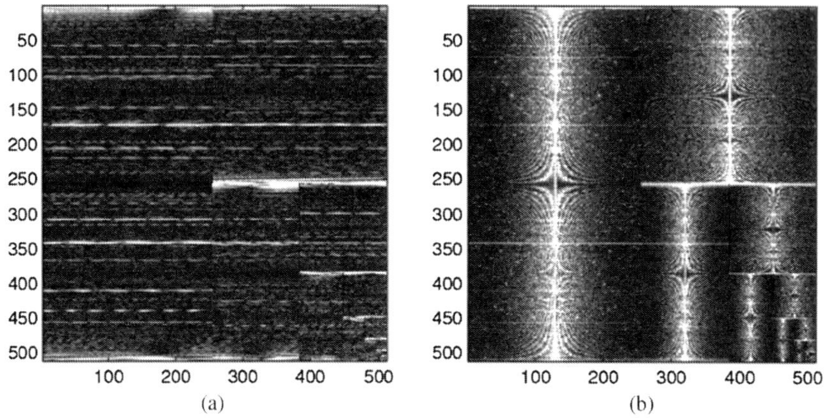

Figure 2.20 (a) 2D paired transform of the image of size 512 $\times$ 512 and (b) the set of 1D DFTs of the modified splitting-signals.

2.6.2 Complete set of the 2D paired transform

The paired transform does not require multiplications. The components of the transform are calculated by

$$f'_{p,s,2^k t} = \chi'_{p,s,2^k t} \circ f = \sum_{n=0}^{N-1}\sum_{m=0}^{N-1} \chi'_{p,s,2^k t}(n,m)\, f_{n,m},$$

where the basis paired functions are defined as

$$\chi'_{p,s,2^k t}(n,m) = \begin{cases} 1, & \text{if } np+ms = 2^k t \bmod N \\ -1, & \text{if } np+ms = (2^k t + N/2) \bmod N \\ 0, & \text{otherwise.} \end{cases} \tag{2.8}$$

Three such functions for the $N = 32$ case are illustrated in Fig. 2.21 in the form of grayscale images of size 32 $\times$ 32 each. The values of 1, -1, and 0

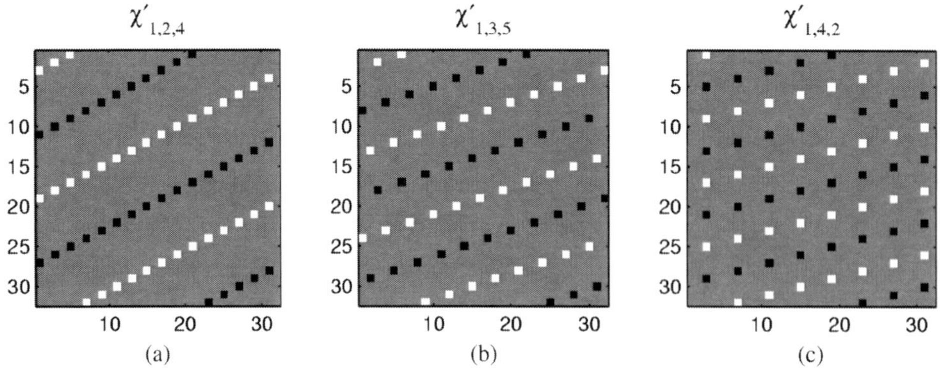

Figure 2.21 The images of 2D paired functions: (a) $\chi'_{1,2,4}(n,m)$, (b) $\chi'_{1,3,5}(n,m)$, and (c) $\chi'_{1,4,2}(n,m)$, where $n,m = 0:31$.

are displayed by intensities of 255 (white), 0 (black), and 128 (gray), respectively. The directions of parallel lines, on which 1 and -1 are located, are defined by the generators as $\varphi(p,s) = \arctan(p/s)$, where $(p,s) = (1,2), (1,3),$ and $(1,4)$.

The complete set of the paired basis functions can be defined by the following set of triplet numbers:

$$U = \{(p,s,2^k t); t = 0 : (N/2^{k+1} - 1), 2^k = \text{g.c.d.}(p,s), (p,s) \in J'_{N,N}\}.$$

All paired basis functions can be calculated from the binary functions of the tensor transform as $\chi'_{p,s,2^k t}(n,m) = \chi_{p,s,2^k t}(n,m) - \chi_{p,s,2^k t+N/2}(n,m)$, where the characteristic function $\chi_{p,s,l}(n,m) = 1$ if $np + ms = l \bmod N$, and 0 otherwise, when $l \in \{0, 1, \ldots, N-1\}$.

2.7 Paired Direction Images

The paired splitting-signals define the corresponding direction image components that together compose the image $\{f_{n,m}\}$. The image is the algebraic sum of direction images; any discrete image can be viewed as the composition of direction images.[78,83,84]

To show this property, consider a generator (p,s) and denote $2^k = \text{g.c.d.}(p,s)$. Let D be the incomplete 2D DFT composed only from the components of the 2D DFT in frequency points of the subset $T'_{p,s}$, i.e.,

$$D = D_{p_1,s_1} = \begin{cases} F_{p_1,s_1}; & \text{if}(p_1,s_1) \in T'_{p,\ s}, \\ 0; & \text{otherwise.} \end{cases} \tag{2.9}$$

The set $J'_{N,N}$ of generators of the paired splitting-signals is divided by groups, or series. Consider the first series of generators, when $\text{g.c.d.}(p,s) = 1$. There are $3N/2$ such generators, and they compose the subset $J_{N,\,N}$. The inverse transform of the above incomplete 2D DFT can be calculated as follows:

$$\begin{aligned} d_{n,m} &= \frac{1}{N^2}\sum_{p_1=0}^{N-1}\sum_{s_1=0}^{N-1} D_{p_1,s_1} W^{-(np_1+ms_1)} = \frac{1}{N^2}\sum_{(p_1,s_1)\in T'_{p,s}} F_{p_1,s_1} W^{-(np_1+ms_1)} \\ &= \frac{1}{N^2}\sum_{k=0}^{N/2-1} F_{\overline{(2k+1)p},\ \overline{(2k+1)s}} W^{-(2k+1)\ \overline{(np+ms)}} \\ &= \frac{1}{2N}\left(\frac{2}{N}\sum_{k=0}^{N/2-1} F_{\overline{(2k+1)p},\ \overline{(2k+1)s}} W_{N/2}^{-kt}\right) W^{-t} \\ &= \frac{1}{2N}(f'_{p,s,t} W^t) W^{-t} = \frac{1}{2N} f'_{p,s,t}, \end{aligned}$$

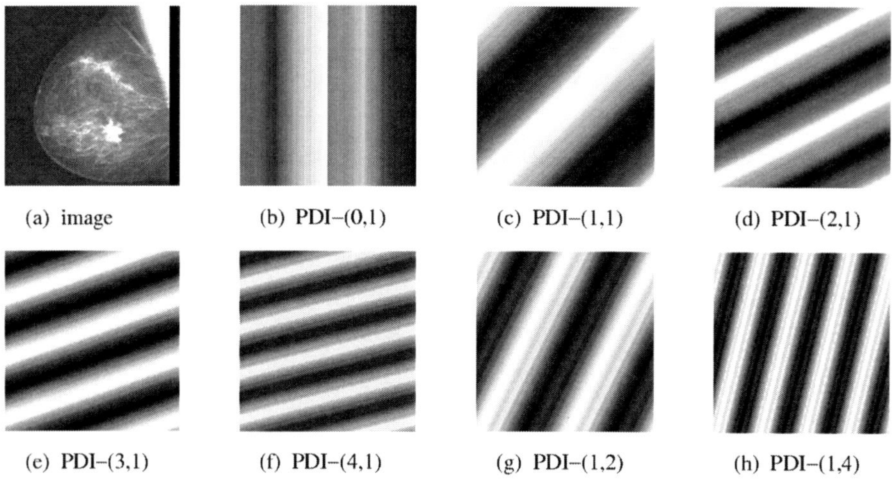

(a) image (b) PDI–(0,1) (c) PDI–(1,1) (d) PDI–(2,1)

(e) PDI–(3,1) (f) PDI–(4,1) (g) PDI–(1,2) (h) PDI–(1,4)

Figure 2.22 (a) Original image and (b)–(h) seven paired direction image (PDI) components of the image. (All images were scaled.)

where $t = (np + ms) \bmod N$. The following direction image of the first series is then obtained:

$$d_{n,m} = d_{n,m;p,s} = \frac{1}{2N} f'_{p,s,(np+ms) \bmod N}, \quad n, m = 0 : (N-1).$$

Now consider that the components $f'_{p,s,t+N/2} = -f'_{p,s,t}$ when $t = 0 : (N/2 - 1)$. In other words, the direction image $N \times N$ is composed of $t = 0 : (N/2 - 1)$ values of the splitting-signal $f_{T'_{p,s}}$, which are placed in the image along the set of parallel lines $np + ms = t \bmod N,\ t = 0 : (N-1)$.

As an example, Fig. 2.22(a) shows an image of size 512 × 512, and Figs. 2.22(b–h) show the seven direction images $d_{n,m;p,s}$ for generators $(p,s) = (0,1), (1,1), (2,1), (3,1), (4,1), (1,2)$, and $(1,4)$, respectively.

Consider the kth series of generators, i.e., when g.c.d.$(p,s) = 2^k k \in \{2, 3, ..., r-1\}$. This set $2^k J'_{N/2^k, N/2^k}$ has $3N/2^{k=1}$ generators. The calculation of the inverse transform of the incomplete 2D DFT results in the following direction image of the kth series:

$$
\begin{aligned}
d_{n,m} &= d_{n,m;p,s} \\
&= \frac{1}{N^2} \sum_{p_1=0}^{N-1} \sum_{s_1=0}^{N-1} D_{p_1,s_1} W^{-(np_1+ms_1)} = \frac{1}{N^2} \sum_{(p_1,s_1)\in T'_{p,s}} F_{p_1,s_1} W^{-(np_1+ms_1)} \\
&= \frac{1}{N^2} \sum_{l=0}^{N/2^{k+1}-1} F_{\overline{(2l+1)p},\ \overline{(2l+1)s}} W^{-(2l+1)\,\overline{(np+ms)}} \\
&= \frac{1}{2^{k+1}N} \left(\frac{2^{k+1}}{N} \sum_{l=0}^{N/2^{k+1}-1} F_{\overline{(2l+1)p},\ \overline{(2l+1)s}} W_{N/2^{k+1}}^{-lt} \right) W_{N/2^k}^{-t} = \frac{1}{2^{k+1}N} \quad f'_{p,s,(np+ms) \bmod N},
\end{aligned}
$$

where $2^k t = (np + ms) \bmod N$. In this case also, the direction image is defined by the corresponding splitting-signal. The last series of generators contains only the frequency $(p, s) = (0, 0)$, and the direction image is the following constant image:

$$d_{n,m} = d_{n,m;0,0} \frac{1}{N^2} F_{0,0} = \frac{1}{N^2} \sum_{n_1=0}^{N-1} \sum_{m_1=0}^{N-1} f_{n_1,m_1} = \frac{1}{N^2} f'_{0,0,0} = E[f].$$

2.7.1 Principle of superposition by direction images

All $(3N - 2)$ subsets $T'_{p,s}$ with generators $(p, s) \in J'_{N,N}$ compose a partition of the Cartesian lattice $N \times N$, which means that the sum of corresponding $(3N - 2)$ incomplete 2D DFTs equals the 2D DFT of the image. The sum of all direction images equals the image $f_{n,m}$.

Principle of Superposition: The discrete image of size $N \times N$, where $N = 2^r$, $r > 1$, can be composed from its $(3N - 2)$ direction images as

$$f_{n,m} = \sum_{(p,s) \in J'_{N,N}} d_{n,m;p,s} = \frac{1}{2N} \sum_{k=0}^{r-1} \frac{1}{2^k} \sum_{(p,s) \in 2^k J_{2^{r-k}, 2^{r-k}}} f'_{p,s,(np+ms) \bmod N} + \frac{1}{N^2} f'_{0,0,0}. \tag{2.10}$$

The splitting-signals of the paired transform can be normalized for each series of generators as

$$f'_{p,s,t} \rightarrow \frac{1}{N 2^{k+1}} f'_{p,s,t}, \quad \text{g.c.d.}(p, s) = 2^k, k \in \{0, 1., \ldots r - 1\},$$

and $f'_{0,0,0} \rightarrow f'_{0,0,0}/N^2$. The above formula of image composition via splitting-signals can then be written as

$$f_{n,m} = \sum_{k=0}^{r-1} \sum_{(p,s) \in 2^k J_{2^{r-k}, 2^{r-k}}} f'_{p,s,(np+ms) \bmod N} + f'_{0,0,0}, \qquad n, m = 0 : (N - 1).$$

This is the formula for the inverse 2D paired transform, i.e., for image composition from its direction images, or paired splitting-signals.[21,78–83] The inverse paired transform requires only operations of addition/subtraction and N^2 divisions by powers of two.

2.7.2 Paired method of image enhancement

Similar to the image enhancement by splitting-signals in tensor representation, the splitting-signals can be processed in paired representation and the image can be enhanced by performing 1D α-rooting enhancement of the signals. The inverse 2D DFT of the modified transform can then be calculated. This is an effective method of image enhancement. However, consider a method that processes the splitting-signals without performing the 2D DFT. The image enhancement has a simple structure: First, the image is represented as a set of

1D signals, or splitting-signals, which is the 2D paired transform of the image. Then, the problem of image enhancement is transformed to processing these signals. In other words, the 2D problem is reduced to processing 1D signals separately. The processed signals are transformed to the 2D DFT and then back to the image by performing the inverse 2D DFT; alternatively, the inverse 2D paired transform can be used with the calculations in accordance with the principle of superposition.

Here, the method of α-rooting is described for processing the splitting-signals, but other methods of signal processing for image enhancement can also be considered. Let (p_0, s_0) be the generator of the splitting-signal that is selected for image enhancement. According to the paired representation, the processing of the splitting-signal

$$f_{T'_{p_0,s_0}} \to \hat{f}_{T'_{p_0,s_0}}$$

with number (p_0, s_0) leads to a change of the 2D DFT of the image at frequency points of the subset T'_{p_0,s_0}. The image is changed to

$$f_{n,m} \to f_{n,m} + \frac{1}{2^{k+1}N}\Delta f'_{p_0,s_0,\,(np_0+ms_0) \bmod N}, \tag{2.11}$$

where

$$\Delta f'_{p_0,s_0,t} = \hat{f}'_{p_0,s_0,t} - f'_{p_0,s_0,t}, \quad t = 0 : (N-1).$$

The processing of the image through its splitting-signals is a process along parallel lines in certain directions. In image enhancement by α-rooting, one or a few paired splitting-signals can be processed by the following scheme:[77]

$$f'_{p_0,s_0,t} \to f'_{p_0,s_0,t}W^t \overset{1-DDFT}{\to} F_m \overset{\alpha-rot}{\to} \{\hat{F}_m = C_m F_m\} \overset{1-DIDFT}{\to} \hat{g}'_t \to \hat{g}'_t W^{-t} = \hat{f}'_{p_0,s_0,t}.$$

Some splitting-signals or direction images are highly expressed, and others are little expressed, by their characteristics, such as energy. The following are the main steps of the paired splitting α-rooting algorithm when processing one paired splitting-signal with number (p_0, s_0) with g.c.d.$(p_0, s_0) = 2k$, $k \geq 0$.

Algorithm II: Image enhancement by one paired splitting-signal.

Step 1: Calculate the splitting-signals $f_{T'_{p_0,s_0}}$.

Step 2: Calculate the 1D DFT F_m of the modified splitting signal:

$$\{g'_{p_0,s_0,2^k t} = f'_{p_0,s_0,2^k t}W^t_{N/2^k};\, t = 0 : N/2^{k+1} - 1\}.$$

Step 3: Calculate coefficients of enhancement $C_m = |F_m|^{\alpha-1}$, $m = 0 : (N/2^{k+1} - 1)$.

Step 4: Change values of the 1D DFT by

$$F_m \to \hat{F}_m = C_m F_m, \quad m = 0 : (N/2^{k+1} - 1).$$

Step 5: Calculate the enhanced splitting-signal $\hat{f}_{T'_{p_0,s_0}}$ by the inverse 1D DFT modified by the twiddle factors $W^{-t}_{N/2^k}$, as follows:

$$\hat{f}'_{p_0,s_0,2^k t} = W^{-t}_{N/2^k} \sum_{m=0}^{N/2^{k+1}-1} \hat{F}_m W^{-mt}_{N/2^{k+1}}, \quad t = 0 : (N/2^{k+1} - 1).$$

Step 6: Calculate the new directional image by

$$\hat{d}_{n,m} = \frac{1}{2^{k+1}N} \hat{f}'_{p_0,s_0,(np_0+ms_0) \bmod N}, \quad n,m = 0 : (N-1).$$

Step 7: Calculate the enhanced image by

$$\begin{aligned}\hat{f}_{n,m} &= f_{n,m} + [\hat{d}_{n,m} - d_{n,m}] \\ &= f_{n,m} + \frac{1}{2^{k+1}N}\left[\hat{f}'_{p_0,s_0,(np_0+ms_0) \bmod N} - f'_{p_0,s_0,(np_0+ms_0) \bmod N}\right].\end{aligned}$$

Figure 2.23(a) shows the splitting-signal $f_{T'_{1,361}}$ of length 256 of the FISH image, (b) shows the 256-point DFT of the modified splitting-signal $g_{T'_{1,361}}$,

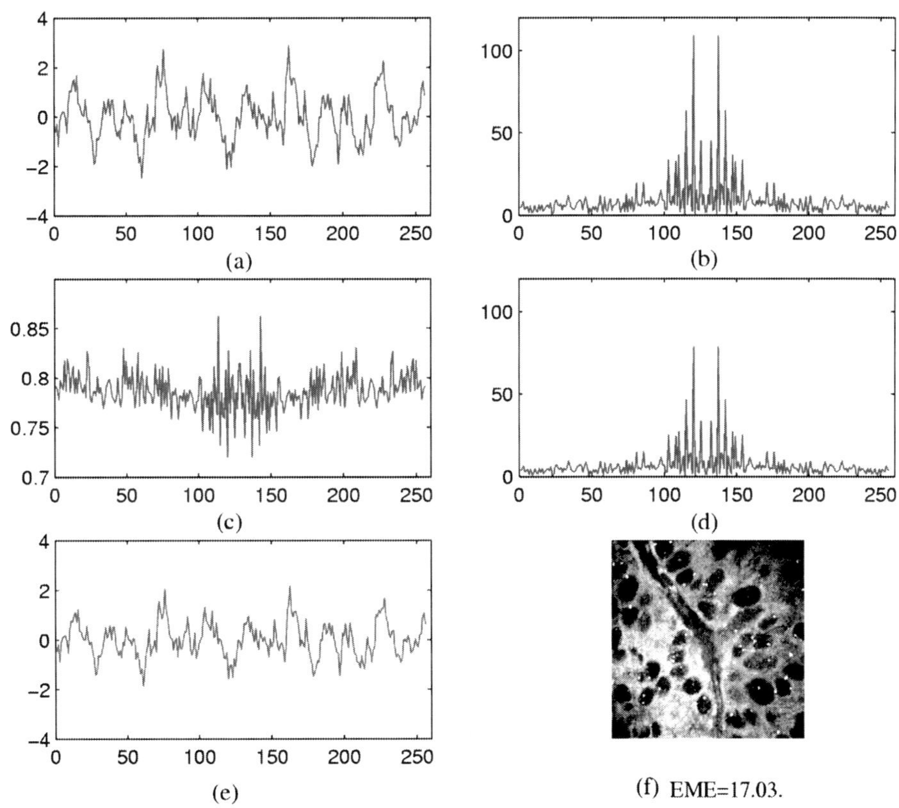

Figure 2.23 (a) The splitting-signal $f_{T'_{1,361}}$. (b) The 256-point DFT of the modified splitting-signal. (c) Coefficients $c_{m,m=0:255}$ of the 1D α-rooting enhancement for $\alpha = 0.97$. (d) The processed 256-point DFT. (e) The splitting-signal after processing by the α-rooting method. (f) The FISH image enhanced by the splitting-signal.

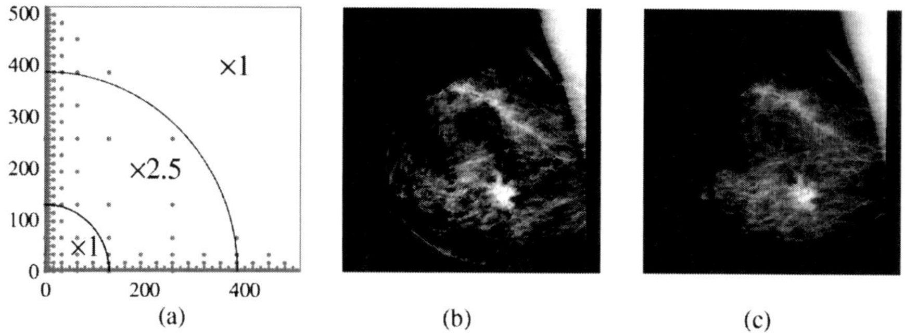

Figure 2.24 (a) All generators for the 512 × 512 case are separated by three zones, (b) the x-ray breast image with amplified splitting-signals of zone 2, and (c) the original image.

(c) shows the coefficients c_m, calculated for $\alpha = 0.97$, and (d) shows the 256-point DFT of the processed signal $g_{T'_{1,361}}$. The splitting-signal $\hat{f}_{T'_{1,361}}$ after processing by the α-rooting method is shown in Fig. 2.23(e), and the image enhanced by the splitting-signal appears in (f). The enhancement of the image equals 17.03.

The splitting-signals can also be processed by dividing the set of all generators by subsets. As an example, Fig. 2.24(a) shows all 1,534 generators $(p, s) \in J'_{256,256}$. This set of frequency points was divided into three zones by circles with radii of 100 and 400. The splitting-signals with generators of the second zone were amplified by a factor of 2.5. The result of processing the splitting-signals is shown in Fig. 2.24(b) together with the original x-ray breast image in (c).

2.8 Enhancement by a Series of Direction Images

Consider the decomposition of the image by its direction images in more detail. The paired splitting-signals have different energies and for image enhancement, the signals with high energies can be selected. Figure 2.25 shows

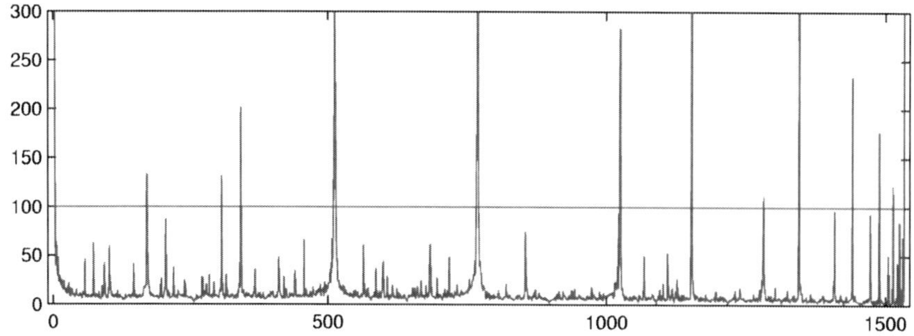

Figure 2.25 Energy graph of paired splitting-signals of the breast x-ray image of size 512 × 512.

a graph of the energies of all splitting-signals for the breast x-ray image. For a selected energy level of 100, there are 28 splitting-signals with energies greater than 100.

The paired transform is a unitary transform, and according to the Parseval's theorem, the energy of all splitting-signals equals the energy of the image, i.e.,

$$E^2[f] = \sum_{n=0}^{N-1}\sum_{m=0}^{N-1} f_{n,m}^2 = \sum_{(p,s)\in J'_{N,N}} E^2[f_{T'_{p,s}}]$$
$$= \sum_{(p,s)\in J'_{N,N}} \frac{1}{2^{k(p,s)}} \sum_{t=0}^{N/2^{k(p,s)+1}-1} [f'_{p,s,t}]^2, \qquad (2.12)$$

where $k = k(p,s) = \text{g.c.d.}(p,s)$, and $k(0,0) = r$.

The image composed of 28 high-energy splitting-signals, or the corresponding direction images, is shown in Fig. 2.26(a), and the image composed of the remaining $1548 - 28 = 1506$ splitting-signals is shown in (b). The sum of these two images equals the original breast x-ray image.

It follows from the definition of the paired representation that from each image, specific periodic structures can be extracted, which together compose the image.[78,81,83] These structures do not have the smooth forms of cosine or sine waves, but the forms that are defined by binary paired basis functions are united by subsets.

To illustrate this property, the sum of direction images corresponding to the subset of generators $2^k J_{N/2^k, N/2^k}$,

$$S_{n,m}^{(k)} = \sum_{(p,s)\in 2^k J_{N/2^k,N/2^k}} d_{n,m;p,s}, \quad k \in \{0,1,\ldots,r-1\},$$

is called the kth *series image*. For the $k = r$ case, such an images is defined by $S_{n,m}^{(r)} = F_{0,0}/N^2$.

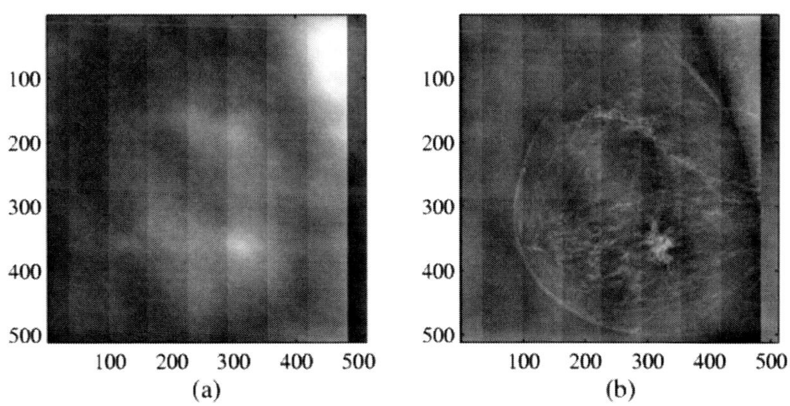

Figure 2.26 (a) Image composed of the splitting-signals with energies greater than 100, and (b) image composed of the remaining splitting-signals.

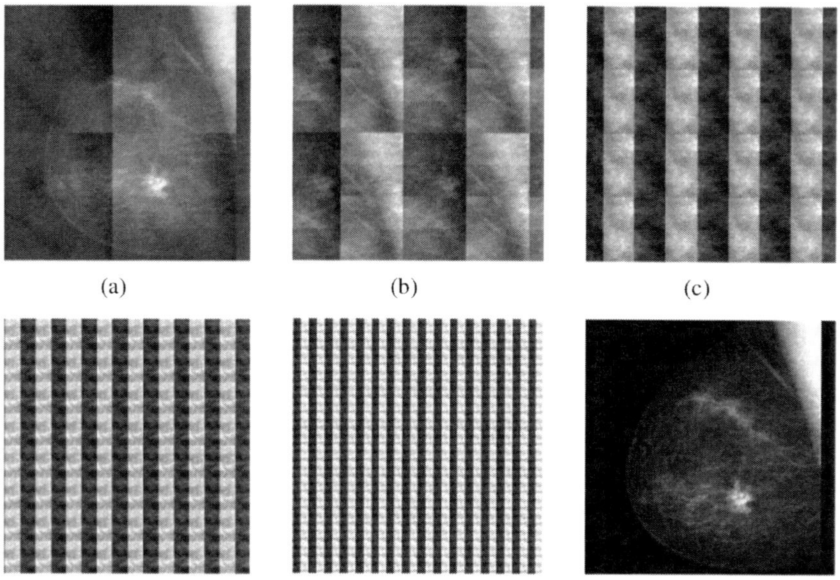

(a) (b) (c)

(d) (e) (f)

Figure 2.27 (a)–(e) The first five series images of the image, and (f) the sum of these series images.

Figure 2.27(a–e) shows the first five series images for the breast x-ray image. The sum of only these series images is shown in Fig. 2.27(f). One can see that each series image, starting from the second one, has a periodic structure with a resolution that increases exponentially with the number of the series. The number 2^k is called the resolution of the kth series image. Each resolution is referred to as a periodic structure of one part of the image. The first series image is the component of the image with the lowest resolution, and the $(r-1)$th series image is the component of the image with the highest resolution. The constant image $S_{n,m}^{(r)}$ has 0 resolution. The sum of all series images equals the original image.

The consequent sum of the three first series images of the breast x-ray image is given in Fig. 2.28; one can see that series images with a resolution of

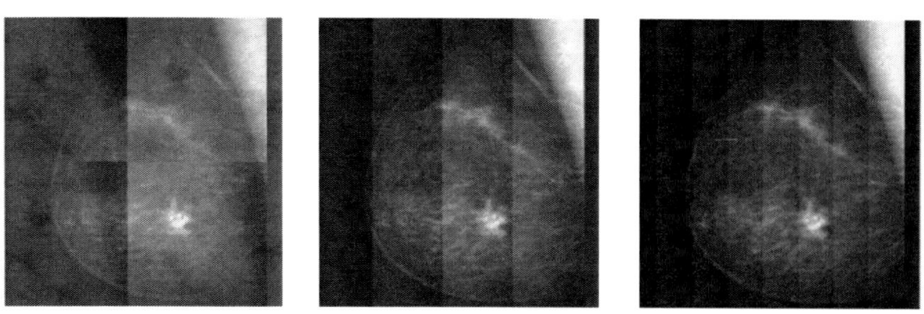

(a) (b) (c)

Figure 2.28 (a) The first series image, (b) the first plus second series images, and (c) the sum of the first three series images.

1, 2, and 4 result in good-quality images. The other four resolutions add more details to the image.

2.9 Compression: Multiresolution Map of the Image

It is important to mention that the first series image is also composed of periodic structures $N/2 \times N/2$. In this image, as well as the rest of the series images, subsets of direction images can be separated in the following way. The set of generators J_{2^r} is divided by three parts as

$$J_{2^r}^{(1)} = \{(1, 2s); s = 0 : (N/2 - 1)\}, \quad J_{2^r}^{(2)} = \{(2p, 1); p = 0 : (N/2 - 1)\},$$

$$J_{2^r}^{(3)} = \{(1, 2s + 1); s = 0 : (N/2 - 1)\}.$$

The division of the first series image S^0 by these subsets is denoted as

$$P_{n,m}^{(0)} = \sum_{(p,s)\in J_{2^r}^{(1)}} d_{n,m;p,s}, \quad N_{n,m}^{0} = \sum_{(p,s)\in J_{2^r}^{(2)}} d_{n,m;p,s}, \quad U_{n,m}^{0} = \sum_{(p,s)\in J_{2^r}^{(3)}} d_{n,m;p,s},$$

so that $S^{(0)} = P^{(0)} + N^{+(0)} + U^{(0)}$.

Figure 2.29(a) shows the image P^0 for the breast x-ray image, and Figs. 2.29(b) and (c) show the images $N^{(0)}$ and $U^{(0)}$ respectively. In these images, one can notice different parts of the original image with their negative versions periodically shifted by $N/2 = 256$ along the horizontal, vertical, and diagonal directions.

Each image is divided by four parts $(N/2 \times N/2)$ with similar structures, which can be used to form the entire series image $S^{(0)}$. Therefore, the series image components $P^{(0)}$, $N^{(0)}$, and $U^{(0)}$ can be defined by their first quarters, which are denoted by P_1, N_1, and U_1 respectively, as follows:

$$P^{(0)} = \begin{bmatrix} P_1 & P_1 \\ -P_1 & -P_1 \end{bmatrix}, \quad N^{(0)} = \begin{bmatrix} N_1 & -N_1 \\ N_1 & -N_1 \end{bmatrix}, \quad U^{(0)} = \begin{bmatrix} U_1 & -U_1 \\ -U_1 & U_1 \end{bmatrix}.$$

Figure 2.30 shows the decomposition of the second series image S^1 for the breast x-ray image.

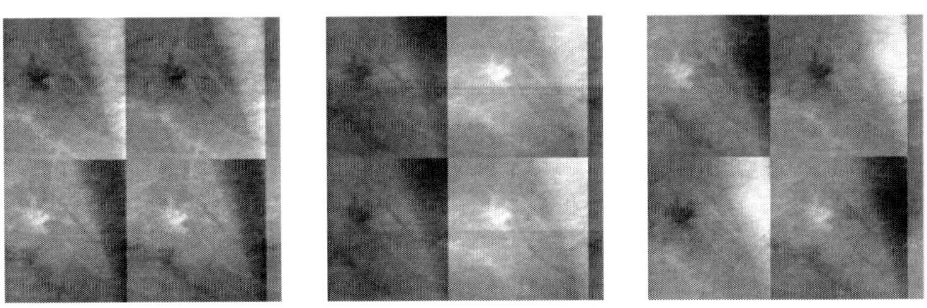

(a) (b) (c)

Figure 2.29 Three components of the first series of the breast x-ray image.

(a) (b) (c)

Figure 2.30 Three components of the second series of the breast x-ray image.

For this series image, as well as the remaining series images $S^{(k)}$, $k = 2 : (r - 1)$ the similar decompositions can be used. Each of these images can be defined by the three quarters P_{k+1}, N_{k+1}, U_{k+1} of their periods $N/2^{k+1} \times N/2^{k+1}$ in a way similar to that for the first series image. As a result, the following resolution map (RM) associates with the image f:

$$RM[f] = \boxed{\begin{array}{cccccc} P_1 & & & U_1 & & \\ & P_2 & & & U_2 & \\ & & P_3 & & & U_3 \\ N_1 & & & & P_4 & & U_4 \\ & N_2 & & & & \\ & & N_3 & N_4 & & \ldots \end{array}} \tag{2.13}$$

The last element in the RM equals the mean of the image, $F_{0,0}/N^2$. This resolution map has the same size as the image and contains all periodic parts of the series images, i.e., all periods by means of which the original image can be reconstructed. Each periodic part is extracted from the direction images, whose directions are given by subsets of generators of $J'_{N,N}$. The RM itself represents the image packed by its periodic structures that correspond to a specific set of directions. The image can be reconstructed exactly from its resolution map; the RM can be used to change the resolution of the entire image by processing selected direction components of the image.[81–93]

As an example, Fig. 2.31(a) shows a FISH image, and its 512×512 resolution map appears in Fig. 2.31(b) For better visualization of the resolution map, the series images have been magnified by factors 1, 2, 4, 8, 16, 32, 64, 128, and 256, respectively. The last coefficient of the RM is the mean of the image; it has been normalized by a factor of $1/(512)^2$.

Figure 2.32(a) shows the grayscale cell image of size 512×512, and the resolution map of the image appears in Fig. 2.32(b).

2.9.1 *A*-series linear transformation

The resolution map can be used to change the resolution of an image by processing direction images along the desired directions or set of directions.

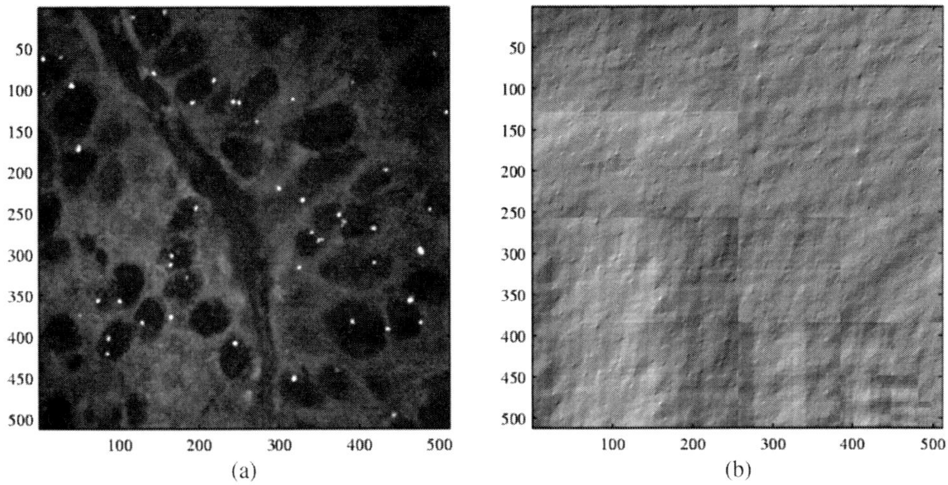

Figure 2.31 (a) FISH image and (b) its resolution map (this image was scaled).

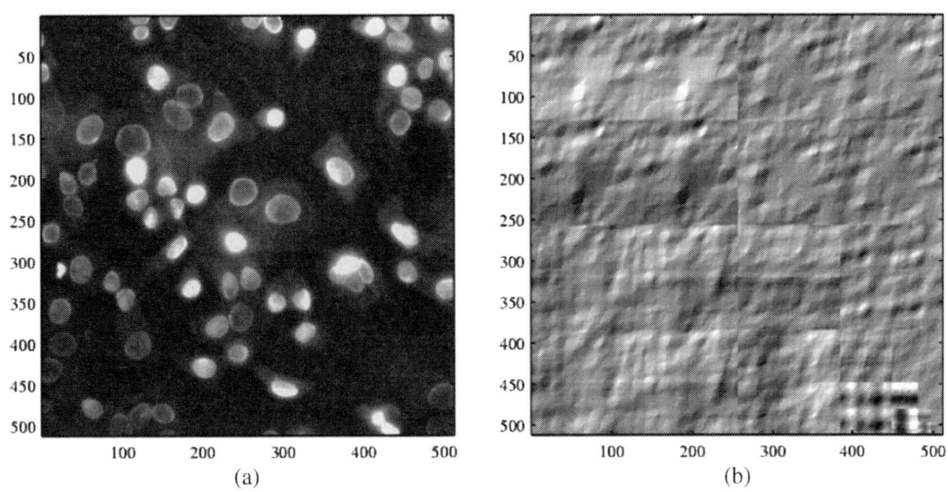

Figure 2.32 (a) The cell image and (b) its resolution map (this image was scaled).

For instance, consider the following simple method of image enhancement.[78,83] Let A be a set of r parameters, $A = \{a_0, a_1, ..., a_{r-1}\}$ that are considered to be the weighed coefficients for the series images. The image $f_{n,m}$ enhanced by the set A is defined as

$$\hat{f}_{n,m} = \sum_{k=0}^{r} a_k S_{n,m}^{(k)}, \quad n, m = 0 : (N-1). \tag{2.14}$$

In the case $a_k = 1$, $k = 0 : (r-1)$ and the image $\hat{f}_{n,m}$ equals $f_{n,m}$. The operation $A : f_{n,m} \rightarrow \hat{f}_{n,m}$ is called the *A-series linear transformation*. The selection of a coefficient a_k greater than 1 means that the resolution 2^k of the image increases, and in the $a_k < 1$ case, the resolution decreases. The $a_k = 0$ case means that the corresponding resolution will be removed from the image.

As an example, Fig. 2.33(a) shows the x-ray breast image of size 512×512 and Fig. 2.33(b) shows the image when the high resolution (series number 8) was removed.

The quality of this image can be improved by manipulating the resolution of the series images in the desired way. For example, for the FISH image, Fig. 2.34(a) shows the result of the A-series linear transformation when the set of parameters A equals $\{1.5, 2, 1.5, 1, 1, 1, 1.5, 1.5, 0\}$, and Figs. 2.34(b) and (c) show the results for $\{1, 2, 1, 1, 1, 2, 4, 2, 1\}$ and $\{1, 3, 2, 1, 1, 1, 1, 1.5, 1\}$, respectively. In (a), the resolutions of 1, 2, 4, 64, and 128 of the series images have been increased, but the high resolution was removed; many small spots were removed from the FISH image. In (b) and (c), the resolutions 2, 32, 64, 128 and 2, 4, 128, respectively, have been increased.

Figure 2.35(a) shows the result of the A-series linear transformation on the breast x-ray image when the set of parameters A equals $\{1, 1, 1, 1, 1, 1, 1, 0.75, 2\}$. The similar result with set of coefficients $\{1, 1, 1, 1, 1, 0.25, 0.25, 0.5, 1\}$ is shown in Fig. 2.35(b) and its negative image appears in (c).

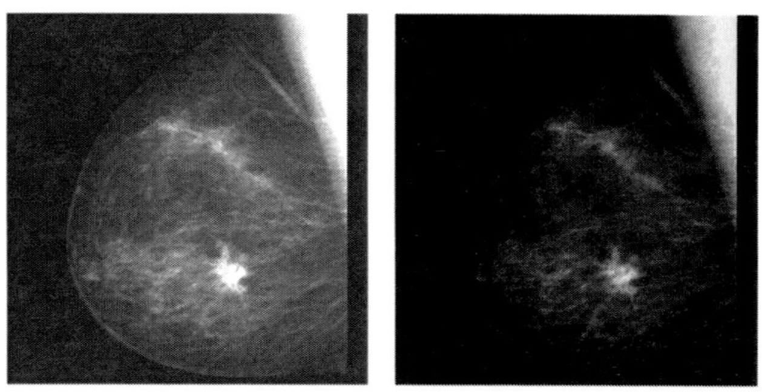

(a) (b)

Figure 2.33 (a) The image and (b) the image without the eighth series.

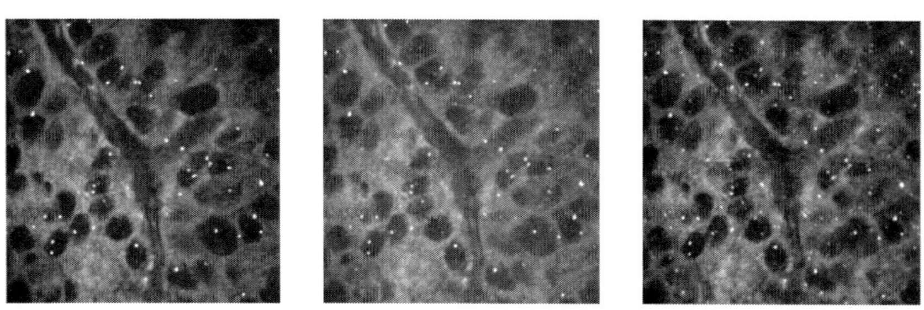

(a) (b) (c)

Figure 2.34 *A*-series linear transformation of the FISH image by the sets of parameters (a) $\{1.5, 2, 1.5, 1, 1, 1, 1.5, 1.5, 0\}$, (b) $\{1, 2, 1, 1, 1, 2, 4, 2, 1\}$, and (c) $\{1, 3, 2, 1, 1, 1, 1, 1.5, 1\}$.

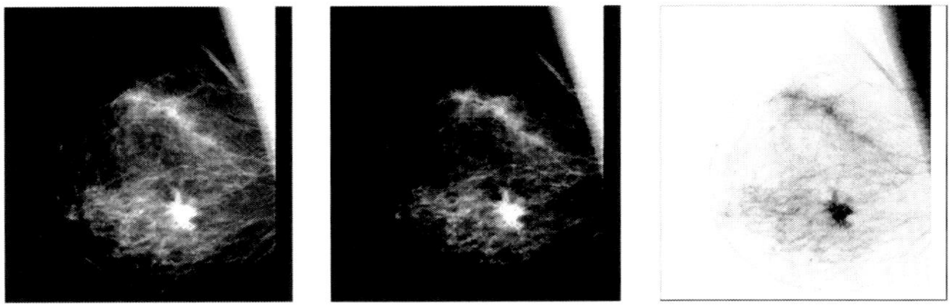

(a) (b) (c)

Figure 2.35 *A*-series linear transformation of the FISH image by the sets of parameters (a) {1, 1, 1, 1, 1, 1, 1, 0.75, 2}, (b) {1, 1, 1, 1, 1, 0.25, 0.25, 0.5, 1}, and (c) the negative image.

The paired form of image representation leads to the splitting of the 2D DFT by the set of 1D DFTs of splitting-signals that define the direction images as components of the image. This representation allows for extracting the periodic structures of the image components, which are defined by direction images united in special subsets of directions, referred to as series images. The image can be packed and described by its resolution map, which can be used for image enhancement. It is clear that each periodic structure in the resolution map can also be represented by its resolution map. In such a recursive way, the resolution map can be crushed into small pieces from which the whole image can be reconstructed. Thus, such a set of resolution maps can be used for image compression.

2.10 Compression by the Tensor Transform

This section briefly describes new methods of image compression via the tensor transform, or splitting-signals. Consider the case where the size of the image is $N \times N$ and N is a prime. It is important to say that the 2D $N \times N$ -point discrete tensor transform is not redundant, and it is composed of $(N + 1)$ splitting-signals that are all independent. Each of these splitting-signals contributes as a basis image component in the specific direction. These direction components cannot therefore be substituted or approximated from other components, even when the directions are at angles with a very small difference range. As an example, Fig. 2.36(a–c) shows three splitting-signals generated by the frequency points $(p, s) = (150, 1), (151, 1)$, and $(152, 1)$, respectively. These signals define the image components along the directions at angles of 89.6180 deg, 89.6206 deg, and 89.6231 deg within the difference range of 0.0051 deg.

Figure 2.37(a–c) shows another three splitting-signals of the same image that are generated by the frequency points $(p, s) = (220, 1), (221, 1)$, and $(222, 1)$, respectively. These signals define the image components along the directions at angles of 89.7396 deg, 89.7407 deg, and 89.7419 deg. The difference range of 0.0023 deg is approximately two times smaller than 0.0051 deg, and again it appears that these splitting-signals are not correlated.

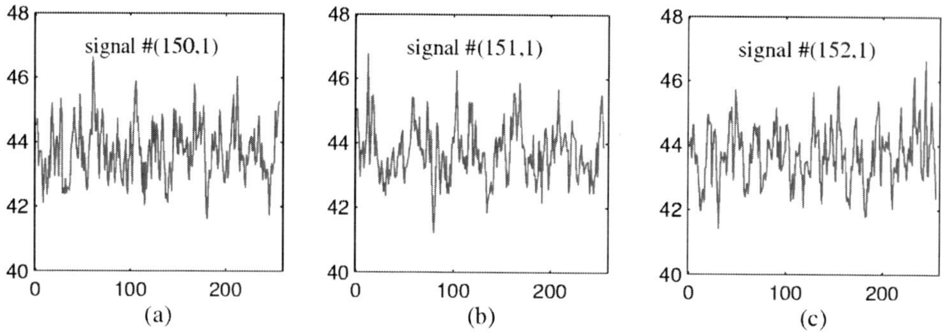

Figure 2.36 The splitting-signals of FISH image that define the direction images at angles of (a) 89.6180 deg, (b) 89.6206 deg, and (c) 89.6231 deg.

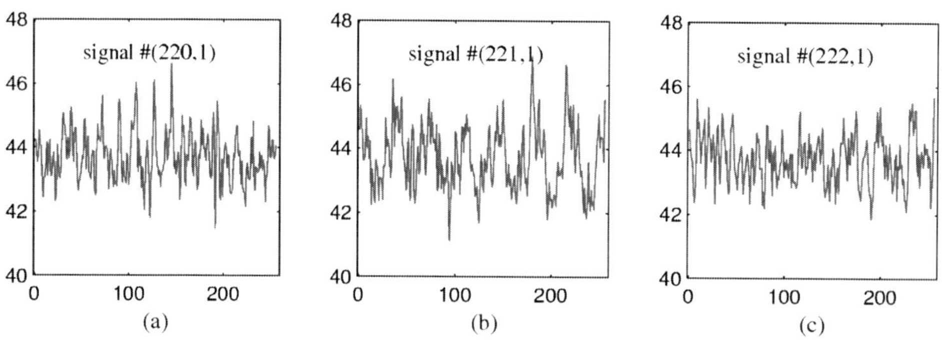

Figure 2.37 The splitting-signals of FISH image that define the direction images at angles of (a) 89.6180 deg, (b) 89.6206 deg, and (c) 89.6231 deg.

The difference of these signals even in such a small difference range of angles can also be seen in the corresponding direction images, which are given in Fig. 2.38. Each direction image has its own structure with a pattern that is periodic, unique, and cannot be seen in other direction images.

Remember that these direction images are on the discrete Cartesian lattice, not in the continuous image plane. If such images are obtained from

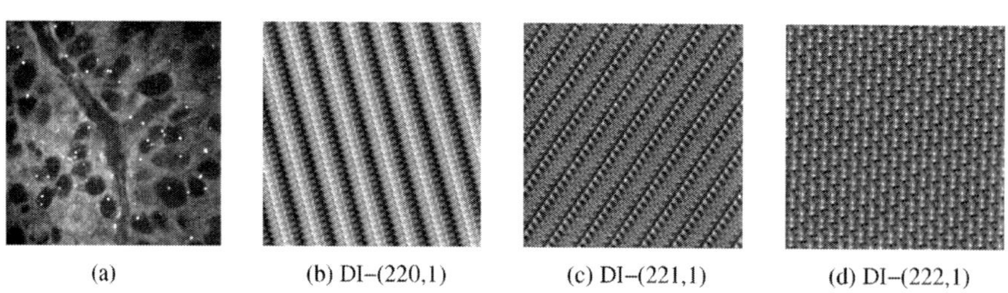

Figure 2.38 (a) FISH image and the direction images calculated from the splitting-signals of FISH image, which are generated by the frequency points (b) (220,1), (c) (221,1), and (d) (222,1). (The direction images were scaled.)

(a) (b) DI–(222,1) (c) DI–(223,1) (d) DI–(224,1)

Figure 2.39 (a) X-ray image and direction images calculated from the splitting-signals generated by the frequency points (b) (220,1), (c) (221,1), and (d) (222,1). (The direction images were scaled.)

the projection data, as a result of transferring geometry from the line integrals to the ray sums,[84] the images will be unique and different. The line integrals may look similar but not the ray sums. For example, the discrete "tomo" image as the result of 257×257 reconstruction in computed tomography is shown in Fig. 2.39(a), along with the three direction images that correspond to the splitting-signals defined by very close angles of 89.7396°, 89.7407°, and 89.7419°. These three direction images have very different periodic textures.

The corresponding splitting-signals of the "tomo" image are shown in Fig. 2.40; the signals were normalized by a factor of $1/N$.

Because each splitting-signal or its direction image is unique and independent of other splitting-signals, the splitting-signals can be processed and encoded for purposes of lossy or lossless image compression. Thus, the problem of compression of the 2D image can be reduced to the compression of the 1D signals, as shown in the block diagram in Fig. 2.41. All splitting-signals are integer, and different encoding methods can be used to process these signals separately in order to achieve an effective image compression. The inverse process of decoding the block tensor transform followed by the inverse 2D tensor transform is also shown in this diagram.

The sample-to-sample difference of components of the splitting-signals may vary more than the actual values. Therefore, to encode the splitting-signals,

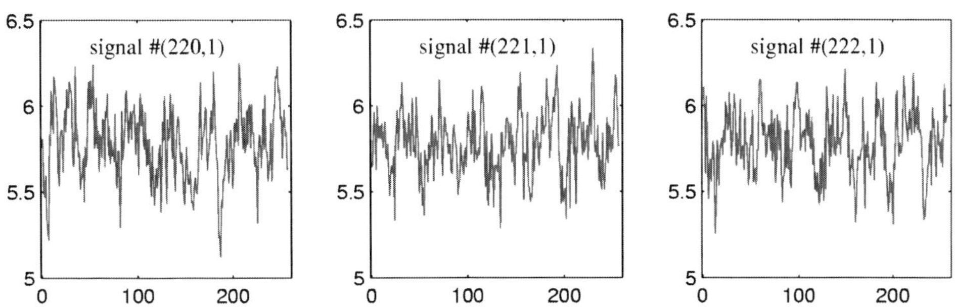

(a) (b) (c)

Figure 2.40 The splitting-signals for the direction images at angles of (a) 89.7396 deg, (b) 89.7407 deg, and (c) 89.7419 deg.

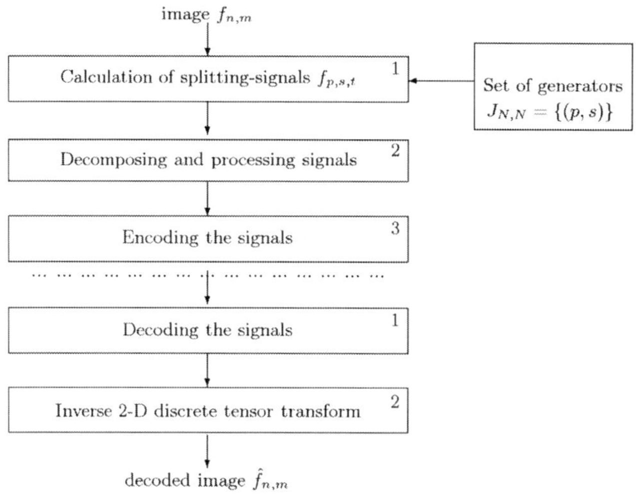

Figure 2.41 Block diagram of image compression by encoding the tensor splitting-signals.

there is the well-known approach of decomposing the splitting-signals into different frequency bands and then applying the best coding methods for these signals. For instance, the components $f_{p,s,t}$, $t = 0 : (N - 1)$ can be decomposed into the average and difference sequences:

$$x_t = \left[f_{p,s,t} + f_{p,s,t+1}\right]/2, \quad \text{and} \quad y_t = \left[f_{p,s,t} + f_{p,s,t+1}\right]/2, \quad t = 0 : ((N-1)/2 - 1).$$

The differential encoding can be used for the sequence x_t, and each value of the sequence y_t can be quantized separately.

In tensor representation, all splitting-signals have the same length N. When N is not prime, the 2D paired transform can be used for image compression in a similar way. For instance, the paired transform is effective for $N \times N$ images, when is a power of 2. In this case, the image is represented by the set of $(3N - 2)$ splitting-signals of different lengths, and the maximal length of signals is $N/2$. As shown previously, the paired splitting-signals can be separated by series, each of which is defined by a different frequency band. Therefore, the coding technique best suited to each splitting-signal can be used to improve the compression performance.

2.10.1 Block-tensor-transform lossy image compression

Consider the traditional approach for lossy image compression, where the image is divided by blocks, and the 2D transform, such as the discrete cosine transform, is calculated for each block; the coefficients of the transform are then quantized and encoded, as in the JPEG standard.[85] For that, there is the case where the size of the image is $N \times N$ and $N > 1$ is a prime.

The image is processed block-wise with calculation of the 2D tensor transform of each block, quantization, and encoding of the transform coefficients. The block diagram of such lossy image compression is given in Fig 2.42. The $N \times N$ image is divided by blocks $f_{[k_1,k_2]}$of size $L \times L$ each,

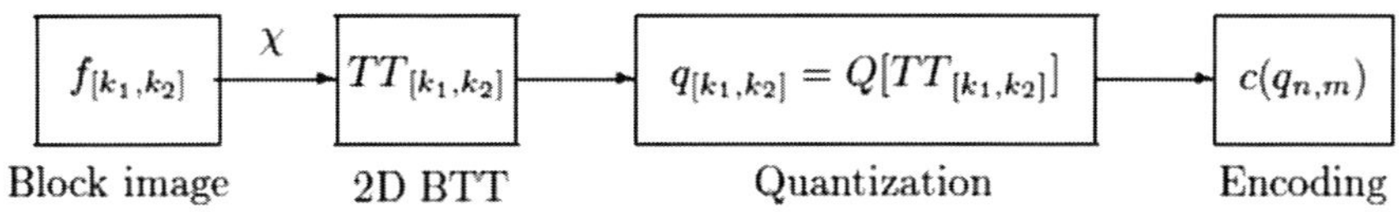

Figure 2.42 Block diagram of the block-tensor-transform-based method of lossy image compression.

where it is common to consider L equal 8 or 16. Here, $[k_1, k_2]$ denotes the block number, and $k_1, k_2 = 1 : \lceil N/L \rceil$ where $\lceil . \rceil$ is the ceil function. For each block, the tensor transform $TT_{[k_1,k_2]}$ is calculated and then quantized by a quantized Q with the subsequent encoding $c(q_{n,m})$, $n, m = 0 : (L - 1)$ of the quantized coefficients $q_{[k_1,k_2]}$ of the tensor transform. During the encoding stage, the Huffman and Art codes can be used.[86,87]

To show the main part of the tensor transform-based lossy image compression, consider the FISH image shown in Fig. 2.43(a) and its block tensor transform (BTT) in (b). The blocks of size 5×5 were used. For simpler processing, the third value of all splitting-signals in the tensor transform of each block was deleted, i.e., 6 coefficients from 25 were removed from the transforms. The result of the inverse BTT is shown in Fig. 2.43(c); the obtained image is of high quality.

For comparison with the known method of the DCT, consider one block in the image, for instance, the block with number [12,12], and its 5×5-point tensor transform,

$$f_{[12,12]} = \begin{bmatrix} 14 & 14 & 0 & 6 & 12 \\ 9 & 9 & 3 & 12 & 3 \\ 0 & 14 & 17 & 23 & 17 \\ 17 & 3 & 3 & 32 & 17 \\ 3 & 14 & 9 & 12 & 17 \end{bmatrix}, \quad TT_{[12,12]} = \begin{bmatrix} 43 & 54 & 32 & 85 & 66 \\ 46 & 36 & 71 & 72 & \underline{55} \\ 57 & 81 & 38 & 57 & \underline{47} \\ 49 & 63 & 46 & 70 & \underline{52} \\ 66 & 52 & 61 & 63 & \underline{38} \\ 89 & 50 & 21 & 60 & \underline{60} \end{bmatrix}.$$

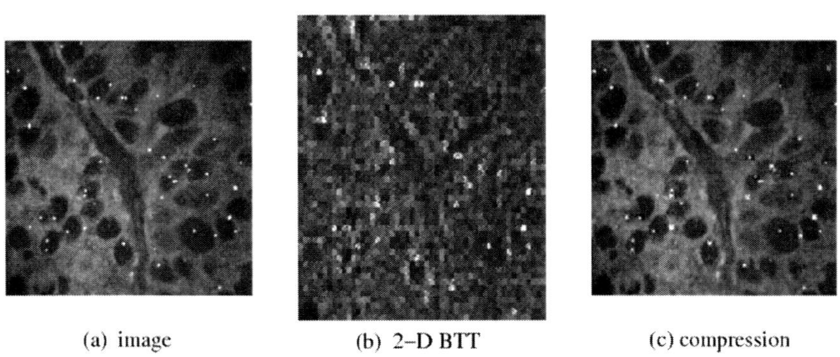

Figure 2.43 (a) The image, (b) the 5×5 block tensor transform of the image, and (c) the compressed image on the first stage (reduction is 24%).

Five underlined coefficients of the tensor transform in the last column can be omitted because the sums of components of all splitting-signals (shown in rows) are equal. The tensor transform is an integer-to-integer transform, and it does not have large DC coefficients, as in the discrete cosine transform. Namely, the concept of DC coefficients does not have a meaning for the tensor transform. In the considered example, the DCT of the block (without subtracting 128 from the values of the block image) is equal to the following matrix with DC coefficient of 56:

$$DCT(f_{[12,12]}) = \begin{bmatrix} 56.0000 & -11.3408 & 3.7419 & 4.5152 & -13.2287 \\ -8.4060 & 12.0652 & 2.8083 & -3.5777 & 0.1927 \\ -6.4101 & 8.4117 & 8.6833 & -8.9848 & 1.6430 \\ 8.1877 & -3.5777 & 8.8183 & 8.9348 & -2.2613 \\ 4.1965 & -5.0272 & -8.6430 & -13.6104 & 3.3167 \end{bmatrix}.$$

The tensor transforms of other blocks can be analyzed and compared with the DCT, as well; the results are similar. The tensor transforms are integer-to-integer transforms, and their coefficients do not fall in a large dynamic range, as shown below for two neighboring blocks with numbers [15,21] and [15,22]:

$$TT_{[15,21]} = \begin{bmatrix} 85 & 82 & 53 & 41 & 62 \\ 95 & 82 & 44 & 51 & \underline{51} \\ 59 & 77 & 52 & 70 & \underline{65} \\ 73 & 56 & 73 & 61 & \underline{60} \\ 65 & 51 & 78 & 59 & \underline{70} \\ 61 & 47 & 76 & 72 & \underline{67} \end{bmatrix}, \quad TT_{[15,22]} = \begin{bmatrix} 71 & 50 & 36 & 70 & 48 \\ 78 & 80 & 51 & 15 & \underline{51} \\ 62 & 53 & 45 & 68 & \underline{47} \\ 59 & 59 & 64 & 60 & \underline{33} \\ 55 & 39 & 54 & 77 & \underline{50} \\ 50 & 48 & 65 & 59 & \underline{53} \end{bmatrix}.$$

Another advantage of using the tensor transform in block-wise image compression is the simplicity of the transform. This transform has the fast algorithm for prime numbers N, and many such prime numbers can be used when selecting the block size. For instance, the main sizes in lossy image compression by the DCT are 8×8 and 16×16. Figures 2.44(a–c) show the block tensor transforms of the FISH image when the block sizes are $7 \times 7, 11 \times 11$, and 17×17, respectively.

With the tensor transform, there are many prime numbers (such as 3, 5, 7, 11, 13, and 19) that can be used for different image sizes to cover the entire image without extending the last columns or rows with zeros if the entire image is not divided exactly by blocks. For instance, for the 257×257 image, the division of $N = 257$ can be written as $257 = 36 \times 7 + 5:$ the image can be divided by $(36)2 + 1$ blocks of size 7×7 each and 100 blocks of size 5×5 each.

The following interesting property of the block tensor transform should be mentioned. For many images, such as the cell image shown in Fig. 2.45(a), the tensor transform can be considered the operation of filtration when working with small-size blocks. The images in Fig. 2.45(b) and (c) illustrate this fact;

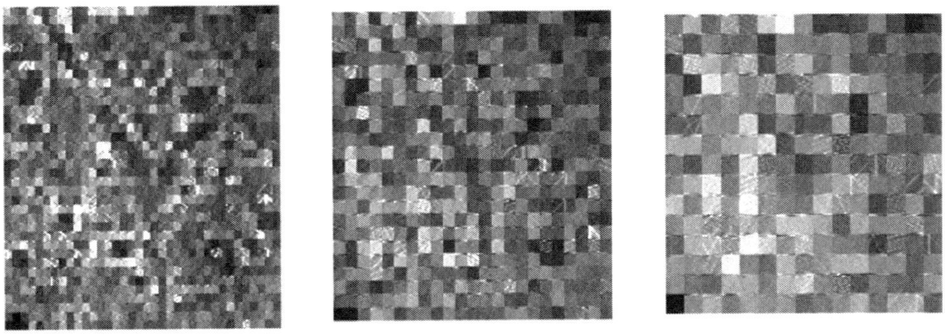

(a) 7x7 2–D BTT (b) 11x11 2–D BTT (c) 17x17 2–D BTT

Figure 2.44 The block TT of the FISH image, when block size is (a) 7 × 7, (b) 11 × 11, and (c) 17 × 17.

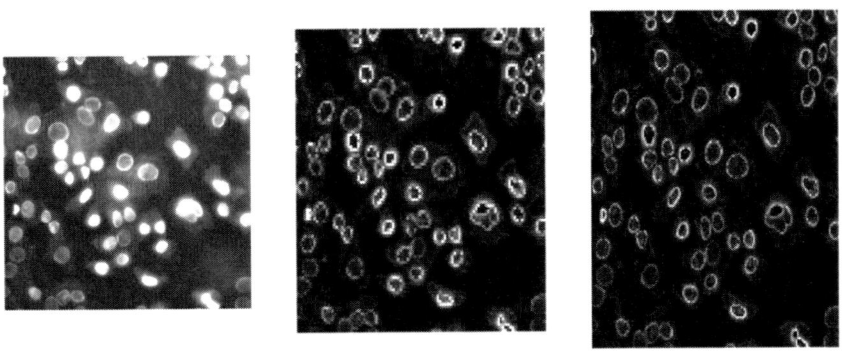

(a) image (b) 5x5 2–D BTT (c) 3x3 2–D BTT

Figure 2.45 (a) The image, and the block TTs of the image, when block size is (b) 5 × 5 and (c) 3 × 3.

the 5 × 5- and 3 × 3-block TT are shown in these images, respectively. One can see the contours of the cells, and these are the images of the block tensor transform, i.e., the images in the 2D frequency and 1D time domain, not the 2D spatial domain.

The histograms of the same images are given in Fig. 2.46. From the histograms of the block tensor transform, it is easy to select the threshold for filtering the images and obtain a high-quality image of contours of all cells. In other words, the image can be visually analyzed in the 2D frequency and 1D time domain.

2.11 Tensor Transform in Image Cryptography

The tensor transform can also be applied for fast and very effective image encryption and decryption without loss of information. The tensor-transform-based image cryptography allows for the effective encryption of images of any

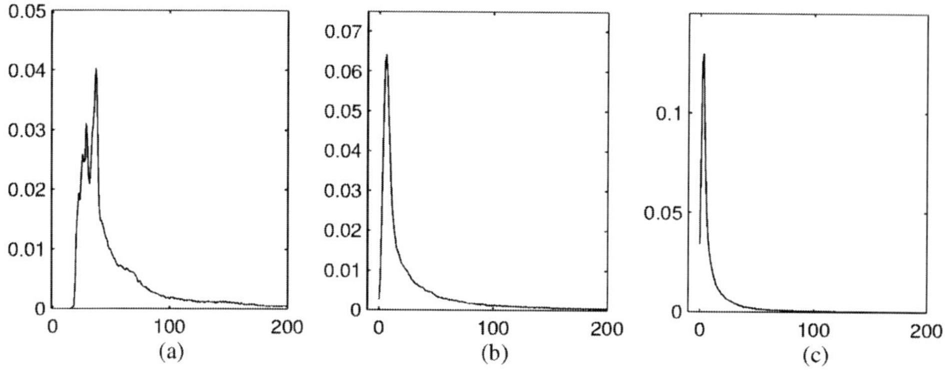

Figure 2.46 The histograms of (a) the original image, and the block TTs of the image, when block size equals (b) 5 × 5 and (c) 3 × 3.

size by distributing the cryptographic key between the sender and recipient, who should keep the key secret. The encryption of the image and its decryption are fast: For instance, the original version of the code written in MATLAB (in other words, not optimized M-code), the CPU working time is 0.0781 s for encrypting a grayscale image of size 256 × 256, and 0.2031 s for an image of size 512 × 512. The decryption uses almost the same amount of time as the encryption.

Consider the method of *mapping the 2D image on the 3D torus*, which was developed for image encryption as well as any other document.[88] For simplicity, let us examine the $N \times N$ case, where $N = 2^r$ and $r > 1$. In the tensor representation of the image, which is in the 3D space, two dimensions of this space are for frequency and one dimension is for time. Let $\{f_{n,m}\}$ be the discrete image defined on the square Cartesian lattice $X_{N,N} = \{(n,m);\ n,m = 0:(N-1)\}$. In tensor representation, the image is described by a set of $3N/2$ splitting-signals of length N each: $\{f_{p,s,0}, f_{p,s,1}, f_{p,s,2}, \ldots, f_{p,s,N-1}\}$ Frequency generators (p,s) are pairs of coprime numbers of types $(p,1)$ and $(1,2s)$, where $p = 0:(N-1)$ and $s = 0:(N/2-1)$. Each component $f_{p,s,t}$ of the splitting-signal is the sum of the image at points of the set $V_{p,s,t}, t = 0:(N-1)$. Thus,

$$f_{p,s,t} = \sum\{f_{n,m}; (n,m) \in V_{p,s,t}\}, \quad t = 0:(N-1).$$

The sets $V_{p,s,t}$ contain N points on the lattice each, and the lattice $X_{N,N}$ can be reordered in such a way that the summation of each component of the splitting-signal will be performed only along the rows (or columns).

In the 3D space, one can identify the opposite sides of boundaries of the square domain $X = [0,N] \times [0,N]$ of the real space R^2 and can consider this square X as a torus and the lattice $X_{N,N}$ as a net traced on the torus. Such a torus and net is called the discrete torus of size $N \times N$. Given a triplet

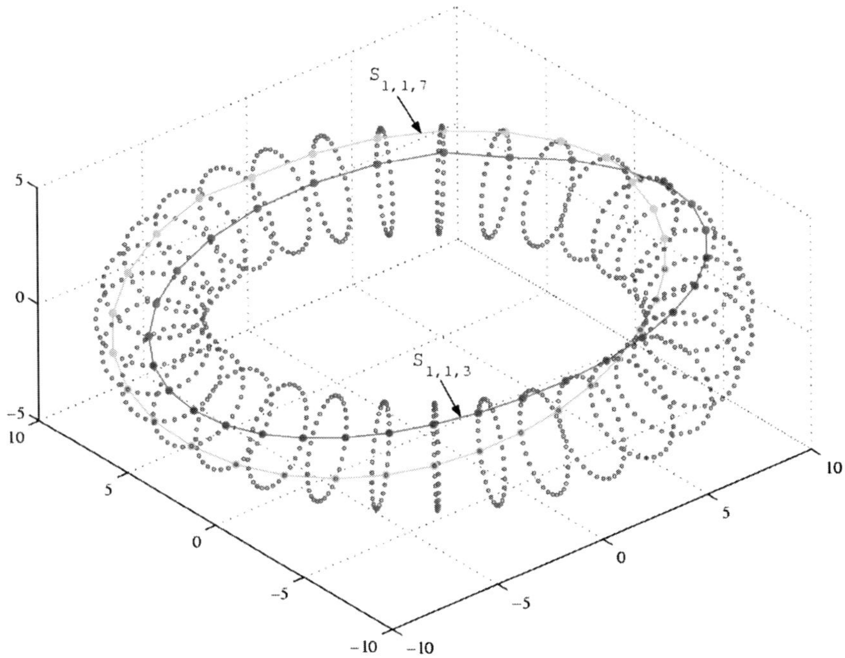

Figure 2.47 The net with knots of the grid of size 32 × 32 in the 3D space with the locus of two spirals $s_{1,1,3}$ and $s_{1,1,7}$.

(p, s, t), the straight lines along which points of the set $V_{p,s,t}$ are located $np + ms = t + kN$, $k = 0 : (p + s - 1)$ compose one discrete spiral $S_{p,s,t}$ on the torus.[21] As an example, for $(p, s) = (1, 1)$, Fig. 2.47 shows the locus of two spirals $s_{1,1,3}$ and $s_{1,1,7}$ on the discrete torus of size 32 × 32.

Spirals have different forms depending on the value of the frequency point (p, s). Fig. 2.48(a) shows the locus of the spiral $S_{2,1,1}$ on the discrete torus of size 128 × 128. The spiral passes through 128 points of the discrete torus. The net of size 128 × 128 on the torus is not shown in the figure. 127 similar, or "parallel," and disjoint spirals $s_{2,1,t}$, $t = 0, 2 : 127$ cover the net of size 128 × 128 on the torus. Other spirals $s_{4,1,2}$, $s_{1,3,1}$, and $s_{1,5,1}$ on the discrete torus of size 128 × 128 are shown in Figs. 2.48(b-d), respectively.

Given the generator $(p, s) \in J_{N,N}$, all values of the image can be reordered and transferred into another matrix $N \times N$. The image seen along these spirals is a permutation of the original image and can be referred to as the *redirected image* or the *image redirected along N spirals on the torus*, which are defined by the frequency points (p, s). In the 2D plane, this image is said to be redirected cyclically along the direction defined by the angle $\varphi(p, s) = \tan(s/p)$ or $\pi - \tan(s/p)$. Such a permutation can be performed with different blocks of the image in a successive way, with different block sizes and in a different order to achieve the "best" encryption.

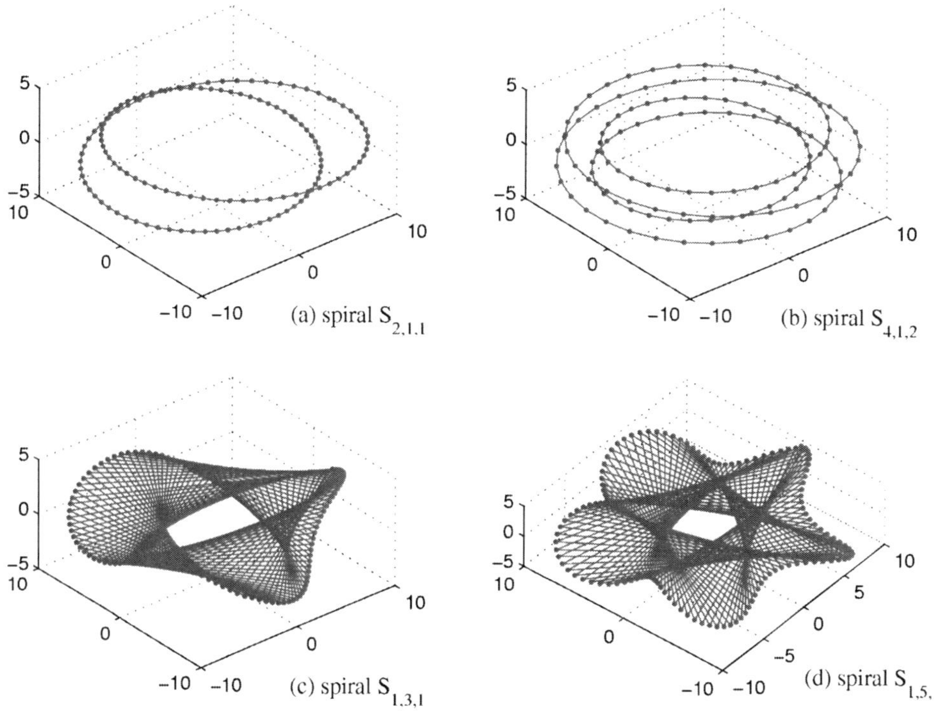

Figure 2.48 Spirals (a) $s_{2,1,1}$, (b) $s_{4,1,2}$, (c) $s_{1,3,1}$, and (d) $s_{1,5,1}$.

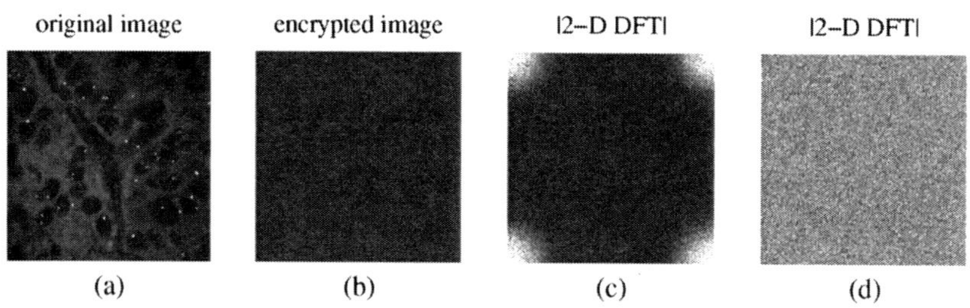

Figure 2.49 (a) FISH image, (b) encrypted image, (c) 2D DFT of the image, and (d) 2D DFT of the encrypted image.

As an example, Fig. 2.49(a) shows the original FISH image, 2.49(b) shows the encrypted image, 2.49(c) shows the 2D DFT of the image in absolute scale, and (d) shows the 2D DFT of the encrypted image.

Likewise, Fig. 2.50(a) shows the x-ray breast image, 2.50(b) shows the encrypted image, and 2.50(c) shows the 2D DFT (in absolute scale) of the encrypted image.

One can see from both examples that the 2D discrete Fourier transforms of the encrypted images have chaotic structures similar to the encrypted image. Correlation coefficients between the pairs of horizontally, vertically,

Figure 2.50 (a) The x-ray breast image, (b) encrypted image, and (c) 2D DFT of the encrypted image.

Table 2.1 Correlation coefficients of two adjacent pixels (among 32620 random pixels).

Correlation	Horizontal Adjacency	Vertical Adjacency	Diagonal Adjacency
Original image	0.94542	0.96855	0.93231
Encrypted image	0.00088	0.00572	0.00019

and diagonal adjacent pixels were calculated in the encrypted images and compared with the original images. Experimental results show that the correlation coefficients are very small for encrypted images. As an example, Table 2.1 shows the correlation coefficients of a pair of horizontally, vertically, and diagonal adjacent pixels of the image of the original tree image and the encrypted image.

Thus, the tensor transform allows for effective encryption of the image. The 2D paired transform can be also used for image encryption with many other methods of image encryption.[89–98]

2.12 Conclusion

The three main preprocessing tools for processing medical images of cancer were considered in this chapter: image enhancement, compression, and encryption. The preprocessing is accomplished from the unique point of view where the 2D image is represented (described) in the form of the 1D splitting-signals that simultaneously carry the spectral information of the image and define the direction components of the image. Such a representation of the image is described in terms of the tensor and paired transforms. This chapter discussed in detail the applications of the tensor and paired forms of image representation for image enhancement by one or a few splitting-signals, via sets of such signals with generators separated by zones and series. The splitting-signals can be processed separately and transferred back to the image. The 1D α-rooting algorithms for image enhancement have been described in the frequency domain and implemented in the spatial domain without calculating the 2D discrete Fourier transform. Emphasis has been placed on the effective implementation of the α-rooting method, but other transform-based methods of image enhancement can be implemented by processing splitting-signals, as well. The methods described previously can be improved and used with other 2D unitary transforms, including the Hadamard and cosine transforms. The tensor transform can be also used for image compression and image encryption; effective methods of such image processing were presented, and preliminary experimental results show that the new forms of the image can be used effectively for preprocessing medical images at different stages of imaging.

References

1. J. Harris, "Constant variance enhancement: a digital processing technique," *Appl. Optics* **16**, 1268–1271 (1977).
2. D. Wang and A. H. Vagnucci, "Digital image enhancement," *Computer Vision, Graphics, Image Processing* **24**, 363–381 (1981).
3. A. Rosenfeld and A. C. Kak, *Digital Picture Processing,* Vol. 1, Academic Press, New York (1982).
4. D. F. Elliot, and K. R. Rao, *Fast Transforms: Algorithms, Analyzes, Applications*, New York, Academic Press, Orlando (1983).
5. W. M. Morrow, R. B. Paranjape, R. M. Rangayyan, and J. E. L. Desautels, "Region-based contrast enhancement of mammograms," *IEEE Trans. Med. Imaging* **11**(3), 392–406 (1992).
6. T. L. Ji, M. K. Sundareshan, and H. Roehrig, "Adaptive image contrast enhancement based on human visual properties," *IEEE Trans. Med. Imaging* **13**(4), 573–586 (1994).
7. P. Zamperoni, "Image enhancement," *Advances in Image and Electron Physics* **92**, 1–77 (1995).
8. P. Suetens, *Fundamentals of Medical Imaging*, Cambridge University Press, Cambridge, UK (2002).
9. B. Chen and R. Ning, "Cone-beam volume CT breast imaging: feasibility study," *Med. Phys.* **29**, 755–770 (2002).
10. A. M. Grigoryan, E. R. Dougherty, J. Kononen, L. Bubendorf, G. Hostetter, and O. P. Kallioniemi, "Morphological spot counting from stacked images for automated analysis of gene copy numbers by fluorescence *in situ* hybridization," *Biomed. Optics* **7**(1), 109–122 (2002).
11. C. L. Andersen, G. H. Hostetter, A. M. Grigoryan, G. Sauter, and A. Kallioniemi, "Improved procedure for fluorescence in situ hybridization on tissue microarrays," *Cytometry* **45**, 83–86 (2001).

12. A. M. Grigoryan and E. R. Dougherty, "Morphological slice-based counting of illuminated 3D bodies," *Proc. SPIE* **4304**, 18–24 (2001) [doi: 10.1117/12.424988].
13. A. M. Grigoryan, E. R. Dougherty, L. Bubendorf, J. Kononen, and O. Kallioniemi, "Rapid analysis of gene copy numbers by FISH and automatic spot counting from stacked confocal images," *Proc. SPIE* **3924**, 96–107 (2000) [doi: 10.1117/12.384245].
14. N. Ahmed and K. R. Rao, *Orthogonal Transforms for Digital Image Processing,* Springer-Verlag, Berlin (1975).
15. R. J. Clarke, *Transform Coding Theory,* Academic Press, New York (1985).
16. O. K. Ersoy and C. H. Chen, "Transform-coding of images with reduced complexity,"*Computer Vision, Graphics, Image Processing* **42**(1), 19–31 (1988).
17. A. K. Jain, *Fundamentals of Digital Image Processing,* Prentice Hal, Englewood Cliffs, NJ (1989).
18. S. Agaian, "Hadamard Matrices and Their Applications,"*Lecture Notes of Mathematics.* **1168**, Springer-Verlag, Berlin (1985).
19. S. S. Agaian, "Advances and problems of the fast orthogonal transforms for signal-images processing applications (part 1),"*Pattern Recognition, Classification, Forecasting* **3**, 146–215, The Russian Academy of Sciences, Nauka, Moscow (1990) (in Russian).
20. S. S. Agaian, "Advances and problems of the fast orthogonal transforms for signal-images processing applications (part 2),"*Pattern Recognition, Classification, Forecasting* **4**, 156–246, The Russian Academy of Sciences, Nauka, Moscow (1991) (in Russian).
21. A. M. Grigoryan and S. S. Agaian, *Multidimensional Discrete Unitary Transforms: Representation, Partitioning, and Algorithms,* Marcel Dekker, New York (2003).
22. T. L. Ji, M. K. Sundareshan, and H. Roehrig, "Adaptive image contrast enhancement based on human visual properties,"*IEEE Trans. Med. Imaging* **13**(4), 573–586, (1994).
23. A. Beghcladi and A. L. Negrate, "Contrast enhancement technique based on local detection of edges,"*Computer Vision, Graphic, Image Processing* **46**, 162–274, (1989).
24. A. Khellaf, A. Beghdadi, and H. Dupoiset, "Entropic contrast enhancement,"*IEEE Trans. Med. Imaging* **10**(4), 589–592, (1991).
25. O. Ersoy, *Fourier-Related Transform, Fast Algorithms and Applications,* Prentice Hall, New York (1997).
26. J. H. McClellan, "Artifacts in alpha-rooting of images,"*Proc. IEEE Int. Conf. Acoustics, Speech, and Signal Process.*, 449–452,(1980).

27. S. Aghagolzadeh and O. K. Ersoy, "Transform image enhancement," *Opt. Eng.*, **31**(3), 614–626 (1992) [doi: 10.1117/12.56095].
28. I. E. Gordon, *Theory of Visual Perception,* John Wiley & Sons, New York (1989).
29. G. T. Fechner, *Elements of Psychophysics,* Vol. 1, Rinehart & Winston, New York (1960).
30. S. S. Agaian, H. G. Sarukhanyan, K. O. Egiazarian, and J. Astola, *Hadamard Transforms,* SPIE Press, Bellingham, WA (2011) [doi: 10.1117/3.890094].
31. S. S. Agaian, K. Tourshan, and J. P. Noonan, "Parametric slant-Hadamard transforms with applications," *IEEE Signal Process. Lett.* **9**(11), 375–377 (2002).
32. K. Panetta, J. Xia, and S. Agaian, "Color image enhancement based on the discrete cosine transform coefficient histogram," *J. Electron. Imaging* **21**(2), 021117 (2012) [doi: 10.1117/1.JEI.21.2.021117].
33. K. Panetta, Y. Zhou, S. Agaian, and H. Jia, "Nonlinear unsharp masking for mammogram enhancement," *IEEE Trans. Information Technol. Biomed.* **15**, 918–928 (2011).
34. S. C. Nercessian, K. A. Panetta, and S. S. Agaian, "Nonlinear direct multiscale image enhancement based on the luminance and contrast masking characteristics of the human visual system," *IEEE Trans. Image Processing* **22**(9), 3549–3561 (2013).
35. K. Panetta, E. Wharton, and S. Agaian, "Human visual system based image enhancement and measure of image enhancement," IEEE transactions on Systems, Man and Cybernetics- Part A: Systems and Humans, vol. 38, no.1, pp.174–188, February 2008.
36. S. S. Agaian, B. Silver, and K. A. Panetta, "Transform coefficient histogram-based image enhancement algorithms using contrast entropy," *IEEE Trans. Image Processing* **16**(3), 741–758 (2007).
37. E. A. Silva, K. A. Panetta, and S. S. Agaian, "Quantifying image similarity using measure of enhancement by entropy," *Proc. SPIE* **6579**, 65790U (2007) [doi: 10.1117/12.720087].
38. C. Gao, K. Panetta, and S. S. Agaian, "A new color contrast enhancement algorithm for robotic applications," *Technologies for Practical Robot Applications (TePRA), 2012 IEEE International Conference*, pp. 42–47 (2012).
39. J. Xia, K. A. Panetta, and S. S. Agaian, "Color image enhancement algorithm based on DCT transforms," *Proc. Man and Cybernetics, SMC 2011,* pp. 1496–1501 (2011).
40. Y. Zhou, K. Panetta and S. S. Agaian, "Parameterized logarithmic image processing model for image enhancement," *IS&T/SPIE Electronic Imaging 2011: Image Processing: Algorithms and Systems IX,* San Jose, CA (2011).

41. J. Xia, K.A. Panetta, and S. Agaian, "Color image enhancement algorithm based on logarithmic transform coefficient histogram shifting," *IEEE Trans. Image Process.* **16**(3), 741–758 (2011).
42. Y. Zhou, K. Panetta, and S. S. Agaian, "Human visual system based mammogram enhancement and analysis," *2010 IEEE International Conference on Image Processing Theory Tools and Applications (IPTA)*, Paris, France, (2010).
43. A. Almuntashri and S. S. Agaian, "Nonlinear filtering for enhancing prostate MR images via alpha-trimmed mean separation," *Proc. Man and Cybernetics, SMC 2010,* pp. 3698–3701 (2010).
44. S. S. Agaian and S. A. McClendon, "Novel medical image enhancement algorithms," *Proc. SPIE* **7532**, 75320W, (2010) [doi: 10.1117/12.839003].
45. Y. Zhou, K. Panetta, and S. S. Agaian, "Non-linear filtering for image enhancement using order statistics decomposition," *SPIE Defense, Security, and Sensing 2010: Mobile Multimedia/Image Processing, Security, and Applications*, Orlando, FL (2010).
46. Y. Zhou, K. Panetta, and S. Agaian, "Human visual system based contrast enhancement for X-ray CT images using alpha weighted quadratic filter," *IS&T / SPIE Electronic Imaging 2010: Image Processing: Algorithms and Systems VIII*, San Jose, CA (2010).
47. Y. Zhou, K. Panetta, and S. S. Agaian, "Mammogram enhancement using alpha weighted quadratic filter," 31st Annual International Conference of the IEEE EMBS, Minneapolis, MN (2009).
48. Y. O. Yesna, D. Q. Abraham, S. S. Agaian, and K. A. Panetta, "Projection image enhancement for explosive detection systems (EDS)," *Proc. SPIE* **6812**, 681208 (2008) [doi: 10.1117/12.766403].
49. E. J. Wharton, K. A. Panetta, and S. S. Agaian, "Human visual-system-based image enhancement," *Proc. SPIE* **6579**, 65790O (2007) [doi: 10.1117/12.719551].
50. E. A. Silva, K. A. Panetta, and S. S. Agaian, "Quantifying image similarity using measure of enhancement by entropy," *Proc. SPIE* **6579**, 65790U (2007) [doi: 10.1117/12.720087].
51. E. Wharton, S. Agaian, and K. Panetta, "A logarithmic measure of image enhancement," *Proc. SPIE* **6250**, 62500P (2006) [doi: 10.1117/12.665693].
52. E. Wharton, S. Agaian, and K. Panetta, "Comparative study of logarithmic enhancement algorithms with performance measure," *Proc. SPIE* **6064**, 606412 (2006) [doi: 10.1117/12.647489].
53. B. Silver, S. S. Agaian and K. A. Panetta, "Contrast entropy based image enhancement and logarithmic transform coefficient histogram shifting," *Proc. ICASSP,* 633–636 (March 2005).

54. S. Agaian and F. Arslan, "Two transform based image enhancement methods," *Int. Signal Process. Conf.,* Dallas, TX (March 2003).
55. A. Grigoryan and S. Agaian, "Reconstruction of images from projections by the tensor transform," *Int. Signal Process. Conf.,* Dallas, TX (March 2003).
56. A. M. Grigoryan, and S. Agaian, "Tensor form of image representation: enhancement by image-signals," *Proc. SPIE* **5014**, 213–220 (2003) [doi: 10.1117/12.473063].
57. S. Agaian, K. A. Panetta, and A. Grigoryan, "A New Measure of Image Enhancement," *IASTED Int. Conf. Signal Process. Comm.,* **5014**, Rhodes, Greece (July 2001).
58. S. Mallat and S. Zhong, "Characterization of signals from multiscale edges," *IEEE Trans. Pattern Anal. Mach. Intel.* **14**, 710–732 (1992).
59. A. Laine, J. Fan, and W. Yang, "Wavelets for contrast enhancement of mammography," *IEEE Eng. Med. Biol.* **14**, 536–550 (1995).
60. G. Cincotti, G. Loi, and M. Pappalardo, "Frequency decomposition and compounding of ultrasound medical images with wavelet packets," *IEEE Trans. Med. Imaging* **20**(8), 764–771 (2001).
61. D.-Y. Tsai, Y. Lee, M. Sekiya, S. Sakaguchi, and I. Yamada, "A method of medical image enhancement using wavelet analysis," *Proc. IEEE ICSP'02*, 723–726 (2002).
62. S. DelMarco and S. Agaian, "The design of wavelets for image enhancement and target detection," *Proc. SPIE* **7351**, 735103 (2009) [doi: 10.1117/12.816135].
63. A. M. Grigoryan, "An algorithm of the two-dimensional Fourier transform," *Radioelectronica* **27**(10), 52–57, Izvestiya vuzov SSSR, Kiev, USSR (1984) [English translation available at http://fasttransforms.com/Art-USSR-papers/Art-IzvuzovSSSR1984.pdf].
64. A. M. Grigoryan, "An optimal algorithm for computing the two-dimensional discrete Fourier transform," *Radioelectronica* **29**(12), 20–25, Izvestiya vuzov SSSR, Kiev, USSR (1986) [English translation available at http://fasttransforms.com/Art-USSR-papers/Art-IzvuzovSSSR1986.pdf].
65. A. M. Grigoryan, "New algorithms for calculating discrete Fourier transforms," *J. Vichislitelnoi Matematiki i Matematicheskoi Fiziki* **26**(9), 1407–1412, AS USSR, Moscow (1986) [English translation available at http://fasttransforms.com/Art-USSR-papers/Art-JVMA-TofASUSSR1986.pdf].
66. A. M. Grigoryan and M. M. Grigoryan, "Two-dimensional Fourier transform in the tensor presentation and new orthogonal functions," *Avtometria* **1**, 21–27, Novosibirsk, USSR (1986).
67. A. M. Grigoryan, "An algorithm for computing the two-dimensional Fourier transform of equal orders," *Proc. IX All-Union Conference on*

Coding and Transmission of Information, Part 2, 49–52, Odessa, USSR (1988) [English translation available at http://fasttransforms.com/Art-USSR-papers/Art-AllUnConf1988.pdf.

68. A. M. Grigoryan, "An algorithm for computing the discrete Fourier transform with arbitrary orders," *J. Vichislitelnoi Matematiki i Matematicheskoi Fiziki* **30**(10), 1576–1581, Moscow (1991) [English translation available at http://fasttransforms.com/Art-USSR-papers/Art-JVMAToFASUSSR1991.pdf.
69. A. M. Grigoryan and S. S. Agaian, "Split manageable efficient algorithm for Fourier and Hadamard transforms," *IEEE Trans. Signal Process.* **48** (1), 172–183 (2000).
70. A. M. Grigoryan, "2D and 1D multi-paired transforms: Frequency-time type wavelets," *IEEE Trans. Signal Process.* **49**(2), 344–353 (2001).
71. A. M. Grigoryan and S. S. Agaian, "Shifted Fourier transform based tensor algorithms for 2D DCT," *IEEE Trans. Signal Process.* **49**(9), 2113–2126 (2001).
72. A. M. Grigoryan and S. S. Agaian, "Efficient algorithms for computing the 2D hexagonal Fourier transforms," *IEEE Trans. Signal Process.* **50**(6), 1438–1448 (2002).
73. S. S. Agaian, K. Panetta, and A. M. Grigoryan, "Transform-based image enhancement algorithms," *IEEE Trans. Image Process.* **10**(3), 367–382 (2001).
74. A. M. Grigoryan and S. S. Agaian, "Transform-based image enhancement algorithms with performance measure," *Advances in Imaging and Electron Physics* **130**, pp. 165–242, Academic Press, San Diego, CA (2004).
75. F. T. Arslan and A. M. Grigoryan, "Fast splitting-rooting method of image enhancement: Tensor representation," *IEEE Trans. Image Process.* **15**(11), 3375–3384 (2006).
76. A. M. Grigoryan and K. Naghdali, "On a method of paired representation: Enhancement and decomposition by series direction images," *J. Math. Imaging and Vision* **34**(2), 185–199 (2009).
77. N. Du and A. M. Grigoryan, "Paired transform method of image reconstruction from projections," in *Object Modeling, Algorithms and Applications,* R. P. Barneva et al., eds., Research Publishing, New York (2010).
78. A. M. Grigoryan and N. Du, "Principle of superposition by direction images," *IEEE Trans. Image Process.* **20**(9), 2531–2541 (2011).
79. A. M. Grigoryan and N. Du, "2D images in frequency-time representation: Direction images and resolution map," *J. Electron. Imaging* **19**(3), 033012 (2010) [doi: 10.1117/1.3483906.

80. A. M. Grigoryan, "Multidimensional Discrete Unitary Transforms," Chapter 19 in *The Transforms and Applications Handbook,* 3rd ed., The Electrical Engineering Handbook Series, Alexander Poularikas, Ed., CRC Press, Boca Raton, FL (2010).
81. A. M. Grigoryan and M. M. Grigoryan, *Brief Notes in Advanced DSP: Fourier Analysis with MATLAB,* CRC Press, Boca Raton, FL (2009).
82. A. M. Grigoryan and M. M. Grigoryan, *Image Processing: Tensor Transform and Discrete Tomography with MATLAB,* CRC Press, Boca Raton, FL (2012).
83. K. Sayood, *Introduction to Data Compression,* 3rd ed., Morgan Kaufmann, Waltham, MA (2006).
84. A. M. Grigoryan, S. Dursun, and E. E. Regentova, "Lossless encoding based on redistribution of statistics," *Proc. IEEE ITCC-2004,* 620–624, Las Vegas, NV (2004).
85. A. M. Grigoryan and E. E. Regentova, "A new method of optimal coding," *Proc. IEEE ITCC-2002,* 378–382, Las Vegas, NV (2002).
86. A. M. Grigoryan, "Image Cryptography on Tensor Representation," U. S. Patent (disclosure submitted to U.T. System, September 23, 2010).
87. Y. Mao, G. Chen, and S. Lian, "A novel fast image encryption scheme based on 3D chaotic Baker maps," *Int. J. Bifurc. Chaos* **14**(10), 3613–3624 (2003).
88. G. Chen, Y. Mao, and C. K. Chui, "A symmetric image encryption scheme based on 3D chaotic cat maps," *Chaos, Solitons, and Fractals* **21**(3), 749–761 (2004).
89. A. A. Yahya and A. M. Abdalla, "A shuffle image-encryption algorithm," *J. Comput. Sci.* **4**(12), 999–1002 (2008).
90. A. Mitra, Y. V. Subba Rao, and S. R. M. Prasanna, "A new image encryption approach using combinational permutation techniques," *Int. J. Comp. Sci.* **1**(2), 13064428 (2006).
91. Y. Zhou, K. Panetta, and S. Agaian, "A lossless encryption method for medical images using edge maps," *Proc. IEEE EMBS 2009,* Minneapolis, Minnesota (2009).
92. Y. Zhou, K. Panetta, and S. Agaian, "Image encryption using discrete parametric cosine transform," *Proc. IEEE ACSSC 2009,* 395–399, Pacific Grove, CA (2009).
93. R. E. L. Metzler and S. S. Agaian, "Selective region encryption using a fast shape adaptive transform," *Proc. IEEE SMC 2010,* 1763–1770 (2010).
94. Y. Wu, J. P. Noonan, P. Joseph, and S. Agaian, "Binary data encryption using the Sudoku block cipher," *Proc. IEEE SMC 2010,* 3915–3921 (2010).

95. Y. Wu, Y. Zhou, K. Panetta, and S. Agaian, "Image encryption using the Sudoku matrix," *Proc. SPIE* **7708**, 77080P (2010) [doi: 10.1117/12.853197].
96. Y. Wu, J. P. Noonan, and S. Agaian, "NPCR and UACI randomness tests for image encryption," *J. Selected Areas Telecomm.*, 31–38 (April 2011).
97. S. C. Nercessian, K. A. Panetta, and S. S. Agaian, "Multiscale image fusion using an adaptive similarity-based sensor weighting scheme and human visual system-inspired contrast measure," *J. Electron. Imaging* **21**(2), 021112 (2012) [doi: 10.1117/1.JEI.21.2.021112].

Artyom M. Grigoryan received M.S. degrees in Mathematics from Yerevan State University (YSU), Armenia in 1978; in Imaging Science from the Moscow Institute of Physics and Technology in 1980; and in Electrical Engineering from Texas A&M University in 1999. He received his Ph.D. in Mathematics and Physics from YSU in 1990. From 1990-1996, he was a senior researcher with the Department of Signal and Image Processing at the Institute for Informatics and Automation Problems (IAAP) at the National Academy of Sciences of the Republic of Armenia (NAS RA). From 1996-2000, he was a Research Engineer with the Department of Electrical Engineering at Texas A & M. In December 2000, he joined the Department of Electrical Engineering at the University of Texas, San Antonio, where he is currently an Associate Professor. He holds two patents for developing an algorithm of automated 3D fluorescent *in situ* hybridization spot counting, and one patent for fast calculation of cyclic convolution. He is the author of three books, three book chapters, two patents, and many journal papers and specializes in the theory and application of fast one- and multidimensional Fourier transforms, elliptic Fourier transforms, tensor and paired transforms, unitary heap transforms, design of robust linear and nonlinear filters, image enhancement, encoding, computerized 2D and 3D tomography, processing biomedical images, and image cryptography.

Sos S. Agaian is the Peter T. Flawn Professor of Electrical and Computer Engineering at the University of Texas, San Antonio, and a professor at the University of Texas Health Science Center, San Antonio. Dr. Agaian received his M.S. degree (summa cum laude) in Mathematics and Mechanics from Yerevan State University, Armenia; his Ph.D. in Mathematics and Physics from the Steklov Institute of Mathematics, Russian Academy of Sciences (RAS), Moscow; and his Eng.Sc.D. from the Institute of Control Sciences, RAS.

He has authored more than 500 scientific papers, seven books, and holds 14 patents. He is a Fellow of SPIE, the American Association for the Advancement of Science (AAAS), and Imaging Sciences and Technology (IS&T). He also serves as a foreign member of the Armenian National Academy. He is the recipient of MAEStro Educator of the Year, sponsored by the Society of Mexican American Engineers. The technologies that Dr. Agaian invented have been adopted by multiple institutions, including the U.S. government, and commercialized by industry. He is an Editorial Board Member for the *Journal of Pattern Recognition and Image Analysis* and an Associate Editor for several journals, including the *Journal of Electronic Imaging* (SPIE, IS&T) and *System* (Elsevier). His research interests include multimedia processing, imaging systems, information security, artificial intelligence, computer vision, 3D imaging sensors, fusion, and biomedical and health informatics.

Chapter 3
Multimodality Imaging for Tumor Volume Definition in Radiation Oncology*

Issam El Naqa
Department of Oncology/Medical Physics Unit, McGill University

3.1 Introduction

Multimodality imaging (or hybrid imaging, as it is sometimes called) is witnessing an extraordinary surge in diagnostic and therapeutic radiology for diagnosing, staging, and monitoring treatment response in complex diseases such as atherosclerosis[1] and cancer.[2,3] This trend has been made possible by the recent evolutionary developments in medical imaging instrumentation technology that allows for the identification of anatomical structures or functional processes *in vivo*. An elegant comprehensive review that follows the historical developments of multimodality imaging instrumentation for clinical use and its future potential has been recently published by Townsend[4] in the first proposal to design and build a combined scanner.

In modern conformal, intensity-modulated image-guided radiotherapy (IGRT) particularly, there is noticeable value in using hybrid multi-modality imaging in treatment planning, diagnosis, and staging of different cancer sites. The goal is to achieve improved target definition by incorporating information from different imaging modalities [computed

* I would like to thank my previous postdoctoral fellow, Dr. Deshan Yang, for his contribution to the development of the MIASYS software. This work was partially supported by the Natural Sciences and Engineering Research Council of Canada under grant NSERC- RGPIN 397711-11.

tomography (CT), magnetic resonance (MR), and 3D ultrasound] as well as improved staging, disease characterization, and localization using functional and molecular imaging [positron emission tomography (PET), single photon emission computed tomography (SPECT), and magnetic resonance spectroscopy (MRS)]. In addition, onboard imaging systems [cone-beam CT (CBCT) and mega-voltage CT (MVCT)] are currently deployed in many clinics to correct patient setup and improve daily delivery of fractionated radiation treatments. Although these advanced imaging modalities have created new opportunities in radiotherapy treatment planning and delivery, they have also presented many technical challenges regarding the integration of different-modalities information, which affects visualization, delineation, and related areas.[5]

Historically, kilovoltage CT (kVCT) has been the standard for treatment planning in 3D conformal or intensity-modulated radiation therapy (IMRT). However, it is well known that there can be significant variability when multiple observers contour tumor target volumes.[6] Several studies indicated that this inter- and intraobserver variability could be reduced by combining information from multiple modalities. For instance, lung-cancer researchers reported reduced variability when using CT with PET for target definition.[7,8] Furthermore, a study of fractionated stereotactic radiotherapy for meningioma patients demonstrated improved target definition by combining physiological information from PET, anatomical structure from CT, and soft-tissue contrasts from MRI, resulting in alterations of the original contour definitions in 73% of the cases.[9] However, this visual approach for integrating multimodality imaging information is prone to observer subjectivity and variations with regard to single-image analysis. Therefore, this chapter discusses methods for integrating multimodality imaging information by extending automated and semi-automated segmentation methods into an interactive multi-modality framework.[10]

3.2 Single versus Multimodality Image Segmentation

Medical image segmentation is a process that separates structures of interest in an image from the background or other neighboring structures. It is a prerequisite step for many medical imaging applications in radiology and radiation therapy. These applications may include automatic organ delineation, quantitative tissue classification, surface extraction, visualization and image registration, etc.[11,12] For instance, Pham and coworkers divided segmentation algorithms into eight different categories: thresholding, region growing, classifiers, clustering, Markov random field models, artificial neural networks, deformable models, and atlas-guided approaches. In our work on PET-guided treatment planning in radiotherapy, we presented a comparative survey of the current methods applied for tumor segmentation.[13,14]

There are several commercial and academic software tools that support different segmentation algorithms. In general, commercial software packages have better implementation (with a user-friendly interface for manual and semi-automatic segmentation methods), but they often lag behind the latest developments in the field. In contrast, academic software packages, such as ITK,[15] BioImage Suite,[16] MIPAV[17] and ImageJ,[18] tend to be more oriented toward single-modality applications and less friendly in handling multi-modality images, as proposed here.

Most automatic algorithms attempt to utilize image intensity variations or image gradient information. However, for low-contrast images, many of these algorithms tend to provide suboptimal solutions that are not clinically acceptable. For such cases, it has been demonstrated that if multiple images are available for the same object (same image modality or different image modalities), all of the available complementary information can be fed into the segmentation algorithms to define the so-called biophysical target.[19] Thus, the segmentation algorithms would benefit from the complementary information from different images, and consequently the accuracy of the final segmentation results could be improved. Similar approaches have been applied to detect the blood-wall interface of heart ventricles from CT, MRI, and ultrasound images using a snake deformable model;[20] to classify coronary artery plaque composition from multiple contrast MR images using the K-means clustering algorithm;[21] and to define tumor target volumes using PET/CT/MR images for radiotherapy treatment planning by using a multivalued, deformable level set approach, as in our previous work. Mathematically, such an approach is a framework that could be thought of as a mapping from the imaging space to the "perception" space identified by radiologists:[19]

$$\text{Biophysical target} = f(CT, PET, MRI, \ldots; \lambda), \tag{3.1}$$

where $f(\cdot)$ is the mapping function from the different imaging modalities to the target space parameterized by λ, which represents the users' defined set of parameters representing prior knowledge. This framework is highlighted in Fig. 3.1.

Despite the opportunities presented by this framework for streamlining the integration of multiple imaging modalities for better tissue classification or target definition, there are several challenges that should be addressed before clinical implementation can be achieved. First, image misalignment is an issue when dealing with images acquired from different scanners. This is partially resolved for PET/CT but not for many other image modalities. Therefore, methods for image registration should be incorporated into the framework. The second and more-challenging issue is the characterization of the mapping in Eq. (3.1) because it relies on translating higher-level human expertise into cues that computer algorithms

Figure 3.1 Biophysical target generated from multimodality imaging by combining anatomical and functional information.

can understand. These challenges have motivated us to develop a software tool to support such a multimodality image segmentation framework that can potentially learn information from the user's interactions. In our recent work, we attempted to resolve some of these problems by developing a dedicated software tool for multimodality image analysis called MIASYS.[10] This software is the first tool to offer a dedicated and comprehensive framework to cope with the emerging needs of therapeutic and diagnostic radiological applications.

3.3 Methods for Multimodality Image Segmentation

There are several methods that have been proposed to integrate multimodality imaging information, primarily by extending automated and semi-automated single- or monomodality segmentation methods into an interactive multimodality framework in which the available complementary information can be fed into the segmentation algorithms to define the biophysical target (as described earlier). Thus, the segmentation algorithms would benefit from the complementary information provided by different images, and consequently the accuracy of the final segmentation results could be improved. Similar approaches have been applied to detect the blood-wall interface of heart ventricles from CT, MRI, and ultrasound images using a snake deformable model (Sebbahi et al., 1997); to classify coronary artery plaque composition from multiple contrast MR images using the K-means clustering algorithm;[22] and to define tumor target volumes using PET/CT/MR images for radiotherapy treatment planning by using a multivalued deformable level set approach.[19] This approach could be applied to several segmentation algorithms that are amenable to such generalization,[10] as discussed in the following sections.

3.3.1 Multiple-image thresholding

Thresholding is one of the most basic image segmentation methods; it is commonly used to delineate objects with high contrast with respect to the surrounding image background. Threshold values can be selected experimentally, e.g., in detecting tumors in PET based on cutoff values of the standardized uptake value (SUV), a threshold value is usually selected as SUV > 2.5 or 40% of maximum SUV.[23] In another example, an optimal threshold image-intensity value could be selected iteratively to separate the lungs from the body and chest wall structures.[24]

One way to expand the thresholding method to support hybrid images involves applying different threshold values to the different images and combining the thresholding results for different images in logical ways to form the final result. Previous work by Yang et al.[10] used the "thresholding conditions" notion to describe such multiple-image thresholding operations. For example, a thresholding condition could be "Im1 < 100 & Im2 > 50 | 20 < Im3 < 150," where Im1, Im2, and Im3 denote the intensity values of images 1, 2, and 3. MIASYS, for instance, is able to interpret the meaning of such a thresholding condition expression and carry out all of the mathematical and logical computations to yield the final combined result. The software tool is implemented with MATLAB, and it accepts any valid MATLAB expression as a thresholding condition. The expressions can contain any arithmetic, logical operators, and parentheses. The new thresholding method, which accepts such mathematical expressions, is flexible, easy to use, and very powerful.

3.3.2 Clustering algorithms

Clustering algorithms are frequently used for different image analysis problems such as image segmentation, object recognition, and image retrieval.[25] For image segmentation, they are used to automatically discriminate different tissue types based on primitive image features such as image intensity. One of the most commonly used algorithms is the K-means algorithm. Another algorithm found to be more robust is the fuzzy C-means (FCM) algorithm.

3.3.2.1 Fuzzy C-means algorithm

Similar to other clustering algorithms, the goal of the FCM algorithm[26] is to divide the image histogram into a few clusters and to iteratively find the center of each cluster by minimizing the following system energy function:

$$J(x,c) = \sum_{i=1}^{N}\sum_{k=1}^{K} \|\mathbf{x}_i - \mathbf{c}_k\|^2, \tag{3.2}$$

where $\mathbf{x}_i$ is the image intensity for pixel i, N is the total number of pixels in the image, K is the total number of clusters, and $\mathbf{c}_k$ is the cluster center intensity value for cluster k.

In the FCM algorithm, a fuzzy membership function is defined and computed as Eq. 3.3, and the cluster center $\mathbf{c}_k^n$ is updated according to Eq. (3.4):

$$\mathbf{u}_{ik}^n = \frac{\|\mathbf{x}_i - \mathbf{c}_k^n\|^{-2}}{\sum_{k=1}^{K} \|\mathbf{x}_i - \mathbf{c}_k^n\|^{-2}}, \tag{3.3}$$

$$\mathbf{c}_k^{n+1} = \frac{\sum_{i=1}^{N} (\mathbf{u}_{ik}^n)^b \mathbf{x}_i}{\sum_{i=1}^{N} (\mathbf{u}_{ik}^n)^b}, \tag{3.4}$$

where $\mathbf{u}_{ik}^n$ is the fuzzy membership probability that image pixel $\mathbf{x}_i$ belongs to cluster k at iteration n, $\mathbf{c}_k^n$ is the updated cluster center intensity value for cluster k at iteration n, and b is a user defined parameter, where $0 < b < 1$.

The user starts the FCM method by setting the value of K, after which the $\mathbf{c}_k^0$ can be automatically and randomly initialized, and the iterations are repeated by computing Eqs. (3.3) and (3.4), respectively. The iterations stop when $\mathbf{c}_k^n$ is stabilized.

3.3.2.2 Extending the fuzzy C-means algorithm to multiple images

The FCM method can be naturally expanded to support multiple images by defining $\mathbf{x}_i$ and $\mathbf{c}_k^n$ as vectors instead of scalar values:

$$\mathbf{x}_i = \langle x_{i,1}, x_{i,2}, \ldots, x_{i,M} \rangle, \tag{3.5}$$

$$\mathbf{c}_k^n = \langle c_{k,1}^n, c_{k,2}^n, \ldots, c_{k,M}^n \rangle, \tag{3.6}$$

where M is the total number of images in the multimodality image data set, $x_{i,M}$ is the image pixel intensity value for the pixel i in the image m ($1 < m < M$), and $\mathbf{c}_{k,m}^n$ is the cluster center intensity value for the cluster k in the image m for iteration n.

With $\mathbf{x}_i$ and $\mathbf{c}_k^n$ defined as vectors, Eqs. (3.3) and (3.4) can be rewritten as

$$\mathbf{u}_{ik}^n = \frac{\left(\sum_{m=1}^{M} \alpha_m (x_{i,m} - c_{k,m}^n)^2\right)^{-2}}{\sum_{k=1}^{K} \left(\sum_{m=1}^{M} \alpha_m (x_{i,m} - c_{k,m}^n)^2\right)^{-2}}, \tag{3.7}$$

$$\mathbf{c}_{k,m}^{n+1} = \frac{\sum_{i=1}^{N} \left(\mathbf{u}_{ik}^{n}\right)^{b} \mathbf{x}_{i,m}}{\sum_{i=1}^{N} \left(\mathbf{u}_{ik}^{n}\right)^{b}}, \quad m = 1, M, \tag{3.8}$$

where α_m is the user-defined weighting parameter for image m.

In this case, the computation of the fuzzy membership value $\mathbf{u}_{ik}^{n}$ is contributed by all of the images in the multimodality image dataset. The new weighting parameters α_m control the contribution from the different images according to the users' prior knowledge.

The FCM algorithm works in the image intensity histogram domain; image pixel spatial information is not considered in the algorithm, and there is also no difference in applying this algorithm to 2D images or to 3D volume images. Better FCM algorithms that consider image pixel spatial information, which may be added to the software tool in the future, have been reported in the literature.[27,28]

3.3.2.3 K-means clustering algorithm

The K-means algorithm[29] is actually a precursor of the FCM algorithm. It uses the hard membership function instead of the fuzzy membership function. The performance of the K-means algorithm is generally comparable to but less robust than the performance of the FCM algorithm. However, both algorithms may suffer from lacking a proper spatial neighborhood definition, which is addressed in the active contour algorithms.

3.3.3 Active contour algorithms

Deformable models are geometric representations of curves (in 2D) or surfaces (in 3D) that are defined explicitly or implicitly in the imaging domain. These models deform under the influence of force-like equations that are computed from the image data.[30,31] Contours of structures in the images are characterized by sharp variations in the image intensity, and therefore the deformable models can be warped to match the contours by means of energy minimization.[31–33]

So-called "snake" algorithms were among the first deformable models developed.[34] Snakes use an explicit parametric representation of the object boundary that deforms by means of energy minimization (or dynamic force equilibrium). Mathematically, if the deformable model is represented by

$$C(s) = \{x(s), y(s), z(s)\}, \ s \in [0, 1], \tag{3.9}$$

then its movement is governed by the following functional:

$$J(C(t)) = \int_0^1 \left(\alpha(s) \left| \frac{\partial C(s,t)}{\partial s} \right|^2 + \beta(s) \left| \frac{\partial^2 C(s,t)}{\partial s^2} \right|^2 \right) ds + \gamma \int_0^1 P(C(s,t)) ds, \tag{3.10}$$

where the first term corresponds to the internal energy and controls the tension and rigidity of the model.† The second term corresponds to the external energy (P represents the potential energy) that could be given as $g(|\nabla I|)$, where g is selected to be a monotonically decreasing function of the gradient of image intensity I. Other examples could include using pressure or balloon representations to represent an expanding object or other diffusing functions. Using calculus of variation techniques, the solution to the equation is obtained by solving the associated Euler–Lagrange PDE:[31,33]

$$\frac{\partial}{\partial s}\left(\alpha\frac{\partial C}{\partial s}\right)+\frac{\partial^2}{\partial s^2}\left(\beta\frac{\partial^2 C}{\partial s^2}\right)+\nabla P(C(s,t))=0. \tag{3.11}$$

However, the formulation in Eq. (3.11) is nonconvex and suffers from several drawbacks such as sensitivity to contour initialization, dependency on parameterization, and an inability to account for topological adaptation (e.g., delineation of a necrotic tumor). To solve the sensitivity problem, the geodesic active contour model was proposed,[35] which in principle is equivalent to Eq. (3.10) if the smoothness constraint is eliminated (i.e., by setting $\beta = 0$). This has led to the development of the flow or curve evolution concept:

$$\frac{\partial C}{\partial t}=\overrightarrow{V}(\kappa), \tag{3.12}$$

where $\overrightarrow{V}$ is the velocity function (of magnitude V) in the normal direction ($\overrightarrow{N}$), and κ is the local contour curvature. However, to resolve the main problem of parameterization and topological adaptation, the level set approach was proposed.[30]

In the level set approach, the curve [in Eq. (3.12)] is embedded in an implicit level set function ϕ. This function defines sets of contour values and positions, including the target boundary at the zero level [$\phi(C) = 0$], as illustrated in Fig. 3.2. In this case, the evolution equation is rewritten as

$$\frac{\partial \phi}{\partial t}=V(\kappa)|\nabla\phi|+F(\Theta), \tag{3.13}$$

where V is defined to be proportional to the curvature and inversely proportional to the image gradient, and $F(\Theta)$ is the external force constraint with vector parameter Θ that could be used to add context-knowledge information such as shape priors. The level set function ϕ is typically selected as a signed distance function. Efficient solutions were developed for Eq. (3.13) by using finite difference and fast marching methods.[30]

† The first-order derivative suppresses stretching and makes the contour behave like an elastic string. The second-order derivative suppresses bending and makes the model behave like a rigid rod.

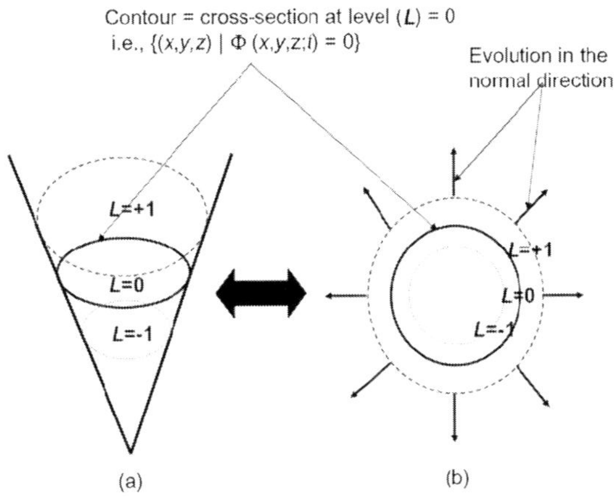

Figure 3.2 Deformable image segmentation by the level set method. (a) Representation of the level set surface at time t of the evolving function ϕ. (b) A projected view showing the evolution direction. Typically, the function ϕ evolves at a velocity proportional to the curvature of the contour and inversely proportional to the image gradient. In the present example, ϕ is represented by a signed Euclidean distance transform of value L. The contour is extracted at $L = 0$, with negative values inside the contour representing the volume of interest and positive values outside representing the background.

3.3.3.1 "Active-contour-without-edge" algorithm

The "active-contour-without-edge" algorithm is one of the frequently used active contour image segmentation algorithms, originally reported by Vese and Chan.[36] Unlike many other algorithms, it is based on regional average intensity value instead of image gradient because gradients tend to be more sensitive to noise and therefore less reliable.

To separate an image into two mutually exclusive partitions, the following system energy equation is to be minimized:

$$F(c_1, c_2, C) = \int_{\Omega_1} (I - c_1)^2 d\Omega_1 + \int_{\Omega_2} (I - c_2)^2 d\Omega_2 + \mu|C| + \nu \cdot \Gamma(C), \quad (3.14)$$

where I is the image intensity; Ω_1 and Ω_2 are the two partitions (Ω_1 is inside the contour defined by the level set, and Ω_2 is outside the contour); c_1 and c_2 are the average image intensity in partition Ω_1 and Ω_2, respectively; C is the partition boundary and $|C|$ is the total length of C for 2D images (or the total surface area of C for 3D images); and ν and μ are user-selected weighting constants. $\Gamma(C)$ is the area inside C for 2D images or the volume inside C for 3D images.

The level set formation for Eq. (3.14) is

$$F(c_1, c_2, C) = \int_{\Omega} \Big[(I - c_1)^2 H(\phi) + (I - c_2)^2 (1 - H(\phi)) + \mu \cdot \delta(\phi)|\nabla\phi| + \nu \cdot H(\phi) \Big] d\Omega, \quad (3.15)$$

where $\Omega = \Omega_1 + \Omega_2$ is the entire image, H is the Heaviside function (a unit step function) of level set ϕ, and δ is the delta function of ϕ.

The iterative solution for the level set equation is given by

$$\frac{\partial \phi}{\partial t} = \delta_\varepsilon(\phi)\left(\mu \text{div}\left(\frac{\nabla \phi}{|\nabla \phi|}\right) - \nu - (I - c_1)^2 + (I - c_2)^2\right), \quad (3.16)$$

where $\delta_\varepsilon(\phi)$ is delta function of level set ϕ with width ε. Note that the constant parameter ν is also often known as the balloon force because it controls the constant inflation or deflation of the level set.

The two average image intensity values c_1 and c_2 are updated per iteration according to

$$c_1(\phi) = \frac{\int_\Omega I \cdot H(\phi) d\Omega}{\int_\Omega H(\phi) d\Omega}, \quad (3.17)$$

$$c_2(\phi) = \frac{\int_\Omega I(1 - H(\phi)) d\Omega}{\int_\Omega (1 - H(\phi)) d\Omega}. \quad (3.18)$$

The implementation here also adds an additional spring force $(c_1 - c_2)^2$ to further balance the distance between c_1 and c_2. This level set equation is given by

$$\frac{\partial \phi}{\partial t} = \delta_\varepsilon(\phi)\left(\mu \text{div}\left(\frac{\nabla \phi}{|\nabla \phi|}\right) - \nu - \lambda_1(I - c_1)^2 + \lambda_2(I - c_2)^2 + \lambda_3(c_1 - c_2)^2\right), \quad (3.19)$$

where λ_1, λ_2, and λ_3 are user-configurable constant parameters that were previously omitted for simplicity.

3.3.3.2 Extension to multiple images

The concept of a multivalued level set used to extend the "active-contour-without-edge" method to multiple images was originally suggested to process multichanneled color images as follows:[37,38]

$$\frac{\partial \phi}{\partial t} = \delta_\varepsilon(\phi)\left(\mu \text{div}\left(\frac{\nabla \phi}{|\nabla \phi|}\right)\right.$$
$$\left. - \nu - \sum_{m=1}^{M} \alpha_m\left(\lambda_1(I_m - c_{1,m})^2 - \lambda_2(I_V - c_{2,m})^2 - \lambda_3(c_{1,m} - c_{2,m})^2\right)\right), \quad (3.20)$$

where α_m is the user-defined relative weighting parameters for image m with image intensity u_m, M is the total number of images, and $c_{1,m}$ and $c_{2,m}$ are the c_1 and c_2 values for image m that are computed using I_m as intensity I in Eqs. (3.17) and (3.18). In this generalization, the level set iterative evolution is contributed by all of the images in the multimodality image dataset. The weighting parameters α_m control the contribution amounts from different images according to the user's perception of importance.

3.4 Examples of Multimodality Tumor Volume Definition

3.4.1 PET/CT target definition in radiotherapy

Radiotherapy is a localized treatment of directed high-energy irradiation toward the tumor target. For example, in the case of radiation therapy treatment planning of lung cancer, CT and PET images are often registered and fused to help the physician more-accurately delineate the "true" tumor volume. CT images contribute to determine the anatomic boundaries, while PET images contribute to determine the tumor metabolic activity boundary.[39–42] Using the MIASYS software tool, Fig. 3.3 shows a pair of co-registered CT and PET images of a lung tumor.

Figure 3.4 shows some demonstrative results obtained by using the generalized "active-contour-without-edge" algorithm, the FCM clustering algorithm, and the thresholding algorithm. These example results demonstrate that different segmentation algorithms could be applied for the

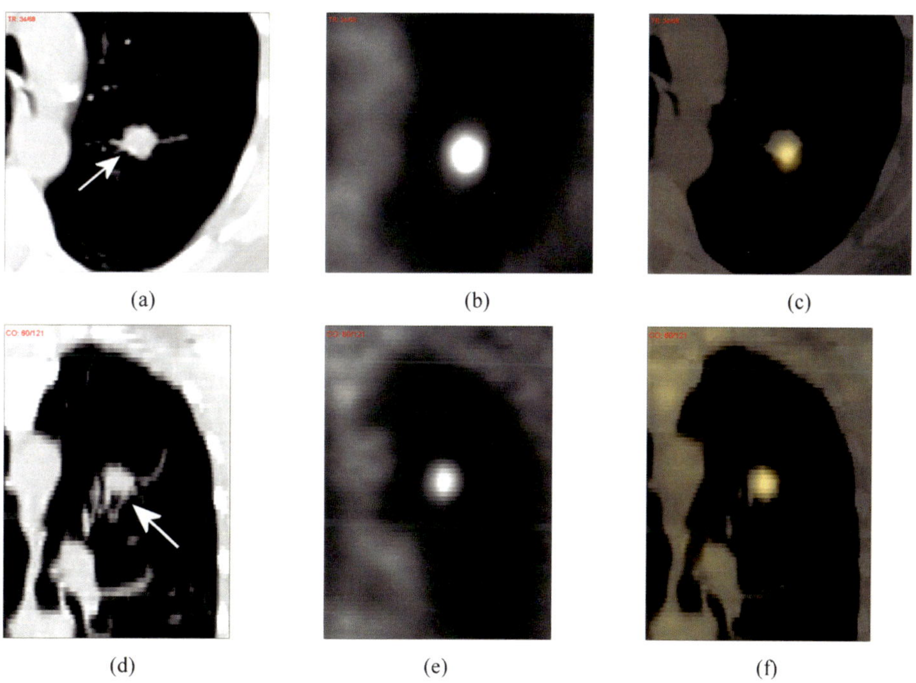

Figure 3.3 Co-registered CT and PET images of a lung cancer tumor. The tumor is indicated by the arrows: (a), (b), and (c) are the transverse views; and (d), (e), and (f) are the coronal views. (a) and (d) are CT images, (b) and (e) are PET images, and (c) and (f) are fused CT and PET images.

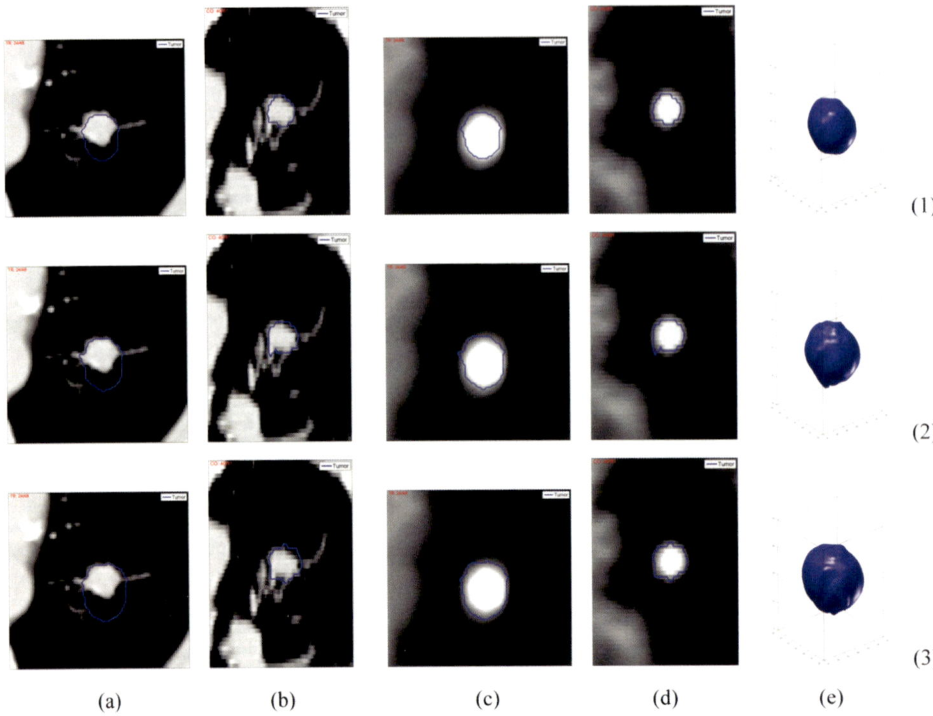

Figure 3.4 Examples of using (row 1) the active contour algorithm, (row 2) the FCM clustering algorithm, and (row 3) the thresholding method, with the condition "(CT > 500 & PET > 1500) | (CT > 400 & PET > 2000) | PET > 2500." Columns (a)–(e) are CT transverse, CT coronal, PET transverse, PET coronal view, and 3D views, respectively.

combined PET/CT images and would generate similar but slightly different results in this case.

3.4.2 PET/CT segmentation of cervix cancer example

In this example PET/CT images were taken from patients with cervix cancer. The PET image was sharpened using a deconvolution approach.[43] The 40% maximum SUV thresholding has been adopted by many institutes to estimate gross tumor volume for cervix cancer patients due to the high target-to-background ratio of these tumors in PET and the difficulty of distinguishing their boundaries in CT. This is different than the lung case, where such criteria suffer from significant variability, as discussed earlier. In Fig. 3.5, the active contour algorithm is initialized with a circle (in white) 15.9 mm in diameter. The evolved contour took ten iterations (in blue), and the final estimated contours (in thick black) are shown. The algorithm converged in just 30 iterations. This fast convergence could be attributed in part to the almost-spherical shape of the tumor and the sharpness of the gradient. It is noticed that the results of the algorithm matches the PET ground truth (99%). Hence,

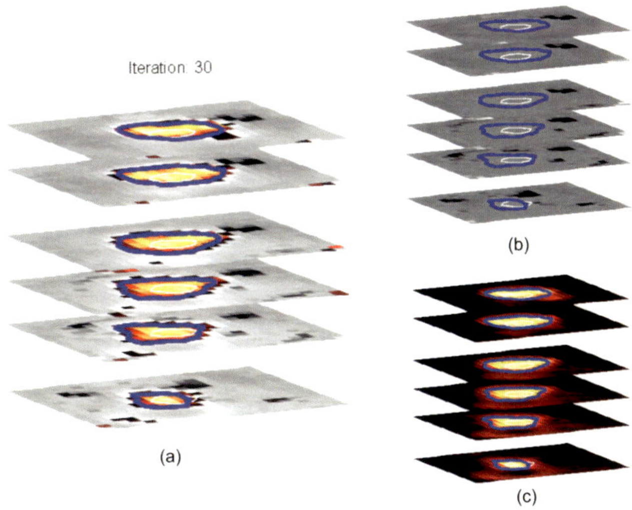

Figure 3.5 A 3D generalization of the MVLS algorithm in the case of PET/CT cervix. (a) The MVLS algorithm is initialized with a sphere (in white) diameter of 15.9 mm, curve evolution in steps of ten iterations (in magenta), and the final estimated contour (in thick blue). The algorithm converged in 30 iterations. (b) MVLS-estimated contour superimposed on CT. (c) MVLS-estimated contour superimposed on PET.

the delineation results were explained mainly by PET in this case, although information from CT could still be used to steer the algorithm, if desired.

3.4.3 MR/CT segmentation of prostate cancer example

A more-challenging case was the analysis of a prostate MRI/CT, shown in Fig. 3.6(a). The images were co-registered using the rigid-body mutual information algorithm. The normalized mutual information (NMI)

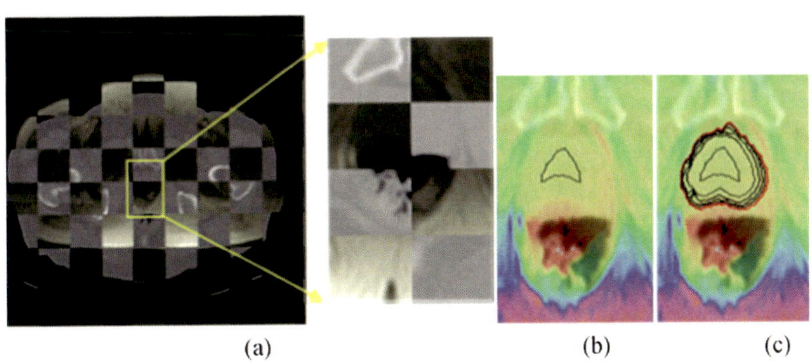

Figure 3.6 Analysis of a prostate MRI/CT. (a) Co-registered MRI/CT and selected ROI, (b) the MVLS algorithm is initialized with a shape prior that roughly resembles prostate (in black), and (c) the curve evolution in steps of ten iterations and the final estimated contour (in thick red).

improved from 1.07 to 1.11. The results of this example seemed to be more dependent on the initial shape; hence, the initial contour (in white) was emphasized in the algorithm as prior knowledge, as shown in Fig. 3.6(b). The evolved contour is shown in steps of 10 iterations (in blue) and the final estimated contour (in thick red) in Fig. 3.6(c). The algorithm converged in 50 iterations (less than 1 s). Note that in this case the delineation of the prostate in the CT image is significantly improved by incorporating the MR. On the other hand, CT also improved the convergence results of MR by offering an additional external force to guide the curve evolution to the desired target's boundary.

3.4.4 Coronary artery plaque MR image analysis

The composition of an atherosclerotic plaque is an important predictor for thromboembolic events. Intraplaque hemorrhage and a lipid core are considered especially high-risk components.[44] Multicontrast MR images can be applied to detect these components. This example used multiple MR images that were acquired by scanning cadavers' coronary artery tissue samples. The *ex vivo* tissue samples were scanned using T1w (T1 weighting), T2w (T2 weighting), and PD (proton density) sequences. The generalized "active-contour-without-edges" algorithm was used to perform automatic image segmentation and compare the results to contours manually drawn by an experienced radiologist for demonstration purposes. The artery lumen, the calcification component, and the lipid core component were contoured, as shown in Fig. 3.7.

3.5 Issues, Problems, and Future Directions

While advances in multimodality imaging have created ample opportunities in diagnostic and therapeutic radiology, they have also presented many technical challenges to integrating information from different modalities, which affects visualization, delineation, and related areas.

A set of applicable algorithms for concurrent segmentation of images from different imaging modalities has been presented here. The underlying principle in this chapter is to combine complementary information from different imaging sources to better understand the nature of the imaged object relevant to the depicted clinical task. This would result in a better target definition in radiotherapy treatment planning of cancer by integrating anatomical and physiological information to decide the target extent, and better classification of different tissue types from multispectral MR images in cardiac diagnosis routines, for instance. The multiple images to be integrated into such a framework need not be from different image modalities; they could be from the same image modality but with different acquisition protocols (e.g., different MRI pulse sequences), or from the same acquisition protocol but at different times (e.g., 4D-CT images). Despite the framework's

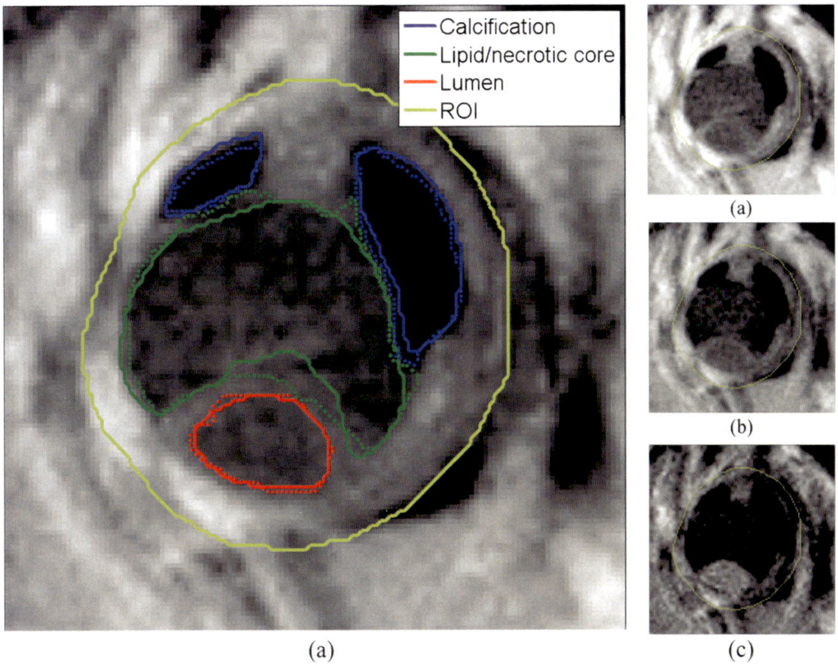

Figure 3.7 Examples of *ex vivo* coronary artery plaque segmentation: (a) the PD slice, (b) the T1w slice, (c) the T2w slice, and (d) the PD slice overlaid with both manual contours and automatically segmented contours for calcification, lipid components, and artery lumen. The solid yellow lines are the manually drawn ROI regions. The other solid lines are manually drawn contours. The dotted lines are the automatic segmentation results. The expert's manual contour and the multimodality automated segmentation results are in good agreement.

strong potential for streamlining multiple-imaging-modality integration for better target definition during treatment planning and better localization during treatment fractions, several challenges must be addressed before clinical implementation.

3.5.1 Image understanding

The biophysical target formalism presented in Eq. (3.1) relies on characterization of higher-level human expertise that must be translated into primitives understandable by a computer algorithm. A new weighting-parameters approach to control the influence of each image in the multimodality image dataset has been introduced.[19] These weighting parameters should be determined according to the user's experience and prior knowledge. However, this procedure may be too simplistic, and this issue is still an open area for future research. A promising direction would be the utilization of human–machine interaction methods to capture radiologists' perception of image similarity. This concept has been previously applied to content-based image retrieval (CBIR) from mammographic databases successfully,[45,46] which could be extended into the multimodality framework by using machine learning algorithms.

3.5.2 Deformable image registration

Image misalignment is a major issue when dealing with images acquired from different scanners, which has been partially resolved for PET/CT but not for many other combined modalities. Further improvements to performance could be further achieved by combining segmentation and deformable image registration algorithms in an iterative framework. Improvement in the robustness and effectiveness of coupling segmentation and registration has been recently demonstrated.[47] The working idea is that prior knowledge of the boundary obtained from segmentation improves the registration, as in finite element registration methods.[48] At the same time, the registration can improve locally weak boundary segmentation via a correspondence to stronger edges in the other image (template), as in atlas-based segmentation.[49]

3.6 Conclusions

This chapter presented a framework for tumor volume definition using multimodality imaging in radiation oncology. It has shown that many different single-modality image segmentation algorithms could be extended to support simultaneous multimodality image analysis tasks. This was demonstrated by using clinical examples from radiation oncology as well as diagnostic radiology. In addition, the current challenges in the field were discussed, as well as the opportunities for medical image scientists to improve tumor definition using the multimodality framework to support better diagnostic and therapeutic interventions of cancer.

References

1. J. H. Rudd, K. S. Myers, J. Sanz, and Z. A. Fayad, "Multimodality imaging of atherosclerosis (magnetic resonance imaging/computed tomography/positron emission tomography-computed tomography)," *Top. Magn. Reson. Imaging* **18**, 379–388 (2007).
2. A. D. King, "Multimodality imaging of head and neck cancer," *Cancer Imaging* **7**(A), S37–46 (2007).
3. A. R. Hsu et al., "Multimodality molecular imaging of glioblastoma growth inhibition with vasculature-targeting fusion toxin VEGF121/rGel," *J. Nucl. Med.* **48**, 445–454 (2007).
4. D. W. Townsend, "Multimodality imaging of structure and function," *Phys. Med. Biol.* **53**, R1–R39 (2008).
5. I. El Naqa, "Radiotherapy informatics: targeted control," *Enterprise Imaging and Therapeutic Radiology Management* **18**, 39–42 (2008).

6. W. L. Smith et al., "Prostate volume contouring: a 3D analysis of segmentation using 3DTRUS, CT, and MR," *Int. J. Rad. Oncol. Biol. Phys.* **67**, 1238–1247 (2007).
7. E. M. Toloza, L. Harpole, and D. C. McCrory, "Noninvasive staging of non-small cell lung cancer: a review of the current evidence," *Chest* **123**, 137S–146S (2003).
8. J. D. Bradley, C. A. Perez, F. Dehdashti, and B. A. Siegel, "Implementing biologic target volumes in radiation treatment planning for non-small cell lung cancer," *J. Nucl. Med.* **45**(1), 96S–101S (2004).
9. S. Milker-Zabel et al., "Improved target volume definition for fractionated stereotactic radiotherapy in patients with intracranial meningiomas by correlation of CT, MRI, and [68Ga]-DOTATOC-PET," *Int. J. Rad. Oncol. Biol. Phys.* **65**, 222–227 (2006).
10. D. Yang et al., "Techniques and software tool for 3D multimodality medical image segmentation," *J. Rad. Oncol. Informatics* **1**(1), 1–21 (2009).
11. D. L. Pham, C. Xu, and J. L. Prince, "Current methods in medical image segmentation," *Ann. Rev. Biomed. Eng.* **2**, 315–337 (2000).
12. J. S. Suri, S. K. Setarehdan, and S. Singh, *Advanced Algorithmic Approaches to Medical Image Segmentation: State-of-the-Art Applications in Cardiology, Neurology, Mammography, and Pathology*, Springer, New York (2002).
13. H. Zaidi, M. Abdoli, C. L. Fuentes, and I. M. El Naqa, "Comparative methods for PET image segmentation in pharyngolaryngeal squamous cell carcinoma," *Eur. J. Nucl. Med. Mol. Imaging* **39**(5), 881–891 (2012).
14. H. Zaidi and I. El Naqa, "PET-guided delineation of radiation therapy treatment volumes: a survey of image segmentation techniques," *Eur. J. Nucl. Med. Mol. Imaging* **37**(11), 2165–2187 (2010).
15. T. S. Yoo et al., "Engineering and algorithm design for an image processing API: a technical report on ITK - The Insight Toolkit," *Proc. Medicine Meets Virtual Reality*, 586–592 (2002).
16. X. Papademetris, M. Jackowski, N. Rajeevan, R. T. Constable, and L. Staib, "BioImage Suite: An integrated medical image analysis suite," *The Insight Journal*, http://hdl.handle.net/1926/209 (2005).
17. M. J. McAuliffe et al., "Medical Image Processing, Analysis and Visualization in Clinical Research," *Proc. Fourteenth IEEE Symp. Computer-Based Medical Systems*, 381–386 (2001).
18. M. D. Abramoff, P. J. Magelhaes, and S. J. Ram, "Image Processing with ImageJ," *Biophoton. Int.* **11**, 36–42 (2004).
19. I. El Naqa et al., "Concurrent multimodality image segmentation by active contours for radiotherapy treatment planning," *Med. Phys.* **34**, 4738–4749 (2007).

20. A. Sebbahi, A. Herment, A. de Cesare, and E. Mousseaux, "Multimodality cardiovascular image segmentation using a deformable contour model," *Comput. Med. Imaging and Graphics* **21**, 79–89 (1997).
21. J. Zheng et al., "Quantitative assessment of coronary artery plaque vulnerability by high-resolution magnetic resonance imaging and computational biomechanics: A pilot study ex vivo," *Magn. Reson. Med.* **54**, 1360–1368 (2005).
22. J. Zheng et al., "Quantitative assessment of coronary artery plaque vulnerability by high-resolution magnetic resonance imaging and computational biomechanics: a pilot study ex vivo," *Magn. Reson. Med.* **54**, 1360–1368 (2005).
23. U. Nestle et al., "Comparison of different methods for delineation of 18F-FDG PET-positive tissue for target volume definition in radiotherapy of patients with non-small cell lung cancer," *J. Nucl. Med.* **46**, 1342–1348 (2005).
24. S. Hu, E. A. Hoffman, and J. M. Reinhardt, "Automatic lung segmentation for accurate quantitation of volumetric X-ray CT images," *IEEE Trans. Med. Imaging* **20**, 490–498 (2001).
25. A. K. Jain, M. N. Murty, and P. J. Flynn, "Data clustering: a review," *ACM Comput. Surv.* **31**, 264–323 (1999).
26. J. C. Bezdek, *Pattern Recognition with Fuzzy Objective Function Algorithms*, Plenum Press, New York (1981).
27. D. L. Pham, "Spatial models for fuzzy clustering," *Comput. Vis. Image Underst.* **84**, 285–297 (2001).
28. C. Weiling, C. Songcan, and Z. Daoqiang, "Fast and robust fuzzy c-means clustering algorithms incorporating local information for image segmentation," *Pattern Recogn.* **40**, 825–838 (2007).
29. M. R. Anderberg, *Cluster Analysis for Applications*, Academic Press, New York (1973).
30. J. A. Sethian, *Level Set Methods and Fast Marching Methods: Evolving Interfaces in Computational Geometry, Fluid Mechanics, Computer Vision, and Material Science*, Cambridge University Press, Cambridge, UK (1999).
31. C. Xu, D. L. Pham, and J. L. Prince, "Image Segmentation Using Deformable Models," Chapter 3 in *Handbook of Medical Imaging: Medical Image Processing and Analysis*, Vol. 2, M. Sonka and J. M. Fitzpatrick, Eds., 129–174, SPIE Press, Bellingham, WA (2002) [doi: 10.1117/3.831079.ch3].
32. G. Aubert, *Mathematical Problems in Image Processing: Partial Differential Equations and the Calculus of Variations*, Springer, New York (2006).

33. S. Osher and R. P. Fedkiw, *Level Set Methods and Dynamic Implicit Surfaces*, Springer, New York (2003).
34. M. Kass, A. Witkin, and D. Terzopoulos, "Snakes: Active contour models," *Int. J. Comput. Vision* **1**, 321–331 (1988).
35. V. Caselles, R. Kimmel, and G. Sapiro, "Geodesic active contours," *Int. J. Comput. Vision* **22**, 61–79 (1997).
36. T. F. Chan and L. A. Vese, "Active contours without edges," *IEEE Trans. Image Process.* **10**, 266–277 (2001).
37. T. F. Chan, B. Y. Sandberg, and L. A. Vese, "Active contours without edges for vector-valued images," *J. Visual Comm. Image Rep.* **11**, 130–141 (2000).
38. J. Shah, "Curve evolution and segmentation functionals: application to color images," *Proc. Int. Conf. Image Processing*, 461–464 (1996).
39. T. Beyer et al., "A combined PET/CT scanner for clinical oncology," *J. Nucl. Med.* **41**, 1369–1379 (2000).
40. J. Bradley et al., "Impact of FDG-PET on radiation therapy volume delineation in non-small-cell lung cancer," *Int. J. Rad. Oncol. Biol. Phys.* **59**, 78–86 (2004).
41. J. Hanley et al., "Deep inspiration breath-hold technique for lung tumors: the potential value of target immobilization and reduced lung density in dose escalation," *Int. J. Rad. Oncol. Biol. Phys.* **45**, 603–611 (1999).
42. B. Zangheri et al., "PET/CT and breast cancer," *Eur. J. Nucl. Med. Mol. Imaging* **31**(1), S135–142 (2004).
43. I. El Naqa, D. A. Low, J. D. Bradley, M. Vicic, and J. O. Deasy, "Deblurring of breathing motion artifacts in thoracic PET images by deconvolution methods," *Med. Phys.* **33**, 3587–3600 (2006).
44. V. C. Cappendijk et al., "Assessment of human atherosclerotic carotid plaque components with multisequence MR imaging: initial experience," *Radiology* **234**, 487–492 (2005).
45. I. El Naqa, Y. Yang, N. P. Galatsanos, R. M. Nishikawa, and M. N. Wernick, "A similarity learning approach to content-based image retrieval: application to digital mammography," *IEEE Trans. Med. Imaging* **23**, 1233–1244 (2004).
46. J. H. Oh, Y. Yang, and I. El Naqa, "Adaptive learning for relevance feedback: application to digital mammography," *Med. Phys.* **37**, 4432–4444 (2010).
47. M. Droske and M. Rumpf, "Multiscale joint segmentation and registration of image morphology," *IEEE Trans. Pattern Anal. Mach. Intell.* **29**, 2181–2194 (2007).
48. J. M. Hensel et al., "Development of multiorgan finite element-based prostate deformation model enabling registration of endorectal coil

magnetic resonance imaging for radiotherapy planning," *Int. J. Rad. Oncol. Biol. Phys.* **68**, 1522–1528 (2007).

49. S. Klein et al., "Automatic segmentation of the prostate in 3D MR images by atlas matching using localized mutual information," *Med. Phys.* 35, 1407–1417 (2008).

Issam El Naqa received his B.S. (1992) and M.S. (1995) in Electrical and Communication Engineering from the University of Jordan, Amman, Jordan; his Ph.D. (2002) in Electrical Engineering from the Illinois Institute of Technology, Chicago; and his M.A. (2007) in Biology Science from Washington University in St. Louis, where he completed a postdoctoral fellowship in medical physics, was subsequently a faculty member in the Departments of Radiation Oncology and Biomedical and Biological Sciences, and served as an adjunct faculty at the Department of Electrical Engineering. He is currently an Associate Professor at McGill University Health Centre/Medical Physics Unit. He is a recognized expert in the fields of image processing, bioinformatics, computational biology, and treatment outcomes modeling and has published extensively in these areas. He is an acting member of several academic and professional societies, which include IEEE, AAPM, and ASTRO. His research has been funded by several federal and private grants and serves as a peer reviewer and associate editor for several international journals in his areas of expertise.

Chapter 4
Nonlinear Unsharp Masking for Enhancing Suspicious Regions in Mammograms

Yicong Zhou and C. L. Philip Chen
Department of Computer and Information Science, University of Macau, Macau, China

Sos S. Agaian
Department of Electrical and Computer Engineering, University of Texas at San Antonio, San Antonio, TX, USA

Karen Panetta
Department of Electrical and Computer Engineering, Tufts University, Medford, MA, USA

4.1 Introduction

Breast cancer is the leading cause of death in women between the ages of 35 and 55. The National Cancer Institute estimates that one out of eight women in the United States will develop breast cancer at some point during her lifetime.[1] The mortality rate of 30% in the U. S. and 45% in Europe has been demonstrated by repeated, randomized, and controlled trials.[2] Currently, there are no effective ways to prevent breast cancer.[3,4] However, treatments of breast cancer in the early stages are more successful; therefore, early detection is an important and effective method to significantly reduce mortality. There are several imaging techniques for breast examination, including magnetic resonance imaging (MRI), ultrasound imaging, positron emission tomography (PET) imaging, computed tomography (CT) imaging, optical tomography/ spectroscopy, and x-ray imaging. Among them, mammography (x-ray image) is the most-common technique that radiologists use to detect and diagnose breast cancer.[5,6] Two types of mammography are currently used: film mammography and digital mammography. Digital mammography is preferred by physicians[7,8] because it has better image quality, requires a lower x-ray dose,[8] provides interpretations with greater confidence in difficult cases, and offers faster diagnosis for routine cases.[9]

Due to the limitations of the x-ray hardware systems, screened mammograms—even when using digital mammography—may present low resolution or low contrast, making it difficult to detect tumors at an early stage. Important indicators of early breast cancer,[10,11] such as irregularly shaped microcalcifications, are very small calcium deposits that appear as bright, granular spots in mammograms.[12,13] The distinction between the tiny malignant tumors and the benign glandular tissue is not readily discernable; misinterpretation results in unnecessary additional examinations and biopsy.[14] The situation becomes worse when radiologists routinely interpret large numbers of mammograms and can misdiagnose a condition.[15]

To improve the visual quality of mammographic images, more image data can be collected at the data acquisition stage, thus improving the image resolution. However, this significantly increases the overall acquisition time, the amount of radiation that a patient is exposed to, and hardware costs.[16] On the other hand, the image visual quality can be enhanced during the post-image-processing stage in medical imaging systems. It utilizes different image enhancement techniques to enhance the contrast of mammograms. In this way, the visual quality of mammograms is improved without affecting the acquisition process or increasing the hardware costs.

The underlying concept of mammogram enhancement involves applying image enhancement algorithms to improve the contrast of suspicious regions and objects in mammograms, and then use a threshold to separate them from their surroundings.[11] To employ it in the medical imaging system, two problems need to be addressed: (1) How to automatically choose the best enhancement algorithm, and (2) how to automatically select the thresholding.

Several algorithms for mammogram enhancement have been developed recently. They can be classified into two main categories: frequency domain methods and spatial domain methods.

- **Frequency domain methods** are based on different transforms or fuzzy logic theory. These transforms include the discrete Fourier transform (DFT),[17–23] discrete cosine transform (DCT),[24–26] discrete wavelet transform (DWT),[27–44] and other transforms.[23,26,45,46]

  The DWT-based enhancement algorithms for mammograms first decompose mammograms into a multiscale subband representation using the contourlet transform[30] or other wavelet transforms such as the discrete dyadic wavelet transform,[31,33–35] integrated wavelets,[36] or redundant discrete wavelet transform.[37] Next, the transform coefficients in each subband of the multiscale representation are modified using different technologies, including nonlinear filtering,[38] regression-based extrapolation,[39] adaptive unsharp masking,[40] the wavelet shrinkage function,[41] or direct contrast modification.[42] Finally, the enhanced mammograms can be obtained from the modified coefficients. However, it has been reported that a wavelet representation does not efficiently show the contours and the geometry of edges in images.[30]

Fuzzy set theory has been used to enhance the contrast of mammograms because it is suitable for dealing with the uncertainty associated with the definition of image edges, boundaries, and contrast.[4,47–49] Fuzzy logic has also been successfully integrated with other techniques such as histogram equalization for enhancing medical images[48] and structure tensor for contrast enhancement of micro-calcifications in digital mammograms.[49] However, the frequency domain methods have limitations. They may introduce artifacts called "objectionable blocking effects"[17] or enhance images globally but not enhance all local details/regions in the image very well. Furthermore, it is very difficult to apply them to automatic image enhancement procedures.[18,19]

- **Spatial domain methods** are based on the human visual system (HVS),[50–53] empirical decomposition,[54–57] histogram equalization,[48,52,58–62] logarithmic framework,[63–66] nonlinear filtering,[43,53,67–70] adaptive neighborhood,[9,15,71,72] or unsharp masking.[73–75]

Because nonlinear filtering is known for its ability to obtain more robust characteristics for suppressing noise and preserving edges and details, it is a desirable technique that can be used to enhance mammographic images and other types of medical images. Examples include utilizing the adaptive density-weighted filter,[67] tree-structured nonlinear filters,[43] and also adaptive anisotropic filtering.[76]

Several other algorithms have been developed for mammogram enhancement using adaptive neighborhood (or region-based) contrast enhancement (ANCE).[9,15,71,72] ANCE is intended to improve the contrast of specific regions, objects, and details in mammograms based on the local-region background and contrast. The region contrast is calculated and enhanced according to the region's contrast, its background, its neighborhood size, and its seed pixel value.[9]

Unsharp masking (UM) is another interesting enhancement technique belonging to spatial domain methods. The traditional UM has good performance to enhance the fine details in the original images. However, it also amplifies noise and overshoots the sharp details at the same time.[75,77] To overcome this problem, several modification schemes have been developed by replacing the high-pass filter with the adaptive filter,[77] quadratic filter,[78] and its derived filtering operators, called rational unsharp masking[75] and cubic unsharp masking.[79] Other algorithms using unsharp masking techniques for mammogram enhancement have been developed.[40,73,74] A set of measure metrics for mammogram enhancement is introduced in Singh and Bovis.[80]

This chapter introduces a new nonlinear unsharp masking (NLUM) scheme for mammogram enhancement by combining the nonlinear filtering and unsharp masking techniques. Leveraging the advantages of these two techniques, the new scheme can enhance the contrast of suspicious regions, objects, and details to achieve better visibility of mammographic images for human observers (radiologists). Furthermore, to address the two questions posed earlier concerning the automatic selection of the best enhancement algorithm and of the threshold, a new enhancement measure called the second-derivative-like measure of enhancement (SDME) is introduced. Different parameters in the enhancement algorithm are varied, and the results are measured automatically to choose the best one to present. The NLUM enhancement performance is demonstrated by comparison with other existing enhancement algorithms, quantitative evaluation using the SDME measure, and receiver operating characteristic (ROC) analysis based on a medical doctor's inspection.

Section 4.2 reviews several existing enhancement algorithms that are to be compared with the new NLUM scheme, and the operations of the parameterized logarithmic image processing (PLIP) to be consistent with the human visual system. Section 4.3 introduces the new NLUM scheme. Section 4.4 introduces the new enhancement measure after reviewing several existing ones for quantitatively evaluating the performance of enhancement algorithms. Section 4.5 shows the parameter design and optimization for the NLUM scheme using the SDME measure, compares the NLUM scheme with three existing enhancement algorithms, and evaluates the NLUM using the thresholding technique and ROC analysis. Section 4.6 reaches a conclusion.

4.2 Background

This section briefly discusses traditional unsharp masking and four existing enhancement algorithms including rational unsharp masking (RUM),[75] adaptive neighborhood contrast enhancement (ANCE),[9] contrast-limited adaptive histogram equalization (CLAHE),[58] and direct image contrast enhancement (DICE).[42] Those algorithms will form the basis for comparison to the new NLUM scheme. The arithmetic operations of parameterized logarithmic image processing (PLIP) are also presented here and will be used as an operator in the presented NLUM scheme to better represent the human visual system response.

4.2.1 Traditional unsharp masking

The foundation of the traditional unsharp masking (UM) technique involves subtracting a low-pass filtered signal from its original. The same results can be achieved by adding a scaled high-frequency part of the signal to its original. This is equivalent to adding the scaled gradient magnitude back to the original signal.[78]

The unsharp masking is used to improve the visual quality of images by emphasizing their high-frequency portions that contain fine details as well as

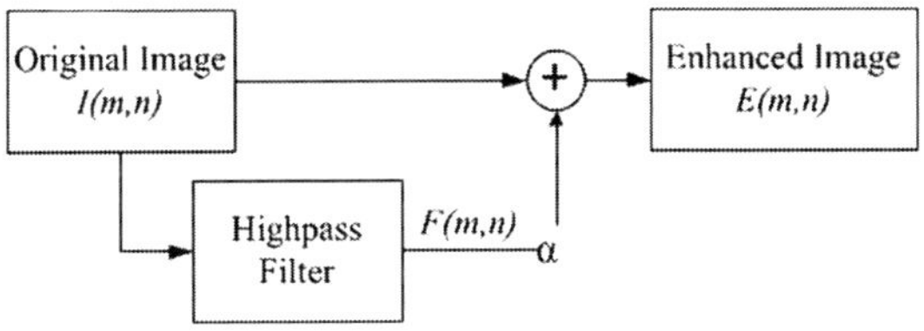

Figure 4.1 The block diagram of the traditional unsharp masking.

noise and sharp details. The scheme for image enhancement is shown in Figure 4.1.

The output enhanced image $E(m,n)$ is defined by

$$E(m,n) = I(m,n) + \alpha F(m,n), \tag{4.1}$$

where the constant α is a scaling factor, and $F(m,n)$ is a high-pass filtered image obtained from the original $I(m,n)$.

The high-pass filter and scaling process in traditional UM amplify those high-frequency portions of original images that contain fine details as well as noise and sharp details. Therefore, due to the fact that traditional UM enhances fine details in images, it also amplifies noise while overenhancing the steep edges.

4.2.2 The RUM algorithm

Rational unsharp masking[75] (RUM) uses a rational function operator to replace the high-pass filter in traditional unsharp masking, shown in Figure 4.2. The rational function is the ratio of two polynomials of the input variables. This scheme is intended to enhance the details in images containing low and medium sharpness without significantly amplifying the noise or affecting the steep edges. The enhanced image is defined by

$$E(m,n) = I(m,n) + \lambda[C_x(m,n)F_x(m,n) + C_y(m,n)F_y(m,n)], \tag{4.2}$$

where λ is the scaling factor, and

$$C_x(m,n) = \frac{[I(m,n+1) - I(m,n-1)]^2}{k[I(m,n+1) - I(m,n-1)]^4 + h}, \tag{4.3}$$

$$C_y(m,n) = \frac{[I(m+1,n) - I(m-1,n)]^2}{k[I(m+1,n) - I(m-1,n)]^4 + h}, \tag{4.4}$$

$$F_x(m,n) = 2I(m,n) - I(m,n-1) - I(m,n+1), \tag{4.5}$$

$$F_y(m,n) = 2I(m,n) - I(m-1,n) - I(m+1,n), \tag{4.6}$$

where k and h are proper positive factors.

4.2.3 The ANCE algorithm

The adaptive neighborhood contrast enhancement (ANCE) method[9] was developed to improve the contrast of objects and features with varying sizes and shapes. In this algorithm, each pixel in an image is considered a seed pixel for a region-growing process. Including those neighboring pixels whose gray values are within a specified gray-level deviation (known as the growth tolerance k) from the seed, a local region—called the foreground—is generated around the seed pixel. Another region—called the background—consists of those neighboring pixels that are outside the range of a specified gray-level deviation. The background, which surrounds the foreground, contains nearly the same number of pixels as the foreground. The region contrast is defined by

$$C = \frac{f - b}{f + b}, \tag{4.7}$$

where f and b are the mean gray-level value of the foreground and background, respectively.

The contrast equation in Eq. (4.7) is similar to Weber's ratio[59] $W = \Delta L/L$, where ΔL is the luminance difference between the central region and the overall image luminance L. The minimum contrast of the region is $C_{\min} = k/2$, and k is the growth tolerance. A Weber's rate of approximately 0.02 for a just-noticeable object under standard light conditions indicates that the growth tolerance should be at most 0.04 if regions or objects are to be distinguishable from their background.

The region's contrast is enhanced by increasing its foreground value when the following conditions are satisfied:

1. The region's contrast is low, i.e., $0.02 \leq C \leq 0.4$; and
2. The pixels in the region's background have a standard deviation normalized by their mean values less than 0.1.

The background in the second condition is defined as a region three pixels thick, molded to the original region in shape. The new foreground value is defined by

$$f' = b\frac{1 + C'}{1 - C'}, \tag{4.8}$$

where C' is the increased contrast based on an empirical look-up table described by Morrow et al.[9]

Therefore, only regions with low contrast are enhanced, whereas the high-contrast regions, such as steep edges, remain unaffected. In order to save computational costs, the redundant pixels in the foreground regions, which have the same values as the seed pixels, are changed to the same new values.

4.2.4 The CLAHE algorithm

The contrast-limited adaptive histogram equalization[58] (CLAHE) is a well-known technique for adaptive contrast enhancement. The normal and adaptive histogram equalizations enhance images using the integration operation. This operation yields large values in the enhanced image if the histogram of the nearly uniform regions of the original image contain several high peaks. As a result, those enhancement methods may overenhance noise and sharp regions in the original images. To solve this problem, the CLAHE algorithm uses a clip level to limit the local histogram in such a way that the amount of contrast enhancement for each pixel can be limited. This clip level is a maximum value of the local histogram specified by users. An interactive binary search process is used to redistribute those pixels that are beyond the clip level. The CLAHE algorithm has the following steps:

1. Divide the original image into contextual regions,
2. Obtain a local histogram for each pixel,
3. Clip the histogram based on the clip level,
4. Redistribute the histogram using binary search, and
5. Obtain the enhanced pixel value by histogram integration.

4.2.5 The DICE algorithm

The direct image contrast enhancement (DICE) algorithm was introduced to enhance screening mammograms in the wavelet domain.[42] It directly amplifies the vertical, horizontal, and diagonal subband components of the original image at different levels of the wavelet decomposition and then reconstructs them to obtain the enhanced image.

4.2.6 The PLIP operations

The parameterized logarithmic image processing (PLIP) model was introduced to provide a nonlinear framework for image processing.[64] The PLIP model can process images as absorption filters using the gray-tone function of images, which is a more-precise approach from a human visual system perspective, while keeping the image pixel values within the range $[0, \mu)$. Operations use the human visual system characteristics that are listed in Table 4.1, where $f(i,j)$ is the original image; $g(i,j)$, g, g_1, and g_2 are the gray-tone functions to generate negative photos of the original images; $\tilde{\oplus}, \tilde{\Theta}, \tilde{\otimes}$, and $\tilde{*}$ are PLIP addition, subtraction, scalar multiplication, and image multiplication, respectively; c and β are constants; and μ, γ, k, and λ are parameters that can be selected as the maximum value of images or other values. Note that the PLIP addition and scalar multiplication use the same parameter γ because the scalar multiplication is an extension of addition, adding the image to itself c times.[63]

Table 4.1 The PLIP operations.

PLIP Operation	Definition
Gray-tone function	$g(i,j) = \mu - f(i,j)$
Addition	$g_1 \tilde{\oplus} g_2 = g_1 + g_2 - \frac{g_1 g_2}{\gamma}$
Subtraction	$g_1 \tilde{\Theta} g_2 = k\frac{g_1 - g_2}{k - g_2}$
Scalar multiplication	$c \tilde{\otimes} g = \gamma - \gamma\left(1 - \frac{g}{\gamma}\right)^c$
Image multiplication	$g_1 \tilde{*} g_2 = \varphi^{-1}\left(\varphi(g_1) \cdot \varphi(g_2)\right)$, where $\varphi(g) = -\lambda \cdot \ln^{\beta}\left(1 - \frac{g}{\lambda}\right)$, and $\varphi^{-1}(g) = \lambda \cdot \left(1 - \exp\left(\frac{-g}{\lambda}\right)^{1/\beta}\right)$

4.3 Nonlinear Unsharp Masking

Integrating the nonlinear filtering operation with the unsharp masking technique, this section introduces a new unsharp masking scheme, called nonlinear unsharp masking (NLUM), for mammogram enhancement. This is a complex unsharp masking scheme. It is good at enhancing suspicious regions in mammographic images.

4.3.1 The new NLUM scheme

The block diagram of the NLUM scheme is shown in Figure 4.2. The original mammogram $I(m,n)$ is filtered by a nonlinear filter. The filtered mammogram $F(m,n)$ is then normalized and combined with the original mammogram using the fusion #1 and #2 to obtain an enhanced mammogram $E(m,n)$.

The nonlinear filtering operation applies a nonlinear operation to the pixels within a 3×3 window. Depending on the different applications, the filtering operation and fusion #1 and #2 can be selected as the arithmetic operations, the PLIP operations, or the nonlinear operations such as the mean square root or logic operations. This property makes the NLUM scheme more general, meeting more-complicated requirements for different objects and applications.

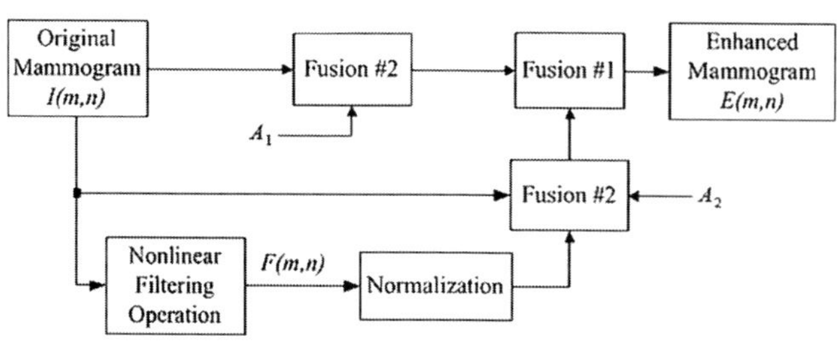

Figure 4.2 Block diagram of the proposed NLUM scheme.

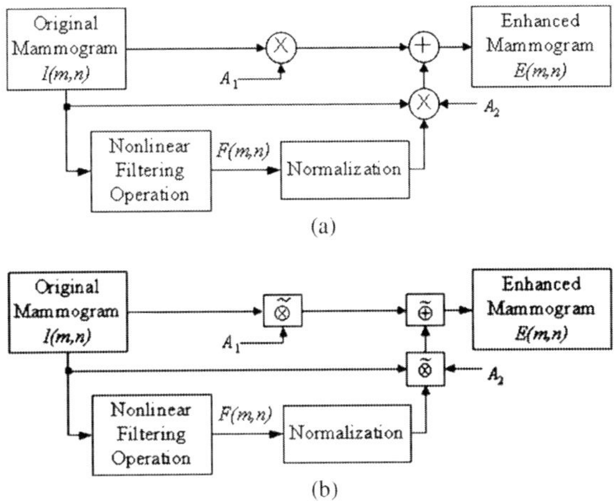

Figure 4.3 Practical examples of the proposed NLUM scheme (a) using the arithmetic operations and (b) using the PLIP operations.

If the NLUM scheme uses the arithmetic operations and fusion #1 and #2 are set to be the arithmetic addition and multiplication, respectively, then the NLUM scheme will resemble the flow shown in Figure 4.3(a). The enhanced mammogram is defined by

$$E(m,n) = A_1 I(m,n) + A_2 \frac{F(m,n)}{|F|_{\max}} I(m,n), \tag{4.9}$$

where A_1 and A_2 are the scaling factors, and $|F|_{\max}$ is the maximum absolute value of the mammogram $F(m,n)$ filtered by a 3×3 nonlinear filter defined by

$$F(m,n) = w_0 I_0 - w_1 I_1 - w_2 I_2, \tag{4.10}$$

where constants $w_0, w_1, w_2 \geq 0$ are weight coefficients, and

$$\begin{aligned} I_0 &= I^{2\alpha_0}(m,n), \\ I_1 &= I^{2\alpha_1}(m-1,n) + I^{2\alpha_1}(m+1,n) + I^{2\alpha_1}(m,n-1) + I^{2\alpha_1}(m,n+1), \qquad (4.11) \\ I_2 &= I^{2\alpha_2}(m-1,n-1) + I^{2\alpha_2}(m+1,n-1) + I^{2\alpha_2}(m+1,n-1) + I^{2\alpha_2}(m+1,n+1), \end{aligned}$$

where $\alpha_0, \alpha_1, \alpha_2$ are exponential coefficients, and $I(\bullet)$ is the image pixel intensity value.

On the other hand, if NLUM chooses the PLIP operations and fusion #1 and #2 are selected as the PLIP addition and multiplication, respectively, then the NLUM scheme follows the flow shown in Fig. 4.3(b). The NLUM output will change to

$$E(m,n) = A_1 \tilde{\otimes} I(m,n) \tilde{\oplus} A_2 \tilde{\otimes} \left(\frac{F(m,n)}{|F|_{\max}} \tilde{*} I(m,n) \right), \tag{4.12}$$

where the filtered mammogram $F(m,n)$ is defined as

$$F(m,n) = w_0 \tilde{\otimes} I_0 \tilde{\Theta} w_1 \tilde{\otimes} I_1 \tilde{\Theta} w_2 \tilde{\otimes} I_2 \tag{4.13}$$

and

$$\begin{aligned} I_0 &= I^{2\alpha_0}(m,n) \\ I_1 &= I^{2\alpha_1}(m-1,n) \tilde{\oplus} I^{2\alpha_1}(m+1,n) \tilde{\oplus} I^{2\alpha_1}(m,n-1) \tilde{\oplus} I^{2\alpha_1}(m,n+1) \\ I_2 &= I^{2\alpha_2}(m-1,n-1) \tilde{\oplus} I^{2\alpha_2}(m+1,n-1) \tilde{\oplus} I^{2\alpha_2}(m+1,n-1) \tilde{\oplus} I^{2\alpha_2}(m+1,n+1), \end{aligned} \tag{4.14}$$

where $\tilde{\oplus}, \tilde{\Theta}, \tilde{\otimes}, \tilde{*}$ are PLIP addition, subtraction, scalar multiplication, and image multiplication, respectively, and $A_1, A_2, w_0, w_1, w_2, \alpha_0, \alpha_1, \alpha_2$ are weight coefficients.

A pseudo-code implementation of the NLUM scheme appears below.

```
Input the original image I(m,n)
Set values for parameters A1, A2, w0, w1, w2, α0, α1, α2
   Switch (operation)
   Case: linear operation
         If (Fusion #1 = arithmetic addition) && (Fusion #2 =
         arithmetic multiplication)
               F(m,n) ⇐ apply Eq. (4.2) to input image I(m,n)
               E(m,n) ⇐ Eq. (4.1)
         End
   Case: PLIP operation
         If (Fusion #1 = PLIP addition) && (Fusion #2 = PLIP
         multiplication)
               F(m,n) ⇐ apply Eq. (4.5) to input image I(m,n)
               E(m,n) ⇐ Eq. (4.4)
         End
   End
Output the enhanced image E(m,n)
```

4.3.2 Discussion

NLUM is a complex unsharp masking scheme because there are eight coefficients to be specified for practical applications. However, more coefficients offer NLUM more power and design flexibility to meet more-complex and specific requirements in real-world applications. The nonlinear filtering operation in the NLUM scheme can be designed as a combination of two different types of filters, which offers NLUM more-robust characteristics. For example, the coefficients w_0, w_1, w_2 can be designed as a high-pass filter, and $\alpha_0, \alpha_1, \alpha_2$ can be chosen as a center-weighted mean filter.

The users can manually/experimentally select all of the NLUM coefficients. However, this is a time-consuming method that makes it difficult to reach the best enhancement results due to a lack of criteria for quantitative evaluation. Alternatively, the NLUM coefficients could be represented by one or two variables based on some reasonable assumptions to simplify the NLUM design and reduce the number of its coefficients in practical applications. An enhancement measure approach could then be used to optimize the coefficients, thus obtaining the best enhancement result. (This method is discussed in Section 4.5.2.)

In summary, the presented new NLUM scheme can be an embodiment of the following scenarios:

1. The fusion operators can be defined as different linear or nonlinear operations.
2. The new nonlinear filtering operator can be designed as a combination of different types of filters.
3. The coefficients allow users to change the NLUM properties to better meet application specific requirements.

These scenarios offer users more design flexibility to adapt the scheme to more specific and complicated requirements in real-world applications. The proposed NLUM can also be applied to other imaging modalities.

4.4 New Enhancement Measure

Developing a good quantitative measure to assess image enhancement is extremely difficult because the improvement in the enhanced images is often subjective and hard to measure. On the other hand, a good quantitative measure is important in order to select the best enhancement results for computer-aided detection (CAD) systems. This section reviews several existing methods of measuring the quality of image enhancement and then introduces a new enhancement measure using the concept of the second derivative.

4.4.1 Discussion

Several measures of image enhancement have been developed by using a contrast measure. The EME (measure of enhancement)[81] and the EMEE (measure of enhancement by entropy) have been developed by Agaian et al.[18] These two measures are based on a Weber-law-based contrast measure. Including the Michelson contrast law,[82] the AME (Michelson–Law measure of enhancement) and AMEE (Michelson–Law measure of enhancement by entropy) were later introduced to improve the measure performance of the EME and EMEE.[19] Because PLIP subtraction has been shown to be consistent with Weber's contrast law and characteristics of the human visual system,[65] the contrast information can be presented and processed more accurately. Including the PLIP operators to further improve these measures, Panetta et al.[51,83] have developed the logAME (logarithmic Michelson contrast measure) and logAMEE (logarithmic AME by entropy). The improved versions of the logAME are the SAME (similarity-based logAME) proposed by Wharton et al.[84] and the Global LogAMEE by Gao et al.[69]

All of these enhancement measures divide an image into $k_1 \times k_2$ blocks and then calculate the average values of the measure results of all blocks in the

Table 4.2 The definition of several enhancement measures.

	Name	Definition
1	Michelson contrast[82]	$C=\dfrac{I_{\max}-I_{\min}}{I_{\max}+I_{\min}}$
2	EME[18,81]	$EME_{k_1k_2}=\dfrac{1}{k_1k_2}\sum_{l=1}^{k_1}\sum_{k=1}^{k_2}\left[20\ln\left(\dfrac{I_{\max,k,l}}{I_{\min,k,l}}\right)\right]$
3	EMEE[18]	$EMEE_{\alpha k_1k_2}=\dfrac{1}{k_1k_2}\sum_{l=1}^{k_1}\sum_{k=1}^{k_2}\left[\alpha\left(\dfrac{I_{\max,k,l}}{I_{\min,k,l}}\right)^{\alpha}\ln\left(\dfrac{I_{\max,k,l}}{I_{\min,k,l}}\right)\right]$
4	AME[19,81]	$AME_{k_1k_2}=-\dfrac{1}{k_1k_2}\sum_{l=1}^{k_1}\sum_{k=1}^{k_2}\left[20\ln\left(\dfrac{I_{\max,k,l}-I_{\min,k,l}}{I_{\max,k,l}+I_{\min,k,l}}\right)\right]$
5	AMEE[19]	$AMEE_{\alpha k_1k_2}=-\dfrac{1}{k_1k_2}\sum_{l=1}^{k_1}\sum_{k=1}^{k_2}\left[\alpha\left(\dfrac{I_{\max,k,l}-I_{\min,k,l}}{I_{\max,k,l}+I_{\min,k,l}}\right)^{\alpha}\ln\left(\dfrac{I_{\max,k,l}-I_{\min,k,l}}{I_{\max,k,l}+I_{\min,k,l}}\right)\right]$
6	logAME[51,83]	$\log AME_{k_1k_2}=\dfrac{1}{k_1k_2}\,\tilde{\otimes}\sum_{l=1}^{k_1}\sum_{k=1}^{k_2}\left[\dfrac{1}{20}\,\tilde{\otimes}\ln\left(\dfrac{I_{\max,k,l}\tilde{\Theta}I_{\min,k,l}}{I_{\max,k,l}\,\tilde{\oplus}I_{\min,k,l}}\right)\right]$
7	logAMEE[51,83]	$\log AMEE_{k_1k_2}=\dfrac{1}{k_1k_2}\,\tilde{\otimes}\sum_{l=1}^{k_1}\sum_{k=1}^{k_2}\left[\left(\dfrac{I_{\max,k,l}\tilde{\Theta}I_{\min,k,l}}{I_{\max,k,l}\,\tilde{\oplus}I_{\min,k,l}}\right)\tilde{*}\ln\left(\dfrac{I_{\max,k,l}\tilde{\Theta}I_{\min,k,l}}{I_{\max,k,l}\,\tilde{\oplus}I_{\min,k,l}}\right)\right]$
8	SAME[84]	$SAME_{k_1k_2}=\dfrac{1}{k_1k_2}\,\tilde{\otimes}\sum_{l=1}^{k_1}\sum_{k=1}^{k_2}\dfrac{\lvert I_{\max,k,l}\rvert\tilde{\Theta}\lvert I_{\min,k,l}\rvert}{\lvert I_{\max,k,l}\rvert\,\tilde{\oplus}\lvert I_{\min,k,l}\rvert}$
9	Global LogAMEE[69]	$\mathrm{Glog}AMEE_{k_1k_2}=\dfrac{1}{k_1k_2}\,\tilde{\otimes}\sum_{l=1}^{k_1}\sum_{k=1}^{k_2}\left[\dfrac{(I_{\max,k,l}\tilde{\Theta}I_{\min,k,l})(I_{\max}\,\tilde{\oplus}I_{\min})}{(I_{\max,k,l}\,\tilde{\oplus}I_{\min,k,l})(I_{\max}\tilde{\Theta}I_{\min})}\tilde{*}\ln\left(\dfrac{(I_{\max,k,l}\tilde{\Theta}I_{\min,k,l})(I_{\max}\,\tilde{\oplus}I_{\min})}{(I_{\max,k,l}\,\tilde{\oplus}I_{\min,k,l})(I_{\max}\tilde{\Theta}I_{\min})}\right)\right]$
10	Region contrast[42]	$C_w(I)=\dfrac{1}{m}\sum_{w}\lvert c(x,y)\rvert\log\left(1+\lvert c(x,y)\rvert\right)$, where $c(x,y)=4I(x,y)-I(x-1,y)-I(x,y-1)-I(x+1,y)-I(x,y+1)$

entire image. The definitions of these measures are listed in Table 4.2, where the image I is divided into $k_1 \times k_2$ blocks, and α is constant. $I_{\max}$ and $I_{\min}$ are the maximum and minimum of the intensity values in these blocks, respectively. However, these enhancement measures only calculate the maximum and minimum values of the small regions or blocks in images. As a result, they are sensitive to noise and to steep edges in images.

Other enhancement measures include region contrast of a region in an image[42] $I(x, y)$ and contrast in the DCT domain.[85] A new enhancement measure is introduced here using the concept of the second derivative because it measures the change ratio of the variation speed of pixel values.

4.4.2 New enhancement measure: SDME

Integrating the idea of the second-derivative-like visibility operator[44] with the strengths of the earlier reviewed measures, a new enhancement measure called the second-derivative-like measure of enhancement (SDME) is introduced here. It is defined by

$$SDME = -\frac{1}{k_1 k_2}\sum_{l=1}^{k_1}\sum_{k=1}^{k_2} 20 \ln\left|\frac{I_{\max,k,l} - 2I_{center,k,l} + I_{\min,k,l}}{I_{\max,k,l} + 2I_{center,k,l} + I_{\min,k,l}}\right|, \tag{4.15}$$

where an image is divided into $k_1 \times k_2$ blocks, $I_{\max,k,l}$, $I_{\min,k,l}$ are the maximum and minimum values of the pixels in each block separately, and $I_{center,k,l}$ is the intensity of the center pixel in each block. Thus, the size of the blocks should be an odd number of pixels, such as 3×3 or 5×5.

Because $I_{center,k,l} \neq \pm\frac{1}{2}(I_{\max,k,l} + I_{\min,k,l})$ according to the SDME definition, the blocks with $I_{center,k,l} = \pm\frac{1}{2}(I_{\max,k,l} + I_{\min,k,l})$ will be discarded when calculating the SDME of an image. Therefore, when $I_{center,k,l}$ approaches $\pm\frac{1}{2}(I_{\max,k,l} + I_{\min,k,l})$, the SDME value will approach infinity; when $I_{center,k,l} = 0$ for all blocks, the minimal SDME value is zero.

4.5 Simulation Results and Evaluations

This section provides experimental results to discuss the SDME measure performance, the NLUM parameter optimization, and the NLUM enhancement analysis, comparison, and evaluation.

4.5.1 Comparison of enhancement measures

The SDME is compared with six existing measure methods. The measure performance of each method is determined by the consistency of the measure results and subjective evaluation of visual quality of mammograms.

The subjective evaluation method uses the mean opinion score (MOS) recommended by the International Telecommunication Union Telecommunication Standardization Sector (ITU-T).[86] The MOS intends to determine which results are the most visually pleasing for a human observer. In this subjective test, seven human observers visually evaluated all original and enhanced mammograms. Each mammogram was given a MOS score of 1–5, where a score of five indicates the best visual quality.

A set of 19 test mammograms was randomly selected from the Internet and the mini-MIAS database of mammograms.[87] They were enhanced using four algorithms: NLUM, RUM, ANCE, and CLAHE. Therefore, including the original and enhanced mammograms, there were 95 test images in total ($19 \times 5 = 95$) for this comparison. They were evaluated according to the subjective method and enhancement measures.

Table 4.3 shows the average subjective evaluation scores of each observer for the test mammograms. The bottom row lists the average evaluation scores of all human observers on enhanced images categorized by enhancement algorithms. Based on the scores, NLUM gives the best overall visual quality with a score of 4.6857, whereas CLAHE obtains the worst quality with a score of 1.9048.

The SDME and six existing measures are then used to measure the quality of all 95 test images. Each individual enhancement measure has its own data range; a good measure method should yield higher measure results for images with higher visual quality, and vice versa.

As shown in Table 4.4, different measures have diverse evaluation results for these enhancement algorithms. For example, the EME evaluates CLAHE-enhanced images as the best, whereas the AME gives the highest value to the

Table 4.3 Subjective evaluation for the enhanced results by different algorithms. The rating scale: 1 = bad, 2 = poor, 3 = fair, 4 = good, 5 = excellent.

Observer	Original	NLUM	RUM	ANCE	CLAHE
#1	3.5556	4.6111	3.3333	2.6111	1.4444
#2	3.6667	4.9444	3.0000	2.5556	1.7222
#3	3.4444	4.2222	3.1111	2.3889	1.6667
#4	3.7778	4.6889	3.0000	2.3333	2.0000
#5	3.3333	4.8889	3.2222	2.6667	2.6111
#6	3.8889	4.7778	3.2778	2.0000	2.1111
#7	4.1111	4.6667	3.4444	2.4444	1.7778
Average	**3.6825**	**4.6857**	**3.1984**	**2.4286**	**1.9048**

Table 4.4 Comparison of measure results based on different algorithms. For each individual enhancement measure, a higher score indicates better enhancement performance.

Enhancement Measure	Original	NLUM	RUM	ANCE	CLAHE
EME	0.9129	1.0833	1.0024	1.0023	2.5425
EMEE	0.0560	0.0688	0.0715	0.0614	0.1961
AME	26.4940	25.1455	26.3165	25.6358	17.4429
AMEE	0.0611	0.0679	0.0619	0.0653	0.1105
logAME	0.0526	0.0485	0.0522	0.0506	0.0316
logAMEE	0.0894	0.0993	0.0894	0.0942	0.1366
SDME	43.6388	47.2091	43.3729	42.2219	35.5386

original images. Comparing the MOS evaluation results in Table 4.4, the SDME is the only measure whose results are consistent with the MOS evaluation results. The rest of this chapter uses the SDME to assess the enhancement performance of different algorithms.

4.5.2 Parameter optimization

To demonstrate how to design and automatically optimize the NLUM parameters using the proposed SDME, one mammogram obtained from the Internet is used as an example; HVS-based image decomposition is then applied for the visualization and analysis of the enhanced results. The SDME is also used to measure and evaluate the performance of the NLUM for mammogram enhancement.

To assess the enhancement performance of the presented NLUM scheme, the users have the flexibility to use any existing measure approach to establish a qualitative metric of mammogram enhancement. The enhancement measure can also be used to optimize all of the NLUM coefficients to achieve the best enhanced results. Here, the SDME is selected to measure and evaluate the performance of NLUM for mammogram enhancement.

There are eight coefficients in NLUM: To reduce the number of parameters, the user can make assumptions according to the practical design requirements, for example, (1) $A_2 = 1/A_1$, $w_0 = 2$, $\alpha_0 = 8h$, $\alpha_1 = \alpha_2 = h$, and $w_1 = w_2 = -0.125$; or (2) $A_2 = 20A_1$, $w_0 = 8h$, $\alpha_0 = 12h$, $\alpha_1 = h$, $\alpha_2 = 2h$, and $w_1 = w_2 = -h$. These assumptions design the nonlinear filter as a combination of a high-pass filter (w_0, w_1, w_2) and a low-pass filter $(\alpha_0, \alpha_1, \alpha_2)$. More weight is given to the filtered image in order to enhance the fine details in images. With these assumptions, all of the NLUM coefficients are correlated with the parameters A_1 and h. Assumption (1) is selected here to demonstrate how to automatically design NLUM.

By automatically changing the parameters A_1 and h, several enhanced mammograms are generated and then measured by the SDME; the measure results are then plotted as a graph. The parameters giving the best enhanced result can be located at the points where the SDME curve reaches the local extrema.

Different fusion operations can be used in the NLUM scheme; compare the arithmetic operation with the PLIP version. Taking Figure 4.4(a) as a test image, the SDME measure results of the enhanced mammograms by NLUM with arithmetic and PLIP operations are plotted in Figure 4.5. The measure results allow one to find the location of parameters A_1 and h that yield the best enhanced result for each operation.

Using the parameters obtained from the measure in Figures 4.5(a) and (b), the original mammogram is enhanced by NLUM with arithmetic addition and PLIP addition, respectively. The enhanced mammograms and their cropped suspicious regions are shown in Figure 4.4. The visual quality and local contrast of the enhanced mammograms are much better than those of the original. Fine details such as microcalcifications and masses in the original

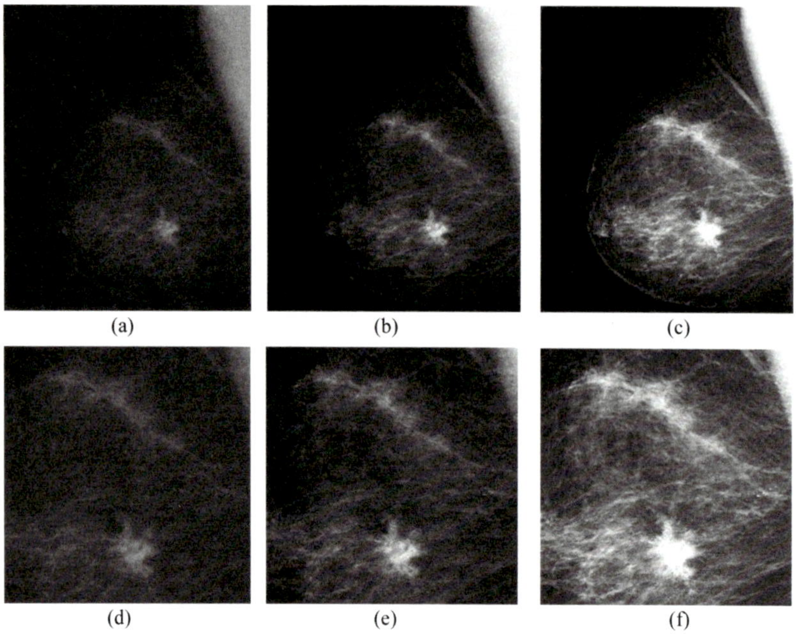

Figure 4.4 Mammogram enhancement: (a) the original mammogram, (b) the NLUM-enhanced mammogram with arithmetic operation, (c) the NLUM-enhanced mammogram with PLIP operation, (d) the cropped region of (a), (e) the cropped region of (b), and (f) the cropped region of (c).

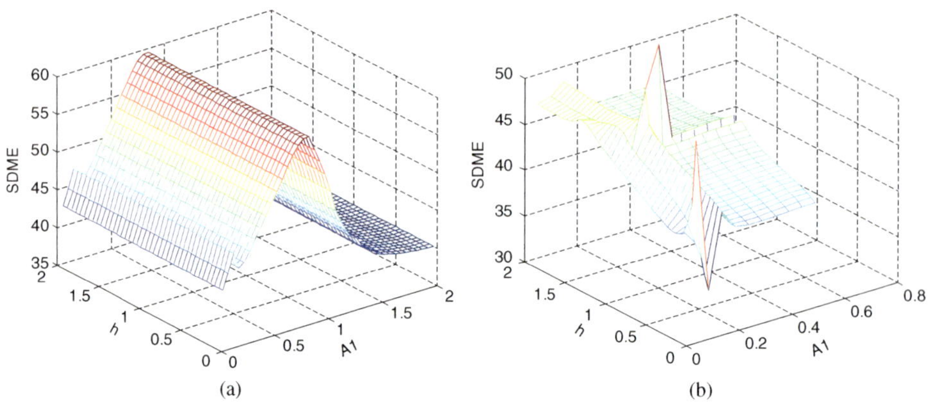

Figure 4.5 The SDME measure plots of mammogram enhancement based on different values of A_1 and h: (a) a SDME measure graph by arithmetic operation and (b) a SDME measure graph by PLIP operation.

mammogram are significantly improved, and the suspicious regions are more recognizable in the enhanced mammograms.

Compared with the enhanced results obtained by using two types of fusion operations in Figure 4.4, the arithmetic operation shows better performance

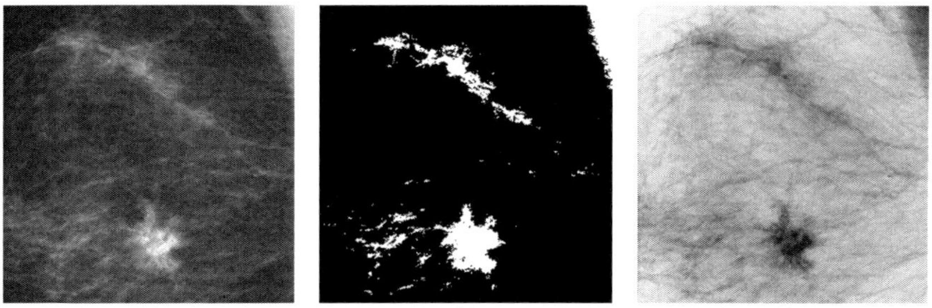

(a) (b) (c)

Figure 4.6 Enhancement analysis: (a) the enhanced region cropped from the mammogram in Figure 4.4(b), (b) the threshold image of (a), and (c) the negative photo of (a).

because the NLUM based on PLIP operation slightly overenhances the mass region, as shown in Figure 4.4(c). Therefore, the arithmetic operation is chosen for NLUM to enhance mammograms in the rest of this chapter.

4.5.3 Enhancement analysis

There are many different methods used to analyze the enhanced images. Figure 4.6 provides two examples: the negative view and thresholding of the specific region of interest (ROI). The shape of the suspicious regions is very clear and easily discernable. This demonstrates that NLUM performs well in improving the contrast of suspicious regions, objects, and details in mammograms.

4.5.4 HVS-based analysis and visualization

While the user can view the entire image's enhanced results, the process would be improved if only the suspicious regions could be emphasized during analysis. Instead of using the segmentation algorithms, HVS-based decomposition can be used as an alternative method to provide visualization of results that isolate ROIs, mainly suspicious regions.

By using the background intensity and the rate of information change, HVS-based decomposition separates images into four subimages based on four defined regions: (1) region 1: the saturation region for overilluminated areas; (2) region 2: the Weber region for properly illuminated areas; (3) region 3: the Devries–Rose region for underilluminated areas; (4) region 4: the fourth region for all pixels containing the least informative pixels.[88,89] This section extends its application to enhancement analysis and visualization.

Figures 4.7 and 4.8 show the HVS-based decomposition results of the enhanced mammogram and its negative (tonal inversion), respectively. In general, the mass regions can be segmented by HVS-based decomposition in one subimage without involving any segmentation algorithms. The results are shown in Figures 4.7(b) and 4.8(d). Therefore, HVS-based decomposition can be used for segmentation and classification of pathological cases in a CAD system.

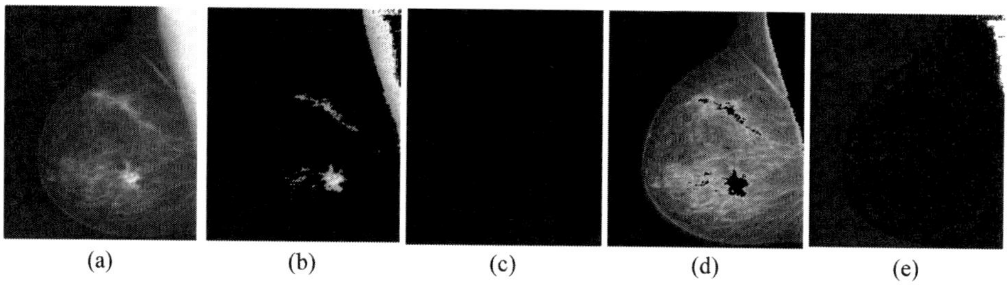

Figure 4.7 HVS-based decomposition of the enhanced mammogram: (a) the enhanced mammogram, (b) the first subimage, (c) the second subimage, (d) the third subimage, and (e) the fourth subimage.

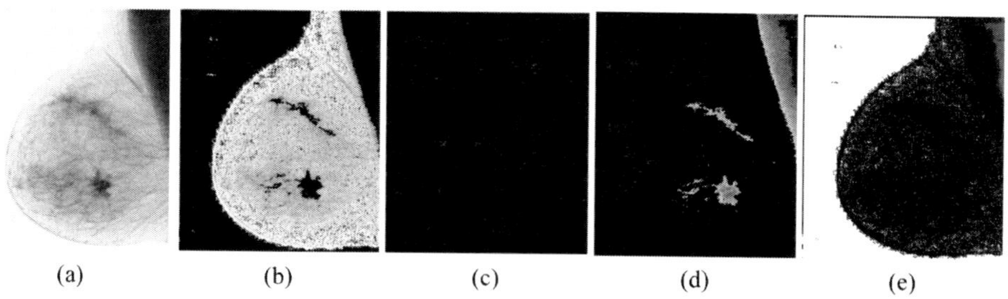

Figure 4.8 HVS-based decomposition of the inversed mammogram: (a) the negative of the image of the enhanced mammogram; (b) the first subimage; (c) the second subimage; (d) the third subimage; and (e) the fourth subimage.

4.5.5 Comparison of enhancement performance

After demonstrating how to automatically optimize the parameters in NLUM, this section applies it to more mammograms and compares it with other well-known enhancement algorithms.

The mammograms for this comparison were obtained from the mini-MIAS database of mammograms.[87] The database consists of 322 mammograms, and the cases of patient records range from fairly dense to extraordinarily dense breast parenchyma. Some cases are completely fatty. Most masses are ill-defined, indistinct, or speculated.

All test mammograms are cropped into smaller-size images for analysis such that the resulting cropped mammographic images contain most of the microcalcifications, masses, and suspicious regions that may be interesting to radiologists. These mammograms have a limited black background, which contains nonobject regions and background project noise.

Six mammograms were used as examples, and the enhanced results are shown in Figures 4.9 and 4.10. They clearly show how the enhancement algorithms change fine details and suspicious regions in images. Their SDME results are shown in Table 4.5.

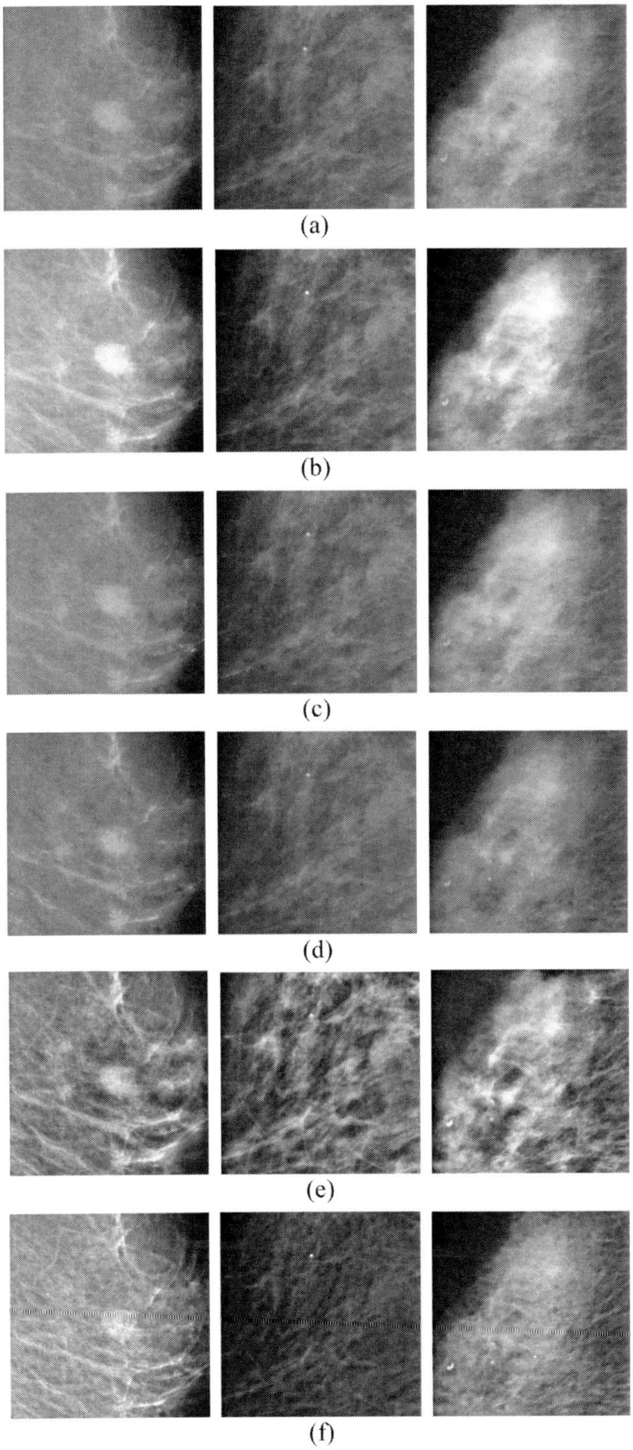

Figure 4.9 Comparison of mammogram enhancement using different algorithms: (a) Original mammograms (Mam_1 to Mam_3), (b) NLUM-enhanced results, (c) RUM-enhanced results, (d) ANCE-enhanced results, (e) CLAHE-enhanced results, and (f) DICE-enhanced results.

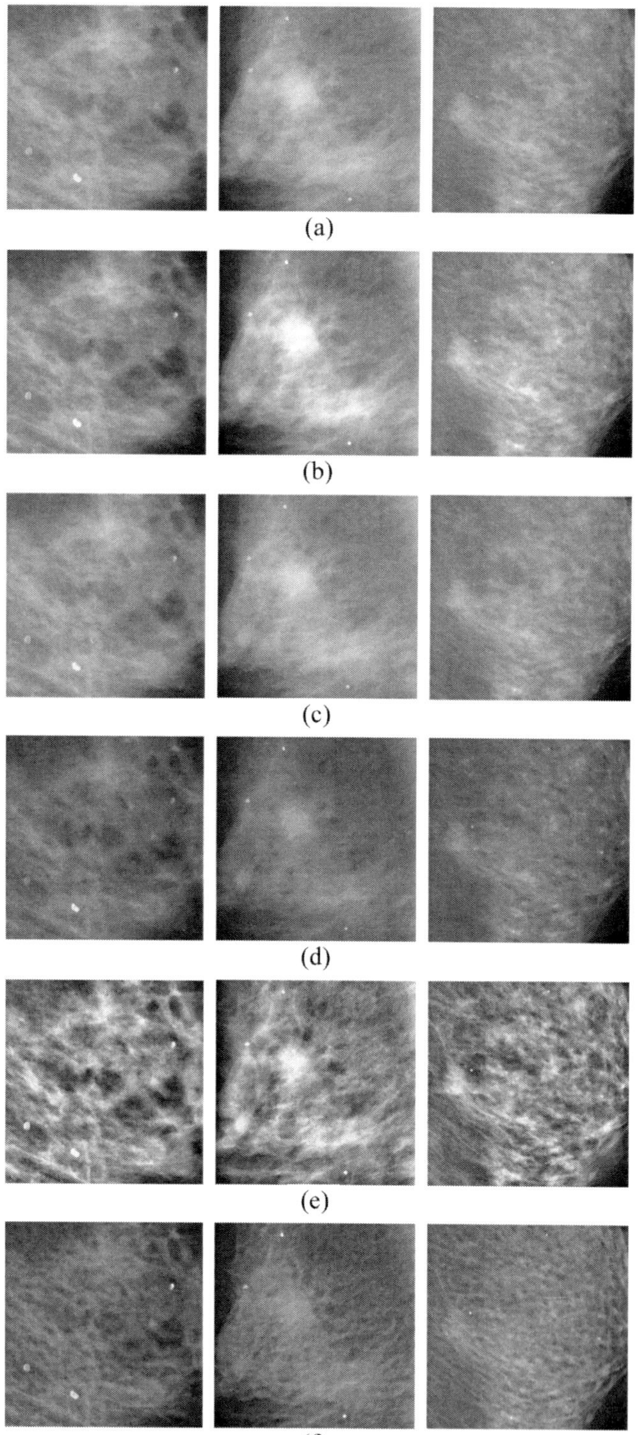

(a) (b) (c) (d) (e) (f)

Figure 4.10 Comparison of mammogram enhancement using different algorithms: (a) Original mammograms (Mam_4 to Mam_6), (b) NLUM-enhanced results, (c) RUM-enhanced results, (d) ANCE-enhanced results, (e) CLAHE-enhanced results, and (f) DICE-enhanced results.

Table 4.5 SDME results of mammograms enhanced by different algorithms. A higher score indicates better enhancement performance.

	Original	NLUM	RUM	ANCE	CLAHE	DICE
Mam_1	44.7378	47.4100	44.5905	41.9886	36.1296	39.748
Mam_2	42.4422	44.7627	42.1725	42.4422	34.2108	37.12654
Mam_3	44.1716	46.9127	44.1132	42.0401	35.8518	39.03196
Mam_4	45.3980	47.6609	45.2934	42.0689	36.2382	39.94165
Mam_5	46.7206	49.7931	46.6719	44.6588	37.5773	41.72588
Mam_6	45.0838	47.6866	44.9040	42.1152	35.9426	39.9139

- RUM slightly improves the visual quality of images, but it generates spot artifacts, as shown in Figures 4.8(c) and 4.9(c).
- ANCE has very limited visual improvement and produces many textile artifacts in the mammograms.
- CLAHE overenhances the background of all mammograms, making microcalcifications and/or masses more unrecognizable than the original ones.
- As evident from Figures 4.9(f) and 4.10(f), DICE improves the contrast of the microcalcifications, but it fails to enhance mass regions. It also generates background noise and textile artifacts.

The measure results in Table 4.5 support these observations. The presented NLUM outperforms the others because it improves the contrast of mammograms and visual quality of suspicious regions such as masses and/or microcalcifications, which is useful for detecting and diagnosing diseases or breast cancer at an early stage. The enhanced mammograms have no detail information loss. The measure results in Table 4.5 verify the excellent enhancement performance of NLUM.

4.5.6 ROC evaluation

The receiver operating characteristic curve was originally developed for signal-detection theory. It is a well-known evaluation methodology used for medical decision making and medical diagnostic imaging systems.[90,91] The ROC curve is a graphical plot of the true positive rate (a fraction of true positives over the positives) versus the false positive rate (a fraction of false positives over the negatives). To determine whether a person has a specific disease in the clinical diagnosis, a true positive case occurs when the person tests positive and actually has the disease. A false positive case, on the other hand, occurs when the person tests positive but does not actually have the disease.[92] The MATLAB® implementation of the ROC analysis is addressed in references.[93,94]

This section uses the ROC curve to evaluate the NLUM enhancement performance. 60 mammograms were selected from the mini-MIAS database. They consist of 30 normal mammograms (which do not contain suspect

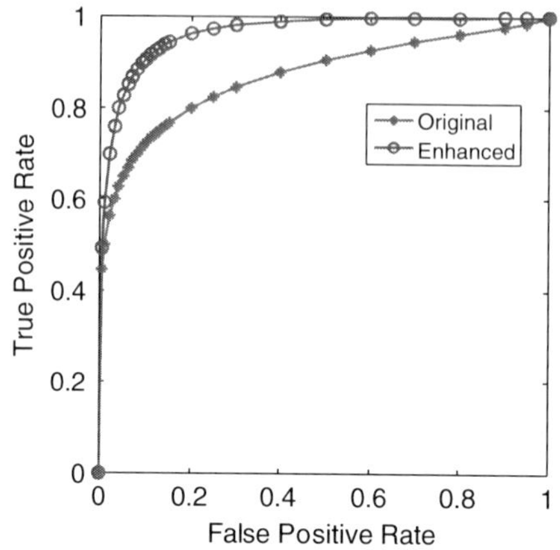

Figure 4.11 The ROC curves of the original and enhanced test mammograms.

regions such as calcifications and masses) and 30 suspicious mammograms. All mammograms were cropped into smaller-size images such that the resulting images have minimal background or contain most of abnormal regions such as microcalcifications and masses.

All mammograms were enhanced by NLUM and then divided in two groups: original and enhanced mammograms. They were inspected by a medical doctor who has a great deal of clinical experience with viewing mammograms. The doctor marked each mammogram with the case type ('0' for the truly negative case indicating a completely normal mammogram and '1' for the truly positive case referring to an abnormal mammogram.) and the confidence rate for each case type. The confidence rate is from 1–5, where '1' indicates a definitely negative case, and '5' means definitely positive.[94]

Using an online code of the ROC analysis developed by Eng,[94] the doctor's inspection results were individually plotted into ROC curves for the original and enhanced mammograms. The results are shown in Figure 4.11.

The area under the ROC curve (AUC) is used to quantitatively evaluate the classification performance of the diagnosis system.[90,91] The AUC value is always between 0 and 1. A higher AUC value indicates better classification performance. The AUC for the enhanced mammogram is 0.957, whereas the AUC for the originals is 0.874. This demonstrates that the NLUM enhancement improves the doctor's diagnosis. It could potentially improve cancer breast diagnosis and detection in the CAD systems.

4.6 Conclusion

This chapter introduced a new nonlinear unsharp masking scheme for mammogram enhancement. NLUM has been shown to provide more design flexibility that makes it possible to meet more specific and complex requirements in real world applications. The simulation results have demonstrated that the NLUM parameters can be optimized by the enhancement measure to obtain the better enhanced result for clinical applications. Enhancement comparison has proven that the NLUM shows better performance for improving the local contrast of suspicious regions and fine details in mammograms. NLUM has potential applications of improving the automatic disease detection and diagnosis in CAD systems.

To quantitatively evaluate NLUM performance for mammogram enhancement, we have introduced a new enhancement measure called the second-derivative-like measure of enhancement. Compared with other existing measure methods, the SDME shows better performance for enhancement measurement and assessment. HVS-based decomposition has been verified to be a useful tool to analyze and display suspicious regions in mammograms.

References

1. L.-M. Wun, R. M. Merrill, and E. J. Feuer, "Estimating lifetime and age-conditional probabilities of developing cancer," *Lifetime Data Analysis* **4**, 169–186 (1998).
2. D. B. Kopans, "The 2009 U. S. Preventive Services Task Force guidelines ignore important scientific evidence and should be revised or withdrawn," *Radiology* **256**, 15–20 (2010).
3. D. E. Stewart et al., "Attributions of cause and recurrence in long-term breast cancer survivors," *Psycho-Oncology* **10**, 179–183 (2001).
4. H. D. Cheng and H. Xu, "A novel fuzzy logic approach to mammogram contrast enhancement," *Information Sciences* **148**, 167–184 (2002).
5. R. Mousa, Q. Munib, and A. Moussa, "Breast cancer diagnosis system based on wavelet analysis and fuzzy-neural," *Expert Systems with Applications* **28**, 713–723 (2005).
6. X. Gao, Y. Wang, X. Li, and D. Tao, "On Combining Morphological Component Analysis and Concentric Morphology Model for Mammographic Mass Detection," *IEEE Trans. Information Techn. Biomed.* **14**, 266–273 (2010).
7. R. A. Smith, V. Cokkinides, and H. J. Eyre, "American Cancer Society guidelines for the early detection of cancer 2006," *CA Cancer J. Clin.* **56**, 11–25 (2006).
8. S. A. Feig and M. J. Yaffe, "Digital mammography," *Radiographics* **18**, 893–901 (1998).
9. W. M. Morrow, R. B. Paranjape, R. M. Rangayyan, and J. E. L. Desautels, "Region-based contrast enhancement of mammograms," *IEEE Trans. Med. Imaging* **11**, 392–406 (1992).

10. L. Wei, Y. Yang, M. N. Wernick, and R. M. Nishikawa, "Learning of perceptual similarity from expert readers for mammogram retrieval," *IEEE J. Selected Topics Signal Process.* **3**, 53–61 (2009).

11. J. Tang, R. M. Rangayyan, J. Xu, I. E. Naqa, and Y. Yang, "Computer-aided detection and diagnosis of breast cancer with mammography: recent advances," *IEEE Trans. Information Tech. Biomed.* **13**, 236–251 (2009).

12. H. Li, K. J. R. Liu, and S. C. B. Lo, "Fractal modeling and segmentation for the enhancement of microcalcifications in digital mammograms," *IEEE Trans. Med. Imaging* **16**, 785–798 (1997).

13. S. Halkiotis, T. Botsis, and M. Rangoussi, "Automatic detection of clustered microcalcifications in digital mammograms using mathematical morphology and neural networks," *Signal Process.* **87**, 1559–1568 (2007).

14. S. Timp, C. Varela, and N. Karssemeijer, "Computer-aided diagnosis with temporal analysis to improve radiologists' interpretation of mammographic mass lesions," *IEEE Trans. Information Tech. Biomed.* **14**, 803–808 (2010).

15. R. M. Rangayyan et al., "Improvement of sensitivity of breast cancer diagnosis with adaptive neighborhood contrast enhancement of mammograms," *IEEE Trans. Information Tech. Biomed.* **1**, 161–170 (1997).

16. I. Larrabide, A. A. Novotny, R. A. Feij´oo, and E. Taroco, "A medical image enhancement algorithm based on topological derivative and anisotropic diffusion," *Proc. XXVI Iberian Latin-American Congress on Computational Methods in Engineering*, Guarapari, Esp´irito Santo, Brazil (2005).

17. S. Aghagolzadeh and O. K. Ersoy, "Transform image enhancement," *Opt. Eng.* **31**(3), 614–626 (1992) [doi: 10.1117/12.56095].

18. S. S. Agaian, K. Panetta, and A. M. Grigoryan, "Transform-based image enhancement algorithms with performance measure," *IEEE Trans. Image Process.* **10**, 367–382 (2001).

19. S. S. Agaian, B. Silver, and K. A. Panetta, "Transform coefficient histogram-based image enhancement algorithms using contrast entropy," *IEEE Trans. Image Process.* **16**, 741–758 (2007).

20. A. M. Grigoryan and S. S. Agaian, *Multidimensional Discrete Unitary Transformations: Representation, Partitioning, and Algorithms*, Marcel Dekker, New York (2003).

21. S. S. Agaian, V. Sridharan, and M. Blanton, "Switching system for image enhancement and analysis of fused thermal and RGBD data," *Proc. SPIE* **8406**, 84060Z (2012) [doi: 10.1117/12.924755].

22. J. Xia, K. Panetta, and S. Agaian, "Color image enhancement algorithm based on logarithmic transform coefficient histogram," *Proc. SPIE* **7870**, 78700Y (2011) [doi: 10.1117/12.872418].

23. A. Grigoryan and S. Agaian, "Image Enhancement," in *Advances in Imaging and Electron Physics* **130**, Academic Press, New York, pp. 165–243 (2004).
24. K. Panetta, J. Xia, and S. Agaian, "Color image enhancement based on the discrete cosine transform coefficient histogram," *J. Electron. Imaging* **21**(2), 021117 (2012) [doi: 10.1117/1.JEI.21.2.021117].
25. X. Junjun, K. Panetta, and S. Agaian, "Color image enhancement algorithms based on the DCT domain," *Proc. IEEE SMC2011*, 1496–1501 (2011).
26. S. Krishnan, S. Agaian, D. Mecke, S. A. Montelongo, and X. Wang, "Method and architecture for quantification of bone structure using microscopic image slices," *Proc. SPIE* **8655**, 865517 (2013) [doi: 10.1117/12.2005120].
27. S. Nercessian, S. S. Agaian, and K. A. Panetta, "Multi-scale image enhancement using a second derivative-like measure of contrast," *Proc. SPIE* **8295**, 82950Q (2012) [doi: 10.1117/12.906494].
28. S. C. Nercessian, K. A. Panetta, and S. S. Agaian, "Multiscale image fusion using an adaptive similarity-based sensor weighting scheme and human visual system-inspired contrast measure," *J. Electron. Imaging* **21**(2), 021112 (2012) [doi: 10.1117/1.JEI.21.2.021112].
29. J. Xia, K. Panetta, and S. Agaian, "Wavelet transform coefficient histogram-based image enhancement algorithms," *Proc. SPIE* **7708**, 770812 (2010) [doi: 10.1117/12.855563]
30. Z. Lu, T. Jiang, G. Hu, and X. Wang, "Contourlet based mammographic image enhancement," *Proc. SPIE* **6534**, 65340M (2007) [doi: 10.1117/12.741340].
31. A. Laine, J. Fan, and W. Yang, "Wavelets for contrast enhancement of digital mammography," *IEEE Eng. Med. Biol. Mag.* **14**, 536–550 (1995).
32. A. Laine, W. Huda, B. G. Steinbach, and J. C. Honeyman, "Mammographic image processing using wavelet processing techniques," *Eur. Radiol.* **5**, 518–523 (1995).
33. A. F. Laine, S. Schuler, F. Jian, and W. Huda, "Mammographic feature enhancement by multiscale analysis," *IEEE Trans. Med. Imaging* **13**, 725–740 (1994).
34. A. Mencattini, M. Salmeri, R. Lojacono, M. Frigerio, and F. Caselli, "Mammographic images enhancement and denoising for breast cancer detection using dyadic wavelet processing," *IEEE Trans. Instrumentation and Measurement* **57**, 1422–1430 (2008).
35. P. Sakellaropoulos, L. Costaridou, and G. Panayiotakis, "A wavelet-based spatially adaptive method for mammographic contrast enhancement," *Phys. Med. Biol.* **48**, 787–803 (2003).

36. P. Heinlein, J. Drexl, and W. Schneider, "Integrated wavelets for enhancement of microcalcifications in digital mammography," *IEEE Trans. Med. Imaging* **22**, 402–413 (2003).
37. S. Skiadopoulos, A. Karahaliou, F. Sakellaropoulos, G. Panayiotakis, and L. Costaridou, "Breast component adaptive wavelet enhancement for soft-copy display of mammograms," *Proc. Digital Mammography*, 549–556, Manchester, UK (2006).
38. C.-M. Chang and A. Laine, "Coherence of multiscale features for enhancement of digital mammograms," *IEEE Trans. Information Technol. Biomed.* **3**, 32–46 (1999).
39. G. Derado, F. DuBois Bowman, R. Patel, M. Newell, and B. Vidakovic, "Wavelet image interpolation (WII): A wavelet-based approach to enhancement of digital mammography images," *Lecture Notes Comp. Sci.* **4463**, 203–214 (2007).
40. F. Y. M. Lure, P. W. Jones, and R. S. Gaborski, "Multiresolution unsharp masking technique for mammogram image enhancement," *Proc. SPIE* **2710**, 830–839 (1996) [doi: 10.1117/12.237989].
41. J. Scharcanski and C. R. Jung, "Denoising and enhancing digital mammographic images for visual screening," *Comp. Med. Imaging and Graphics* **30**, 243–254 (2006).
42. J. Tang, X. Liu, and Q. Sun, "A Direct Image Contrast Enhancement Algorithm in the Wavelet Domain for Screening Mammograms," *IEEE J. Selected Topics Signal Process.* **3**, 74–80 (2009).
43. W. Qian, L. P. Clarke, M. Kallergi, and R. A. Clark, "Tree-structured nonlinear filters in digital mammography," *IEEE Trans. Med. Imaging* **13**, 25–36 (1994).
44. S. DelMarco and S. Agaian, "The design of wavelets for image enhancement and target detection," *Proc. SPIE* **7351**, 735103 (2009) [doi: 10.1117/12.816135].
45. S. C. Nercessian, K. A. Panetta, and S. S. Agaian, "Multiresolution decomposition schemes using the parameterized logarithmic image processing model with application to image fusion," *EURASIP J. Adv. Signal Process 2011*, 1–17 (2011).
46. S. S. Agaian, K. A. Panetta, and S. C. Nercessian, "Image Fusion Using A Parameterized Logarithmic Image Processing Framework," Chapter 8 in *Image Fusion,* Osamu Ukimura, Ed., Intech, New York (2011).
47. F. Sahba and A. Venetsanopoulos, "Contrast enhancement of mammography images using a fuzzy approach," *Proc. IEEE EMBS 2008*, 2201-2204 (2008).
48. I. Stephanakis, G. Anastassopoulos, A. Karayiannakis, and C. Simopoulos, "Enhancement of medical images using a fuzzy model for segment dependent local equalization," *Proc. ISPA 2003*, 970–975 (2003).

49. J. Jiang, B. Yao, and A. M. Wason, "Integration of fuzzy logic and structure tensor towards mammogram contrast enhancement," *Comp. Med. Imaging and Graphics* **29**(1), 83–90 (2005).
50. Y. Zhou, K. Panetta, and S. Agaian, "Human visual system based mammogram enhancement and analysis," *Proc. IEEE Int. Conf. Image Process. Theory Tools Applications*, Paris, France, 229–234 (2010).
51. K. A. Panetta, E. J. Wharton, and S. S. Agaian, "Human Visual System-Based Image Enhancement and Logarithmic Contrast Measure," *IEEE Trans. Systems, Man and Cybernetics, Part B* **38**, 174–188 (2008).
52. E. Wharton, K. Panetta, and S. Agaian, "Adaptive multi-histogram equalization using human vision thresholding," *Proc. SPIE* **6497**, 64970G (2007) [doi: 10.1117/12.704474].
53. Y. Zhou, K. Panetta, and S. Agaian, "Human visual system based contrast enhancement for x-ray CT images using alpha weighted quadratic filter," *Proc. IS&T / SPIE Electronic Imaging 2010: Image Processing: Algorithms and Systems VIII*, San Jose, CA (2010).
54. S. Bakhtiari, S. Agaian, and M. Jamshidi, "An enhanced Empirical Mode Decomposition based method for image enhancement," *Proc. IEEE SMC 2011*, 2681–2686 (2011).
55. S. Bakhtiari, S. Agaian, and M. Jamshidi, "A novel empirical mode decomposition based system for medical image enhancement," *Proc. IEEE SysCon 2011*, 145–148 (2011).
56. S. Bakhtiari, S. Agaian, and M. Jamshidi, "Complex ensemble empirical mode decomposition and alpha-rooting for image contrast enhancement," *Proc. SPIE* **8063**, 806313 (2011) [doi: 10.1117/12.887424].
57. S. Bakhtiari, S. Agaian, and M. Jamshidi, "A color image enhancement method based on Ensemble Empirical Mode Decomposition and Genetic Algorithm," *World Automation Congress 2012*, 1–6 (2012).
58. S. M. Pizer, E. P. Amburn, J. D. Austin, R. Cromartie, A. Geselowitz, T. Greer, B. ter Haar Romeny, J. B. Zimmerman, and K. Zuiderveld, "Adaptive histogram equalization and its variations," *Comp. Vision, Graphics, Image Process.* **39**, 355–368 (1987).
59. R. C. Gonzalez and R. E. Woods, *Digital Image Processing*, 3rd ed., Prentice Hall, New York (2007).
60. E. Wharton, K. Panetta, and S. Agaian, "Human Visual System Based Multi-Histogram Equalization for Non-Uniform Illumination and Shoadow Correction," *Proc. IEEE ICASSP 2007*, I-729–I-732 (2007).
61. L. Lu, Y. Zhou, K. Panetta, and S. Agaian, "Comparative study of histogram equalization algorithms for image enhancement," *Proc. SPIE* **7708**, 770811 (2010) [doi: 10.1117/12.853502].

62. Y. Zhou, K. Panetta, and S. Agaian, "3D CT baggage image enhancement based on order statistic decomposition," *Proc. IEEE Technol. Homeland Security 2010*, Waltham, MA (2010).

63. K. Panetta, S. Agaian, Y. Zhou, and E. J. Wharton, "Parameterized Logarithmic Framework for Image Enhancement," *IEEE Trans. Systems, Man and Cybernetics, Part B* **41**, 460–473 (2011).

64. K. Panetta, E. Wharton, and S. Agaian, "Logarithmic Edge Detection with Applications," *J. Comp.* **3**, 11–19 (2008).

65. M. Jourlin and J. Pinoli, "A model for logarithmic image processing," *J. Microscopy* **149**, 21–35 (1988).

66. M. Jourlin and J.-C. Pinoli, "Image dynamic range enhancement and stabilization in the context of the logarithmic image processing model," *Signal Process.* **41**, 225–237 (1995).

67. N. Petrick, H.-P. Chan, B. Sahiner, and D. Wei, "An adaptive density-weighted contrast enhancement filter for mammographic breast mass detection," *IEEE Trans. Med. Imaging* **15**, 59–67 (1996).

68. E. J. Wharton, K. A. Panetta, and S. S. Agaian, "Edge Preserving Image Enhancement Using Anisotpic Diffusion," *Proc. SPIE* **6812**, 681218 (2008) [doi: 10.1117/12.766893].

69. C. Gao, K. Panetta, and S. Agaian, "A new color contrast enhancement algorithm for robotic applications," *Proc. IEEE Technol. Practical Robot Applications 2012*, 42–47 (2012).

70. R. G. Kogan, S. S. Agaian, and K. Panetta, "Visualization using rational morphology and zonal magnitude reduction," *Proc. SPIE* **3304**, 153–163 (1998) [doi: 10.1117/12.304595].

71. A. P. Dhawan, G. Buelloni, and R. Gordon, "Enhancement of Mammographic Features by Optimal Adaptive Neighborhood Image Processing," *IEEE Trans. Med. Imaging* **5**, 8–15 (1986).

72. V. H. Guis, M. Adel, M. Rasigni, G. Rasigni, B. Seradour, and P. Heid, "Adaptive neighborhood contrast enhancement in mammographic phantom images," *Opt. Eng.* **42**(2), 357–366 (2003) [doi: 10.1117/1.1534846].

73. G. Ramponi, N. K. Strobel, S. K. Mitra, and T.-H. Yu, "Nonlinear unsharp masking methods for image contrast enhancement," *J. Electron. Imaging* **5**(3), 353–366 (1996) [doi: 10.1117/12.242618].

74. Z. Wu, J. Yuan, B. Lv, and X. Zheng, "Digital mammography image enhancement using improved unsharp masking approach," *Proc. IEEE CISP 2010*, 668–672 (2010).

75. G. Ramponi and A. Polesel, "Rational unsharp masking technique," *J. Electron. Imaging* **7**(2), 333–338 (1998) [doi: 10.1117/1.482649].

76. J. George and S. P. Indu, "Fast Adaptive Anisotropic Filtering for Medical Image Enhancement," *Proc. IEEE ISSPIT 2008*, 227–232 (2008).

77. A. Polesel, G. Ramponi, and V. J. Mathews, "Image enhancement via adaptive unsharp masking," *IEEE Trans. Image Process.* **9**, 505–510 (2000).
78. N. Strobel and S. K. Mitra, "Quadratic filters for image contrast enhancement," *Proc. Signals, Systems and Computers 1994*, 208–212 (1994).
79. G. Ramponi, "A cubic unsharp masking technique for contrast enhancement," *Signal Process.* **67**, 211–222 (1998).
80. S. Singh and K. Bovis, "An evaluation of contrast enhancement techniques for mammographic breast masses," *IEEE Trans. Information Technol. Biomed.* **9**, 109–119 (2005).
81. S. Agaian, K. Panetta, and A. Grigoryan, "A new measure of image enhancement," *Proc. Int. Conf. Signal Process. Comm.*, 19–22, Marbella, Spain (2000).
82. S. S. Agaian, "Visual morphology," *Proc. SPIE* **3646**, 139–150 (1999) [doi: 10.1117/12.341081].
83. E. Wharton, S. Agaian, and K. Panetta, "Comparative study of logarithmic enhancement algorithms with performance measure," *Proc. SPIE* **6064**, 606412 (2006) [doi: 10.1117/12.647489].
84. E. Wharton, K. Panetta, and S. Agaian, "Human visual system based similarity metrics," *Proc. IEEE Systems, Man and Cybernetics 2008*, 685–690 (2008).
85. J. Tang, E. Peli, and S. Acton, "Image enhancement using a contrast measure in the compressed domain," *IEEE Signal Process. Lett.* **10**, 289–292 (2003).
86. ITU-T, "Methods for subjective determination of transmission quality," *Recommendation*, International Telecommunication Union, p. 800 (1996).
87. J. Suckling, J. Parker, D. Dance, S. Astley, I. Hutt, C. Boggis, I. Ricketts, E. Stamatakis, N. Cerneaz, and S. Kok, "The mammographic image analysis society digital mammogram database," *Proc. 2nd Int. Workshop Digital Mammography*, 375–378 (1994).
88. M. K. Kundu and S. K. Pal, "Thresholding for edge detection using human psychovisual phenomena," *Pattern Recognition Lett.* **4**, 433–441 (1986).
89. G. Buchsbaum, "An analytical derivation of visual nonlinearity," *IEEE Trans. Biomed. Eng.* **27**, 237–242 (1980).
90. M. J. P. Castanho, L. C. Barros, A. Yamakami, and L. L. Vendite, "Fuzzy receiver operating characteristic curve: an option to evaluate diagnostic tests," *IEEE Trans. Information Technol. Biomed.* **11**, 244–250 (2007).
91. M. Zweig and G. Campbell, "Receiver-operating characteristic (ROC) plots: a fundamental evaluation tool in clinical medicine," *Clin. Chem.* **39**, 561–577 (1993).

92. Wikipedia, "Receiver operating characteristic," http://en.wikipedia.org/wiki/Receiver_operating_characteristic.
93. A. Slaby, "ROC analysis with MATLAB," *Proc. 29th Int. Conf. Information Technol. Interfaces*, Cavtat, Croatia, 191–196 (2007).
94. J. Eng, "ROC analysis: web-based calculator for ROC curves," available at http://www.jrocfit.org (2007).

Yicong Zhou received his B.S. degree from Hunan University, Changsha, China, and his M.S. and Ph.D. degrees from Tufts University, Medford, MA, USA, all in Electrical Engineering. He is currently an Assistant Professor in the Department of Computer and Information Science at the University of Macau, China. Dr. Zhou is a member of the IEEE and SPIE. His research interests include multimedia security, image and signal processing, pattern recognition, and medical imaging.

Sos Agaian is the Peter T. Flawn Professor of Electrical and Computer Engineering at the University of Texas, San Antonio, and a professor at the University of Texas Health Science Center, San Antonio. Dr. Agaian received his M.S. degree (summa cum laude) in Mathematics and Mechanics from Yerevan State University, Armenia; his Ph.D. in Mathematics and Physics from the Steklov Institute of Mathematics, Russian Academy of Sciences (RAS); and his Eng.Sc.D. from the Institute of Control Sciences, RAS. He has authored more than 500 scientific papers, seven books, and holds 14 patents. He is a Fellow of SPIE, the American Association for the Advancement of Science (AAAS), and Imaging Sciences and Technology (IS&T). He also serves as a foreign member of the Armenian National Academy. He is the recipient of MAEStro Educator of the Year, sponsored by the Society of Mexican American Engineers. The technologies that Dr. Agaian invented have been adopted by multiple institutions, including the U.S. government, and commercialized by industry. He is an Editorial Board Member for the *Journal of Pattern Recognition and Image Analysis* and an Associate Editor for several journals, including the *Journal of Electronic Imaging* (SPIE, IS&T) and *System* (Elsevier). His research interests include multimedia processing, imaging systems, information security, artificial intelligence, computer vision, 3D imaging sensors, fusion, and biomedical and health informatics.

Karen Panetta received her B.S. degree in Computer Engineering from Boston University, Boston, MA, and her M.S. and Ph.D. in Electrical Engineering from Northeastern University, Boston. Dr. Panetta is currently a Professor in the Department of Electrical and Computer Engineering at Tufts University, Medford, MA, and director of the Simulation Research Laboratory. She is the Editor-in-Chief of the IEEE *Women in Engineering Magazine*. Dr. Panetta is the IEEE-USA Vice President of Communications and Public Affairs; she also served as the 2011 chair of the IEEE Boston Section, which has over 8800 members. During 2007–2009, she served as the worldwide director for IEEE Women in Engineering, overseeing the world's largest professional organization supporting women in engineering and science. She is a Fellow of the IEEE, and she is also the recipient of the 2012 IEEE Ethical Practices Award and the Harriet B. Rigas Award for Outstanding Educator. In 2011, Dr. Panetta was awarded the Presidential Award for Engineering and Science Education and Mentoring by U.S. President Obama. Her research focuses on developing efficient algorithms for simulation, modeling, signal, and image processing for biomedical and security applications.

C. L. Philip Chen received his M.S. degree from the University of Michigan, Ann Arbor, Michigan, USA and his Ph.D. from Purdue University, West Lafayette, Indiana, USA, both in Electrical Engineering. Dr. Chen is currently a Chair Professor and the Dean of Faculty of Science and Technology at the University of Macau, China. He was a Professor and Chair of the Department of Electrical and Computer Engineering, as well as the Associate Dean for Research and Graduate Studies of the College of Engineering at the University of Texas in San Antonio after he served as an assistant, an associate, and a full professor in the Department of Computer Science and Engineering at Wright State University, Dayton, OH. Dr. Chen has served as a member of organizing committees for many IEEE conferences under different capacities. He is currently the President of the IEEE SMC Society. He is an Accreditation Board of Engineering and Technology Education (ABET) Program Evaluator for computer engineering, electrical engineering, and software engineering programs. He is a fellow of the IEEE and AAAS. His research interests include computer networking, neural networks, fuzzy neural systems, intelligent systems, robotics, and CAD/CAM.

Chapter 5
Skin Lesion Extraction Based on Distance Histogram and Color Information

Yanliang Gu and Jinshan Tang
Medical Informatics Program & Computer Network and Systems Administration Program, School of Technology, Michigan Technological University, Houghton, MI, USA

5.1 Introduction

As is well known, skin is the largest organ in human body, making up 12–15% of body weight, with a surface area of 1–2 square meters.[1] There are two main layers of skin: the top layer is called the epidermis, and the inner layer is called the dermis. Malignancies are always found in the epidermis, and it contains three kinds of cells:[2] squamous, basal cells, and melanocytes. Squamous is on the surface and looks flat and scaly; basal cells are round cells; and melanocytes give color to skin.[2] The dermis contains blood vessels and sweat glands. It is used to keep skin from drying out.[3]

People whose skin is exposed to sunlight for longer amounts of time are more prone to develop skin cancer.[3] Australia and New Zealand are the two countries with the highest rates of skin cancer incidence in the world, with rates that are almost four times that of the United States.[4] Based on the statistics in 2008, 59,695 people in the US were diagnosed with melanomas of the skin, and 8,623 people died from it.[5]

Skin cancers are named after three kinds of cells in the epidermis from which they arise. Basal cell cancer arises on basal cells and is the most-common skin cancer. Squamous cell cancer is found on squamous cells and is less common. Basal and squamous cell cancers are called nonmelanoma skin cancer and are less dangerous than melanoma, which develops on melanocyte cells. Melanoma is less common than nonmelanoma skin cancer, but it is more likely to spread and become fatal.[6–8]

Basal cell cancer appears as a raised, smooth bump on the sun-exposed skin of the head, neck, or shoulder. It grows slowly and painlessly. Even though this kind of skin cancer hardly spreads, it can extend below the skin to the bone and cause further damage. Squamous cell cancer appears as a red, crusted patch on the face, lips, and mouth. It is usually found in elderly people, and it grows rapidly in size to form a large mass.[7] Melanoma appears in an asymmetrical zone, with an irregular border, with a diameter frequently greater than 6 mm. Most melanomas display color between brown and black, but a few appear pink or red. This type of cancer grows rapidly and sometimes displays different colors and shapes, which can be warning signs.[8] Fortunately, even melanoma can be cured if it is detected early. Thus, the detection of skin cancer in its early stage has attracted the attention from different fields.[8]

Computer technologies play an important role in the early detection of skin cancer, and different methods for detecting skin cancer have been proposed. A skin cancer detection system was developed by F. Ercal;[9] in the proposed system, the border of the skin lesion was extracted by doctors, and the features were constructed using the "ABCDE rule." A commercial neural network classifier was adopted to classify the skin lesions into benign or malignant. This system can achieve 80% accuracy rate. Another computer program, named MoleSence, was developed by Opticom Data Research,[10] which also used the ABCDE rule to analyze skin cancer images. An automated melanoma recognition system was developed by H. Ganster.[11] In this system, automated image segmentation was accomplished by fusing the output of 3 algorithms, and the segmentation accuracy is 96%. A 24-NN (nearest neighbor) classifier was used for skin cancer classification, and it achieved a 73% melanoma-recognition rate. The sensitivity and specificity are 87% and 92% for the "not benign" class in a two-class scenario.

Handyscope was developed by the company called Fotofinderin Germany;[12] it merges a dermatoscope with an iPhone and can extract lesion features. However, it does not incorporate lesion classification in the same device. A multi-direction gradient vector flow (GVF) has been proposed for skin cancer border extraction.[13] In the proposed method, a new anisotropic diffusion filter employing new gradient computation and an adaptive threshold selection method was used to remove the noise. After noise reduction, the system used a multidirectional GVF snake that was extended from the single-direction GVF snake to segment the skin cancer image.[13] An automatic detection method employing a radial search algorithm to detect irregular borders in skin tumor images was proposed.[14] A method based on domain-specific algorithms was used to extract the boundary of a skin lesion, and this method can achieve a high success rate.[15] An automatic approach for skin cancer segmentation was proposed by the modification of the region-based active contour model (RACM).[16] In the proposed approach, the initial threshold was set automatically, and a function named Courant–Friedreichs–Lewy (CFL) was used to enforce the stability of curves. Compared with other state-of-the-art methods, this approach achieves better accuracy. A method based on a principal component transform algorithm was developed for automatic color segmentation.[17] This method can detect the borders of skin cancer in six different color spaces. Many images were obtained with low contrast and lack color calibration; thus, an approach to improve the accuracy of skin cancer segmentation for these cases was provided.[18] This approach enhanced color information by a preprocessing step, and then skin cancer segmentation was performed on the preprocessed images. A segmentation algorithm based on an evolution strategy (EV) was provided to detect the skin cancer area.[19] The segmentation results were compared with those obtained by a dermatologist, and the comparison showed that the proposed method can get high accuracy for images with either high noise level, less prominent edges, or very small size lesions. Using region fusion and narrow-band energy graph partitioning technology, a novel multimodal skin lesion segmentation method was developed.[20] The proposed method can process skin cancer images with blurred and irregular edges well and can achieve a better performance compared with other state-of-the-art segmentation methods. A new automatic skin cancer segmentation method was introduced,[21] which first converted a color image into a gray image before segementing it using intensity thresholding. The segmentation is refined using image edges.

This chapter proposes a new method for skin lesion detection. The proposed system uses a distance histogram and color information to segment the skin lesions.

5.2 Color-Based Skin Lesion Segmentation

The proposed skin lesion segmentation system consists of three main stages, shown in Fig. 5.1. In the first stage, noise reduction technology is used to remove the noise and hairs in the skin cancer image. In the second stage, the color image is converted into a color space and an adaptive color model is constructed. In the third stage, color information and a distance histogram are used to extract the border of the lesions from the skin cancer image.

5.2.1 Noise reduction

Because skin cancer images obtained usually include noise and hairs, noise reduction technology is used in the system so that the segmentation results will be good. Many noise reduction methods have been proposed in the past, one of the most popular being a Gaussian filter. However, a Gaussian filter generally blurs the edges of the images, and thus a bilateral filter is investigated in this chapter. Compared with a Gaussian filter, a bilateral filter has good performance in edge preservation.[22] A bilateral filter was first

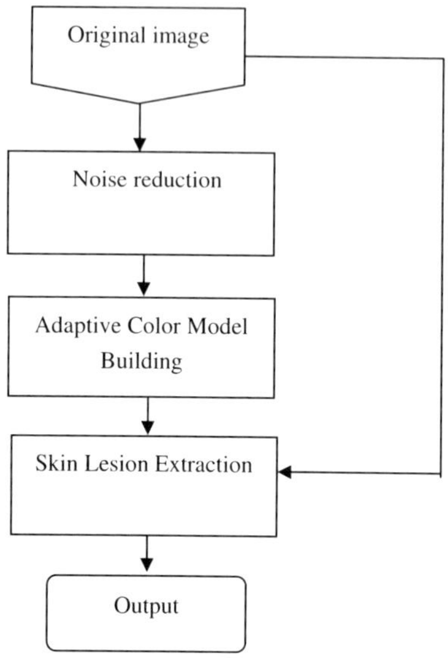

Figure 5.1 System architecture.

introduced by Tomasi and Manduchi[22] and has been found to have wide application in noise reduction.

There are different types of bilateral filters.[22,23] One type of bilateral filter is called the Gaussian bilateral filter, which can be expressed mathematically as[22,24]

$$\bar{I}(X) = \frac{1}{C} \sum_{Y \in N(X)} e^{\frac{-\|Y-X\|^2}{2\sigma_d^2}} e^{\frac{-\|I(Y)-I(X)\|^2}{2\sigma_r^2}} I(Y), \tag{5.1}$$

where $\bar{I}(X)$ is the output pixel value vector, $I(Y)$ is the input pixel value vector, X and Y are the coordinate vectors, σ_d^2 and σ_r^2 are the parameters controlling the fall-off of weights in spatial and intensity domains, respectively, $N(X)$ is a spatial neighborhood of pixel $I(X)$, $\|\|$ is Euclidean distance, and C is used for the normalization and can be expressed as

$$C = \sum_{Y \in N(X)} e^{\frac{-\|Y-X\|^2}{2\sigma_d^2}} e^{\frac{-\|I(Y)-I(X)\|^2}{2\sigma_r^2}}. \tag{5.2}$$

5.2.2 Adaptive Color Model Building

5.2.2.1 Color spaces

Several color spaces are investigated in this chapter: RGB, YUV, HSV, and YCbCr. YUV color space consists of two components: one is luminance (Y), which carries the brightness information of the images, and the others are chrominance components U and V, which represent the color information of the images.[27] HSV color space uses two cylinders to represent the pixels in an RGB color space. In HSV color space, H means "hue," which is the degree of the angle around the center vertical axis in each cylinder; S means "saturation," which indicates the degree to which the hue differs from a neutral gray; and V means "value," which stands for the illumination level.[28] YCbCr, Y′CbCr, or Y Pb/Cb Pr/Cr, also written as YCBCR or Y′CBCR, is a family of color spaces used as a part of the color image pipeline in video and digital photography systems. Y is the luminance component, and CB and CR are the blue-difference and red-difference chroma components.[29] The relationship between RGB color space and other color spaces can be obtained by different mathematics. For RGB to YUV, the following equation is adopted:

$$\begin{aligned} Y &= 0.299R + 0.587G + 0.114B, \\ U &= -0.14713R - 0.28886G + 0.436B, \\ V &= 0.615R - 0.51499G - 0.10001B. \end{aligned} \tag{5.3}$$

The conversion from RGB to YCbCr can be expressed as

$$\begin{aligned} Y &= 0.299R + 0.587G + 0.114B, \\ Cb &= 0.564(B - Y) + 128, \\ Cr &= 0.713(R - Y) + 128, \end{aligned} \tag{5.4}$$

and the conversion from RGB to HSV can be obtained by

$$M = \max(R, G, B),$$
$$m = \min(R, G, B),$$
$$V = M/255,$$

$$S = \begin{cases} 1 - m/M & \text{if } M > 0 \\ 0 & \text{if } M = 0, \end{cases}$$

$$H = \begin{cases} \cos^{-1}\left[(R - 0.5G - 0.5B)/\sqrt{R^2 + G^2 + B^2 - RG - RB - GB}\right] & \text{if } G \geq B \\ 360 - \cos^{-1}\left[(R - 0.5G - 0.5B)/\sqrt{R^2 + G^2 + B^2 - RG - RB - GB}\right] & \text{if } G < B. \end{cases} \tag{5.5}$$

5.2.2.2 Adaptive color model building

In order to use color information to detect the lesions, the color model must be built. This chapter investigates adaptive color modeling by building the color model for background, consisting of two steps. The first step builds a general background color model for color detection, and the second step, wherein the general model is used to detect the color information in a specific image, uses the color information from the four corners of the image detected by the

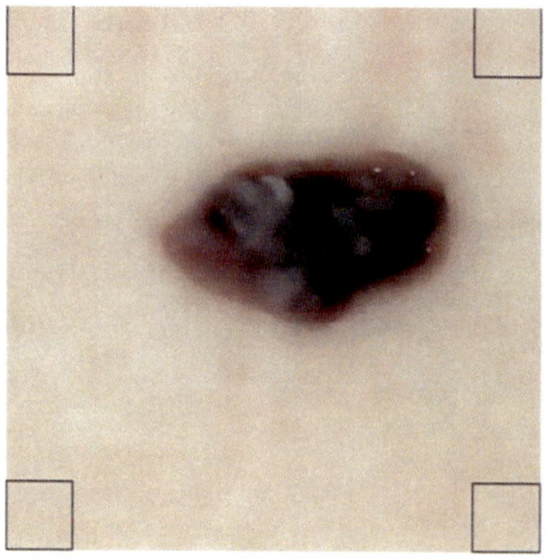

Figure 5.2 Adaptive color modeling.

general model (Fig. 5.2) to rebuild a color model specified for the image; the new color model is then used to detect the lesions in the image. Specifically, in the building processing, the mean and co-variance of color components is calculated for each color space, using 20 images as the general color model. The following is an example of mean and co-variance computations in YUV color space:

$$M = (M_y, M_u, M_v), \tag{5.6}$$

$$C = \begin{pmatrix} C_{yu}(1,1) & C_{yu}(1,2) & C_{yv}(1,2) \\ C_{yu}(1,2) & C_{yu}(2,2) & C_{uv}(1,2) \\ C_{yv}(1,2) & C_{uv}(1,2) & C_{uv}(2,2) \end{pmatrix}, \tag{5.7}$$

where M is the mean value, and C is the co-variance. After the mean and co-variance of the general color model is built, and when color detection begins for a specific image, the general color model is used to detect the color from four corners of the image. The color regions found using the general color model will be used to rebuild a color model for the specific image for which the specific mean and co-variance are obtained.

5.2.3 Distance-histogram-based lesion extraction

The proposed lesion extraction is composed of three steps: skin color detection, adaptive thresholding, and morphological post-processing.

5.2.3.1 Skin color detection

In this step, we will use color information to find the lesion regions based on the assumption that lesion regions and background have different color distributions, which can be characterized using mean and co-variance. Thus, we can classify a pixel as the part of the background or part of the skin lesion using the Mahalanobis distance

$$d(x, y) = (I(x, y) - M)^T \times C^{-1} \times (I(x, y) - M), \tag{5.8}$$

where $I(x, y)$ is a vector constructed from the color components of pixel (x, y), and $d(x, y)$ is the Mahalanobis distance at pixel (x, y). M and C are the mean and the co-variance of the adaptive color model obtained using Eqs. (5.6) and (5.7). A $d(x, y)$ can be obtained for each pixel in the original image. A pixel at location (x, y) is determined to be background by the following rules: for a predetermined threshold T, if $d(x, y) < T$, then either the pixel belongs to background regions or it is classified into skin lesions. The pixel values are set to 1 if the pixel is classified as a skin lesion and 0 if it is classified as background, thus producing a binary image. For convenience, the binary image is denoted as **B**, which has the same size as the original image.

5.2.3.2 Adaptive thresholding

One of the important steps in classifying a pixel as skin lesions or background involves the threshold T. Several methods have been proposed in the literature. In this paper, we will develop a new adaptive thresholding method. The proposed method is a distance-histogram-based method. The basic idea is to use $d(x, y)$ to generate a histogram and determine the threshold based on the histogram. However, the histogram generated directly from $d(x, y)$ is sticky and noisy, it is difficult to find a suitable T. The proposed method first smooths the histogram directly obtained from $d(x, y)$ and then the smoothed histogram is used to get the threshold. Figure 5.3(a) shows the original histogram obtained directly from $d(x, y)$, and Fig. 5.3(b) shows the histogram smoothed using a Gaussian filter ($\sigma = 3$). In order to find the threshold, the lowest point between the first peak and the second peak of the smoothed histogram must be found. The first peak stands for the number of the pixels

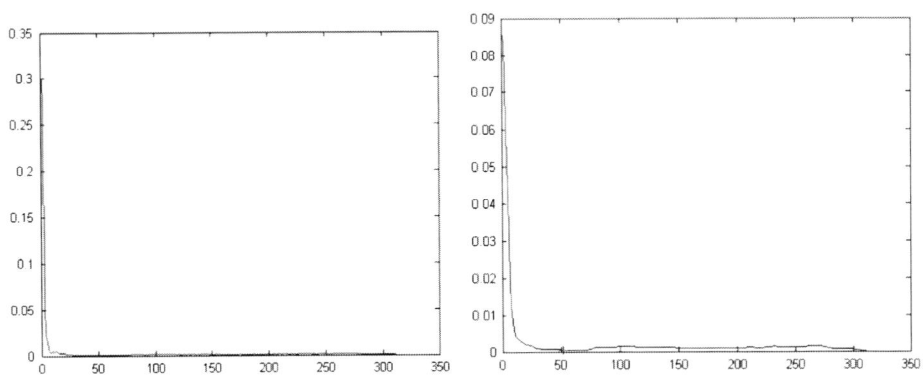

(a) (b)

Figure 5.3 (a) Distance histogram without smoothing. (b) Smoothed histogram from Fig. 5.3(a). The asterisk in the second graph is the threshold.

that are closest to the adaptive color model (normal skin), so the lowest point splits the normal skin and the lesion skin. The parts whose distances are larger than the threshold are the lesion skin.

5.2.3.3 Morphological processing

In this step, morphological technology is used because the image obtained via the method in Section 5.2.3.2 may have noise in the normal skin and the lesion skin. Two morphological operators are used to postprocess the binary image **B**.

The first operator used is called "connected-component labeling technology," which is used to remove the small objects. The basic idea of connected-component labeling technology is to separate the images into several objects and label them. During processing, the background is labeled as 0, the first object is labeled as 1, the second object is labeled as 2, and so on. After all of the objects are labeled, the size of these objects is compared with a preset threshold. If the size of an object is smaller than the preset threshold, the object will be labeled as 0 (the background). Figure 5.4(a) shows the original binary image, and Fig. 5.4(b) shows the processed one.

After the image is processed using the first morphological operator, the second operator is used in the postprocessing, known as a closing operation. Closing is defined simply as a dilation followed by an erosion *using the same structuring element for both operations*; it is useful to remove small holes. Figure 5.5 shows the images after the closing operation.

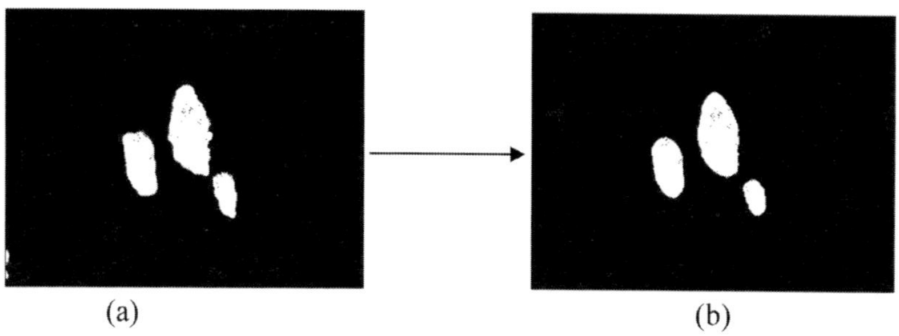

Figure 5.4 (a) The original binary image, and (b) the processed image.

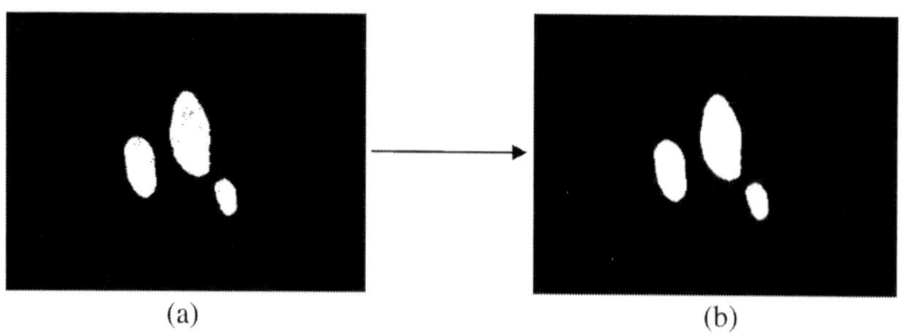

Figure 5.5 (a) The original binary image, and (b) the processed image.

5.3 Experimental Results

Experiments were performed to verify the effectiveness of the proposed method. The experiments are divided into two parts. The first part is used to test the noise reduction method, and the second part is used to test the proposed skin-lesion-extraction method.

5.3.1 Noise reduction on synthetic images

The first experiment was used to perform noise reduction on synthetic images. A red square image was created and different levels of noise were added to the image. In this chapter, Gaussian noise was added to the images. Figure 5.6 shows the original images and the corresponding images after noise reduction.

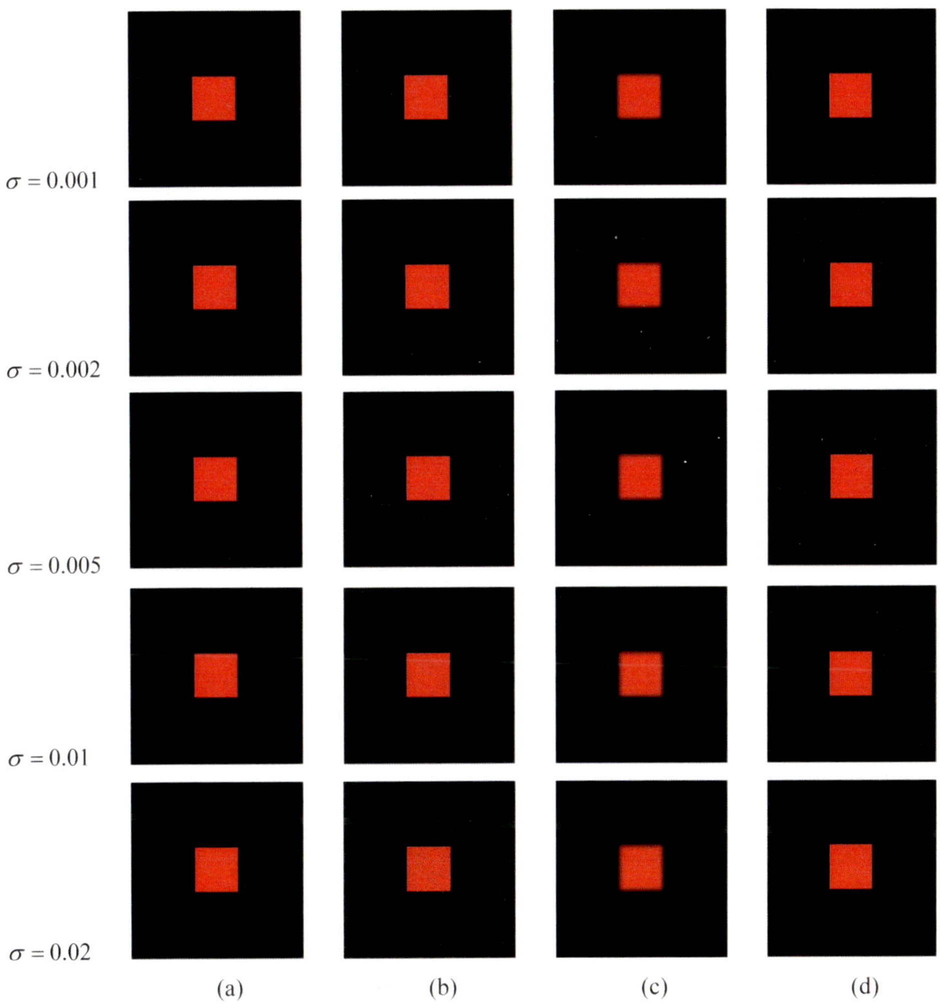

Figure 5.6 Filtered results of different Gaussian-noised synthetic images. Column (a) shows the original synthetic images; column (b) shows the images after Gaussian noise with different variances are added; column (c) shows the filtered images obtained by a Gaussian filter; and column (d) shows the filtered images obtained by a bilateral filter.

Table 5.1 MSE values obtained by Gaussian and bilateral filter.

Variance	Gaussian filter (MSE)	Bilateral filter (MSE)
0.001	6.752	3.183
0.002	7.773	4.455
0.005	9.933	7.034
0.01	12.469	10.232
0.02	16.177	15.636

To evaluate the performance of the two filters, this section uses the quantitative measure mean square error (MSE), defined as follows:[25]

$$M = \sqrt{\sum_{(x,y)\in image} (I_0(x,y) - I(x,y))^2}, \tag{5.9}$$

$$MSE = (M_R + M_G + M_B)/3, \tag{5.10}$$

where I_0 and I are the original synthetic image and the filtered image, respectively. When noise reduction methods are compared, the smaller the MSE value is, the better the filtered result will be. Table 5.1 shows the comparison results of the two noise reduction methods; it shows that under different variances of Gaussian noise, the MSE values obtained by a bilateral filter are always smaller than the MSE values obtained by a Gaussian filter. It means that a bilateral filter has a better performance than a Gaussian filter when they are applied to reduce the noise in the synthetic images.

5.3.2 Noise reduction on skin lesion images

How do the two filters work on the real images? In order to demonstrate their ability, they were applied to 20 different skin lesion images from different sources.[26] Each image has one or more lesions, and some of them have hairs and noise. Figure 5.7 shows a sample of the original image and the corresponding image filtered by a Gaussian filter and bilateral filter respectively. Figure 5.7(a) shows the original skin lesion image, 5.7(b) shows the image filtered by a Gaussian filter, and 5.7(c) shows the image filtered by

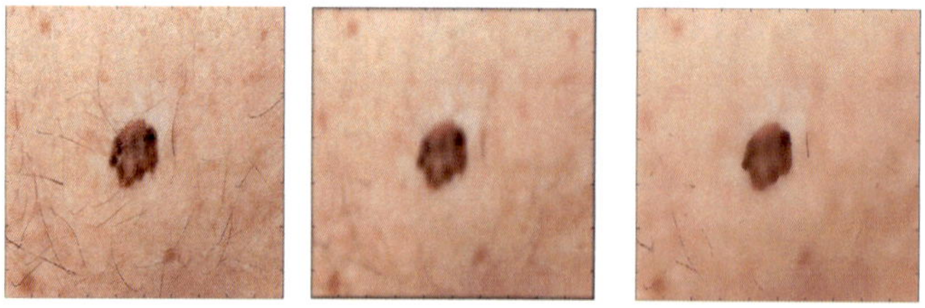

(a) (b) (c)

Figure 5.7 (a) Original skin lesion image; (b) the image filtered by a Gaussian filter; and (c) the image filtered by a bilateral filter.

bilateral filter. From Fig. 5.7, it can be seen that a bilateral filter is much better than a Gaussian filter.

In order to evaluate the performance of the two filters working on the real images quantitatively, two things must be evaluated: the performance for noise reduction, and the performance for edge preservation. Because this chapter focuses on border detection, the performance of edge preservation is much more important. The local contrast of the homogeneous region and edge region in the image is computed to evaluate the performance. The contrast measure used is defined in[13]

$$C_w(I) = \frac{1}{m}\sum_{w} |c(x,y)|\log(1 + |c(x,y)|), \tag{5.11}$$

where $c(x,y)$ is the local contrast at pixel (x, y) and is defined as

$$c(x,y) = 4I(x,y) - \{I(x-1,y) + I(x,y-1) + I(x+1,y) + I(x,y+1)\}, \tag{5.12}$$

where $I(x, y)$ is the pixel intensity value at pixel (x, y) of an image, w is a region or a set of edge points, and m is the number of the pixels in the region or edge points.

Three images were used to compare the performance of the two filtering methods. One homogenous region and one set of edge points for each image were selected for evaluation. Figure 5.8 shows the selections.

When comparing the two filters, this chapter considers the edge preservation in the set of edge points and noise reduction in homogenous regions selected. A good filter should remove the noise while preserving the edges in the image. Thus, in the comparison of the two filters, the edge preservation in the set of selected edge points was compared and the selected homogenous regions were made to have similar smoothness. Table 5.2 shows the average contrasts in the homogenous regions obtained by a Gaussian filter and bilateral filter, respectively, which corresponds to Fig. 5.8(a). Table 5.3 shows the average contrasts in the set of edge points obtained by a Gaussian filter and bilateral filter, respectively, which corresponds to Fig. 5.8(b). The two tables show that the contrast of the set edge points obtained by a the bilateral filter is much bigger than those obtained by a Gaussian filter even though they have the similar contrast in the selected homogenous regions, which means that a bilateral filter is much better than a Gaussian filter.

5.3.3 Experimental results on skin lesion segmentation

For skin lesion extraction, 20 images including one or more skin lesions were used in the experiments. Different color spaces were tested and compared. For each color space, the same procedure developed in Section 5.2.3 was used to extract skin lesions. Figure 5.9 shows some samples of lesion extraction using different color spaces, from which it can be seen that RGB, YUV, and YCbCr color spaces almost achieved the same performance, whereas HSV is much worse than the other three color spaces. To evaluate the performance for skin

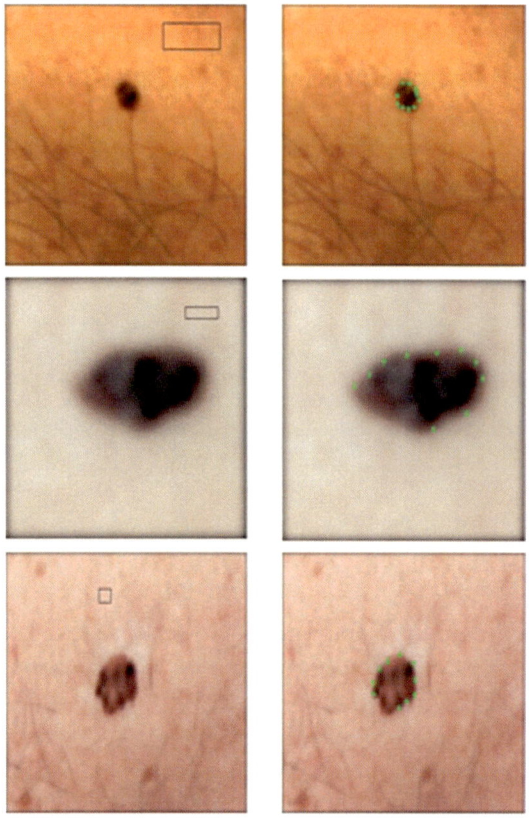

(a) (b)

Figure 5.8 Homogenous regions and edge points selected for evaluating the performance. Column (a) shows the homogenous regions selected, and column (b) shows edge-point sets selected.

Table 5.2 The contrasts of selected homogenous regions.

Image	Gaussian filter (C)	Bilateral filter (C)
Image 1	0.000016	0.000015
Image 2	0.000012	0.000011
Image 3	0.000196	0.000194

Table 5.3 The contrasts of a selected edge-points set.

Image	Gaussian filter (C)	Bilateral filter (C)
Image 1	0.0001	0.0004
Image 2	0.0002	0.0004
Image 3	0.0013	0.0035

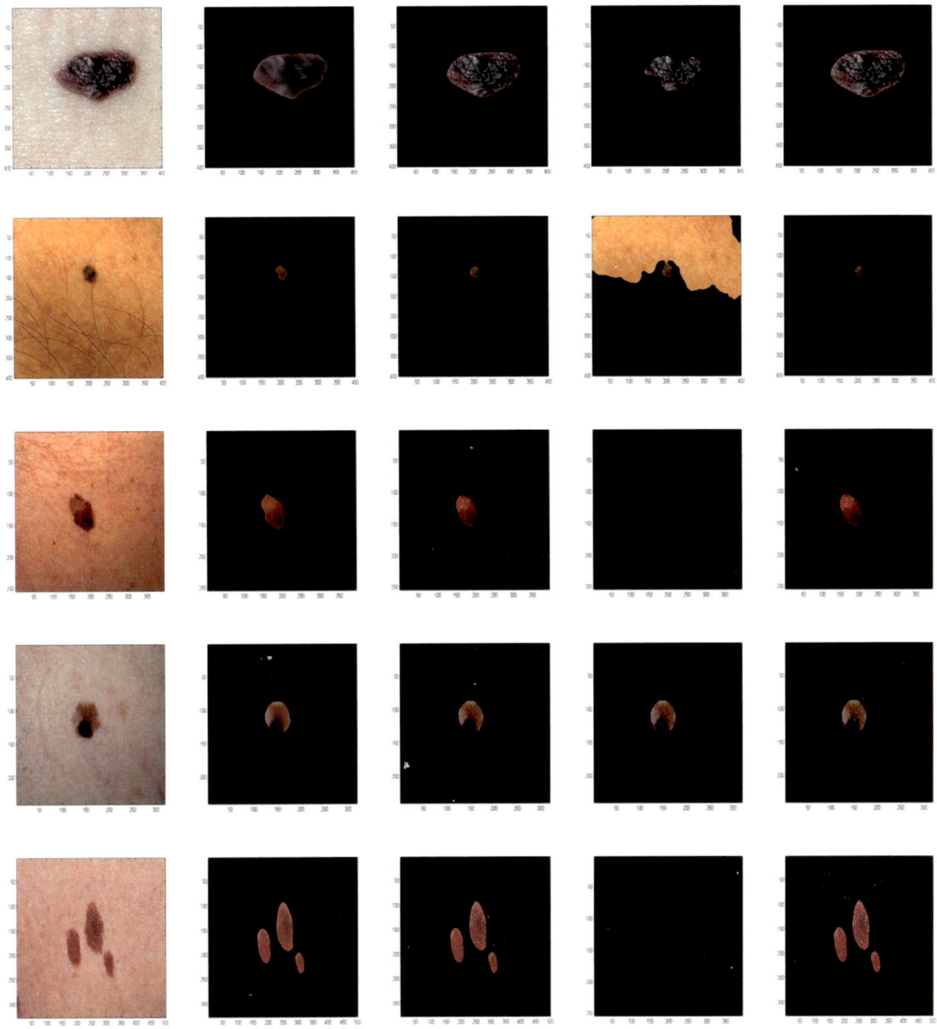

(a) (b) (c) (d) (e)

Figure 5.9 Segmentation results of skin lesions using four different color spaces on five skin lesion images. Column (a): original skin lesion images; column (b): segmentation obtained by RGB color space; column (c): segmentation obtained by YUV color space; column (d): segmentation obtained by HSV color space; and column (e): segmentation obtained by YCbCr color space.

lesion extraction using different color spaces, Pratt's quality measurement metric (FOM) was used. The definition of the FOM is defined as

$$F = \frac{\sum_{i=1}^{I_A} \frac{1}{1+\alpha(d(i)^2)}}{\max(I_A, I_I)}, \tag{5.13}$$

where I_M is the number of boundary pixels that are delineated by dermatologists, I_N is the number of boundary pixels delineated by this system,

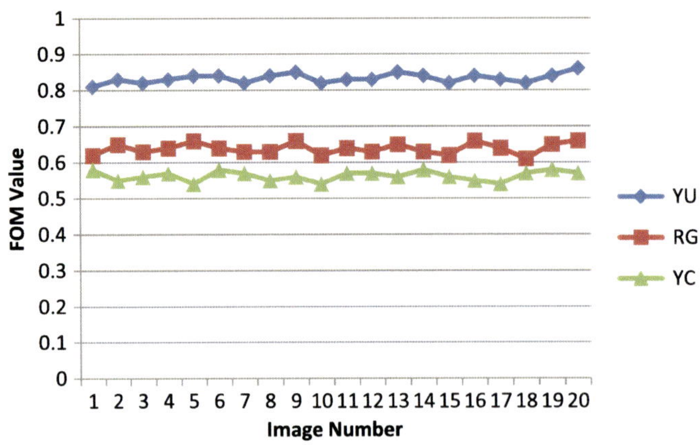

Figure 5.10 The FOM values of 20 segmentation images with RGB, YUV, and YCbCr color spaces

$d(i)$ is the Euclidean distance between a pixel from I_N and the nearest pixel from I_M, and α is a scaling constant (0.05 in this example). The complete matching between the boundary pixels delineated by computer-aided segmentation method and the boundary pixels delineated by the technicians has $F = 1$ in Eq. (5.13).

In experiments, one technician was used to segment the skin lesions manually, and the results were used to compute FOMs using Eq. (5.13). Figure 5.10 shows the FOM values obtained using different color spaces. From Fig. 5.10, it can be seen that YUV color space achieved the highest FOM values, and thus it performed better than other three color spaces using the proposed method.

5.3.4 Speeding up using a GPU

The previous experiments were done using Matlab and many matrix operations. The program is generally very slow when many matrix operations are involved. In order to speed-up the processing, graphics processing unit (GPU) technology was investigated to perform matrix arithmetic instead of the CPU for skin lesion extraction. The typical function of a GPU is to assist with the rendering of 3D graphics and visual effects;[30] it is highly optimized to perform advanced calculations such as floating-point arithmetic, matrix arithmetic, and so on. The design of a GPU is more of data processing than data caching and flow control than a CPU.[31] The following graph (Fig. 5.11[31]) shows the structural difference between CPUs and GPUs. In a CPU [see Fig. 5.11(a)], cache is designed for data caching, control is designed for flow control, and the ALU

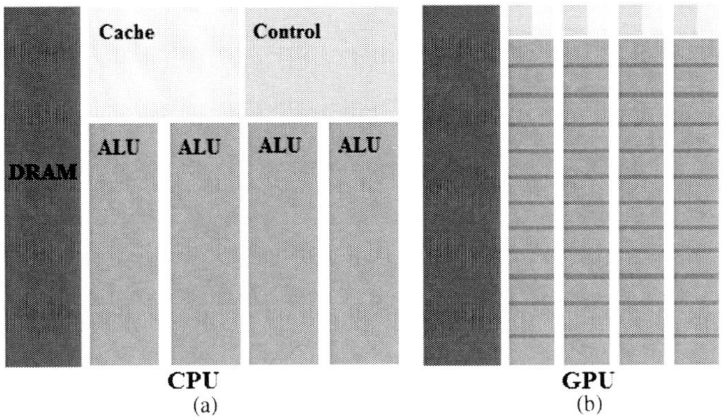

Figure 5.11 CPU and GPU structure.

(arithmetic logic unit) is designed for data processing. Thus, the major difference between GPUs and CPUs is that a GPU has highly parallel structure (many more ALUs), which makes it more effective than a CPU for some operations.

In matrix operations, the size of the matrix will affect the performance. There is a threshold to decide whether or not to use a GPU. If the matrix size is below this threshold, the computing speed between a CPU and GPU are nearly the same. Otherwise, a GPU is faster than a CPU. The bigger the matrix size is, the faster the GPU will be. In this chapter's skin lesion extraction example, GPU technology was used to perform the matrix operations in Eq. (5.8), and the results are shown in Fig. 5.12: border extraction involving GPU technology can improve from 30–40% in running speed according to the matrix size.

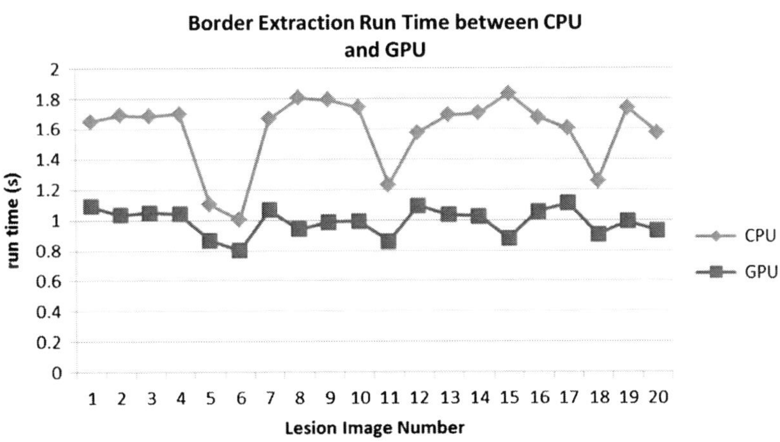

Figure 5.12 Comparison of CPU and GPU technologies for border extraction.

5.4 Conclusion

In this chapter, we provided an overview of skin lesion segmentation and proposed a new skin border extraction based on distance histogram and color information. The new distance-histogram-based skin lesion border extraction can achieve high accuracy.

In order to remove the noise, two noise reduction technologies were investigated. A Gaussian filter and bilateral filter were compared in the noise reduction part. The experimental results show that a bilateral filter is better at edge preservation than a Gaussian filter.

We also compared four color spaces (RGB, YUV, HSV, and YCbCr) for skin lesion extraction, and we found that YUV color space is more suitable for the proposed method. Because YUV color space separately stores the brightness information (Y) and chrominance information (UV) of images, it can reduce the effect by brightness for color images.

We also investigated GPU techniques for skin lesion extraction and the results show a GPU has potential applications in skin lesion extraction.

References

1. "Skin care," http://health-cares.net (2007).
2. P. Lindquist, "Malignant melanoma detection using NIR-Spectroscopy," Research report, University of Maryland, Baltimore, MD (1995).
3. National Cancer Institute, "Skin cancer," http://physioweb.org (2005).
4. H. W. Rogers, M. A. Weinstock, and A. R. Harris, "Incidence estimate of nonmelanoma skin cancer in the United States," *Arch. Dermatol.* **146**(3), 283–287 (2006).
5. Centers for Disease Control, "Skin cancer," http://www.cdc.gov/cancer/skin/ (May 8, 2013).
6. D. L. Kasper, E. Braunwald, and A. Fauci, *Harrison's Principles of Internal Medicine*, 17th ed., McGraw-Hill, New York (2008).
7. R. R. Anderson and A. Parrish, "The optics of human skin," *J. Investigative Dermatol.* **77**, 13–19 (1981).
8. T. DiChiara, "Pictures of Moles and Melanoma skin cancer – Learn to tell the difference with pictures," http://skincancer.about.com/od/symptoms/ss/mole_9.html (Apr. 23, 2011).
9. Sun Protection and Products Guide, "Learning the ABCDE rule for skin cancer can save your life!" http://www.sun-protection-and-products-guide.com/ABCDE-rule-for-skin-cancer.html (2009).
10. F. Ercal, A. Chawla, W. V. Stoecker, H. C. Lee, and R. H. Moss, "Neural network diagnosis of malignant melanoma from color images," *IEEE Trans. Biomed. Eng.* **41**(9), 837–845 (1994).

11. H. Ganster et al., "Automated melanoma recognition," *IEEE Trans. Med. Imaging.* **20**(3), 233–239 (2001).
12. http://www.handyscope.net/
13. J. Tang, "A multi-direction GVF snake for the segmentation of skin cancer images," *Pattern Recognition* **42**(6), 1172–1179 (2009).
14. J. E. Golston, W. V. Stoecker, R. H. Moss, and I. P. S. Dhillon, "Automatic detection of irregular borders in melanoma and other skin tumors," *Comp. Med. Imaging Graphics* **16**(3), 199–203 (1992).
15. J. E. Golston, R. H. Moss, and W. V. Stoecker, "Boundary detection in skin tumor images: An overall approach and a radial search algorithm," *Pattern Recognition* **23**(11), 1235–1247 (1990).
16. Q. Abbas, I. Fondón, and M. Rashid, "Unsupervised skin lesions border detection via two-dimensional image analysis," *Comp. Methods Programs Biomed.* **104**(3), e1–e15 (2011).
17. S. E. Umbaugh, R. H. Moss, and W. V. Stoecker, "An automatic color segmentation algorithm with application to identification of skin tumor borders," *Comp. Med. Imaging Graphics* **16**(3), 227–235 (1992).
18. G. Schaefer, M. I. Rajab, M. E. Celebi, and H. Iyatomi, "Colour and contrast enhancement for improved skin lesion segmentation," *Comp. Med. Imaging Graphics* **35**(2), 99–104 (2011).
19. X. Yuan, N. Situ, and G. Zouridakis, "Automatic segmentation of skin lesion images using evolution strategies," *Biomed. Signal Process. Control* **3**, 220–228 (2008).
20. X. Yuan, N. Situ, and G. Zouridakis, "A narrow band graph partitioning method for skin lesion segmentation," *Pattern Recognition* **42**(6), 1017–1028 (2009).
21. L. Xu et al., "Segmentation of skin cancer images," *Image Vision Comp.* **17**(1), 65–74 (1999).
22. C. Tomasi and R. Manduchi, "Bilateral Filtering for Gray and Color Images," *Proc. IEEE ICCV 1998*, 839–846 (1998).
23. I. Butt and N. Rajpoot, "Multilateral Filtering: A Novel Framework for Generic Similarity-based Image Denoising," *Proc. IEEE ICIP 2009*, 2981–2984 (2009).
24. M. Zhang, "Bilateral Filter in Image Processing," Master's Thesis, Louisiana State University, Baton Rouge, LA (2009).
25. S. H.-S. Hung, "Lower bound on average mean-square error for image restoration," *IEEE Trans. Signal Process.* **39**(2), 497–499 (1991).
26. http://www.meddean.luc.edu/lumen/meded/medicine/dermatology/melton/content1.htm
27. N. Vandenbroucke, L. Macaire, and J. Postaire, "Color image segmentation by pixel classification in an adapted hybrid color space: application to

soccer image analysis," *Comp. Vision Image Understanding* **90**(2), 190–216 (2003).

28. J. L. Liu and D. G. Kong, "Image retrieval based on weighted blocks and color feature in: International Conference on Mechatronic Science," *Proc. Int. Conf. Mechatronic Sci. Electric Eng. Comp.*, Jilin, China, 921–924 (2011).
29. H. Noda and M. Niimi, "Colorization in YCbCr color space and its application to JPEG images," *Pattern Recognition* 40(12), 3714–3720 (2007).
30. M. B. Kolundžija and P. D. Zoric, "Efficient evaluation of MoM matrix elements using CPU and/or GPU," *Proc. IEEE EUCAP 2012*, 702–706 (2012).
31. NVIDA CUDA, "Non-graphic computing with graphics processors," http://ixbtlabs.com/articles3/video/cuda-1-p1.html (Oct, 21, 2008).

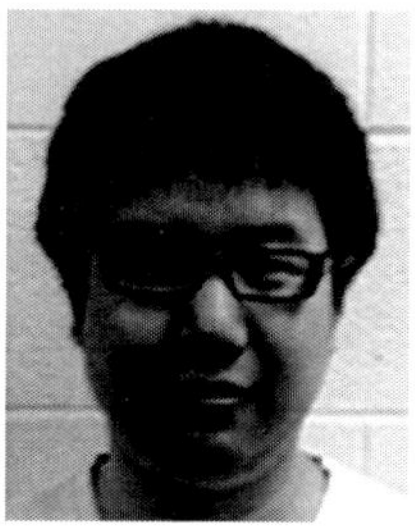

Yanliang Gu was born in Suzhou, Jiangsu, China. He received the BS degree in computer science from Michigan Technological University, Houghton, MI in 2011 and is currently a master's student of Medical Informatics there. His research areas include computer-aided cancer detection and mobile image processing and applications.

Jinshan Tang was born in Chenzhou, Hunan, China. He received his Ph.D. degree from the Beijing University of Posts and Telecommunications, Beijing, China in 1998. He joined ATR Media Integration and Communication Research Laboratories (Japan) after he graduated. In 2000, he joined Harvard Medical School as a Postdoctoral Researcher. In 2001 he joined the University of Virginia, where he worked for three years. He also worked as a senior engineer in Intel and a visiting fellow at NIH. From 2006, he worked as an Assistant Professor at Alcorn State University, Alcorn, MS and currently he is an associate professor in Michigan Technological University, Houghton, MI. He has published more than 90 journal and conference papers, and has served as a guest editor for five journals. He is a senior member of the IEEE and a member of the technical committee on information assurance and intelligent multimedia-mobile communications (IEEE SMC society).

Chapter 6
Geometric Incremental Support Vector Machine for Object Detection from Capsule Endoscopy Videos

Xiaohui Yuan and Mohamed Abouelenien
University of North Texas, Department of Computer Science and Engineering, Denton, TX, USA

Balathasan Giritharan
CityGrid Media, West Hollywood, CA, USA

Jianguo Liu
University of North Texas, Department of Mathematics, Denton, TX, USA

Xiaojing Yuan
University of Houston, Department of Engineering Technology, Houston, TX, USA

6.1 Introduction

Capsule endoscopy (CE) is a method used to visualize the entire small intestine. It is a widely adopted procedure for diagnosing gastrointestinal diseases including obscure bleeding, Crohn's disease, gastric ulcers, and colon cancer. The CE videos used in this research were produced with the Pillcam® by Given Imaging. The imaging component of this system is a vitamin-sized capsule that comprises a color CMOS camera, a battery, a light source, and a wireless transmitter. The device captures two images per second for approximately eight hours and generates approximately 55,000 color images with a size of 256 × 256 pixels during the life of its usage.

Reviewing CE videos to make diagnostic decisions is a tedious task and is achieved by watching the video playback and marking suspicious frames and anatomical landmarks. It usually takes more than one hour to annotate a full-length video, and a typical mid-size hospital produces an average of twelve CE videos per day.

Given the large amount of training data, computer algorithms are in great demand to reduce the review time by identifying frames that contain signs of lesion, bleeding, and polyps, as well as segment videos into gastrointestinal sections. Many existing learning algorithms require all training data to be present in memory to achieve the best generalization performance. Limited by the computing power and memory size, it is usually difficult to implement such a learning scheme. Incremental learning has great potential to accommodate the inclusion of examples that become available over time or represent a change of perception. The initial data set can be used to create a model; when new data becomes available, it is integrated to update the classifier. In practice, clinical videos are acquired over time. Furthermore, knowledge of the visual appearance of the diseases in CE video changes over time due to the relatively shorter practice time. It would be practical to build a classifier based on initial data and revise the classifier as new examples arrive.

A key question of incremental methods is how to retain knowledge from the training examples in each repetition to maximize the unbiased representation of underlying data distribution. Retaining some key examples, e.g., support vectors in a support vector machine (SVM), works well in cases where the existing examples closely represent the topography of the class boundary. However, if a new instance dramatically changes the topography and hence the decision hyperplane, some previously removed examples could become the margin mover.

This chapter presents an incremental learning method that extends the geometric SVM to multiclass classification with large training data. The proposed method identifies important examples and models the data such that when new examples become available, a classifier is built without revisiting all of the past data available but with generalization accuracies, which are comparable to those obtained in the batch-learning setting.

6.2 Related Work

6.2.1 Related work on CE video analysis for automatic object detection

Among efforts in computer-aided CE video analysis, color and texture features are used in many applications,[24] particularly for detecting heterogeneous objects, e.g., ulcers and polyps.[3,10,28,39] Many classification algorithms have been applied to video analysis including neural networks,[39] SVMs,[16] and thresholding. Despite improvements, many previous studies were evaluated with a small number of examples, and to the best of our knowledge no performance was reported with respect to the entire videos.

Table 6.1 Experimental data and detection outcomes. "–" indicates not reported in the paper. The sensitivity, specificity, and accuracy are in percentages.

Reported Studies	Data Set Size			Performance		
	Total	Abnormal	Normal	Sen.	Spe.	Acc.
Kodogiannis and Boulougoura[24]	140	35	35	–	–	95.7
Kodogiannis and Lygouras[25]	140	35	35	–	–	97.1
Vilarino et al.[39]	400	100	300	–	–	95.5
Coimbra and Cunha[10]	1000	–	–	–	–	87
Lau and Correia[26]	1705	577	1128	88.3	–	–
Li and Meng[27]	60	30	30	65.2	82.5	–
Li and Meng[28]	400	200	200	91	93	–
Jung et al.[18]	2000	1000	1000	92.8	89.5	–
Barbosa et al.[3]	204	100	104	98.7	96.6	–
Karargyris and Bourbakis[19]	–	20	30	75	73.3	–
Karargyris and Bourbakis[20]	50	10	40	100	67.5	–

Table 6.1 summarizes the characteristics of experimental data sets and performances in recent related work on automatic detection using CE videos. Despite the use of different features and classification methods, the experimental data and performances vary greatly. Among these studies, results in eight studies were generated from experiments using 1000 examples or less. Two studies used a moderately larger number of examples. Compared to the number of frames available in a CE video (approximately 50,000), however, the training data set size is small. Ideally, if the training set is well selected and comprehensive, the classifier can achieve satisfactory generalization performance. It is unclear if the formed cohort represents the true data distribution. An important question awaits investigation: "Given the relatively small number of positive examples from CE videos, how does one train learning algorithms to achieve minimal false negative detections?"

6.2.2 Related work on incremental learning using SVMs

Although a large number of training examples helps reduce the generalization error, the learning process can become computationally expensive, if not infeasible. Efficient and scalable approaches are needed that can modify the knowledge structure in an incremental fashion without having to revisit all of the previously processed data.

Attempts at an incremental SVM started by retaining the support vectors. The method in Syed et al.[38] keeps only the support vectors at each incremental step. The model obtained via this strategy will be the same or very similar to what would have been obtained by using all training examples. Mitra et al.[32] used an error-driven technique in the incremental SVMs. In addition to the support vectors, this method keeps a number of non-support-vector examples. Given a trained $SVM^{(t)}$ at iteration t, the SVs of $SVM^{(t)}$ (along with a certain number of correctly classified and misclassified instances) are used to train the new model $SVM^{(t+1)}$. Alternatively, Domeniconi and Gunopulos[13] proposed a method that keeps only the misclassified examples. When a given number of misclassified examples is collected, the update occurs. The support vectors of the last-trained SVM, along with the misclassified instances, are used as training data to obtain the new model. The assumption of minimum change in the hyperplane serves as the foundation of the previous methods.

Katagiri and Abe[21] proposed using one-class SVMs to select support vectors, which reduces the possibility of support vectors being deleted when the hyperplane is rotated. A hypersphere is generated for each class, and only the instances lying close to the boundary of the hypersphere are retained as candidate support vectors for future updates. Although this method handles the rotation of the decision boundary, the assumption of a hypersphere to model data distribution is unrealistic in many real-world applications.

To manage the space complexity and size of the representative data set, Hernandez et al.[14] employed a multiresolution approach. Agarwal et al.[40] demonstrated that the concept of the span of support vectors can be used to build a classifier that performs reasonably well while satisfying space and time constraints, thus making it suitable for online learning. Mitra et al.[33] presented probabilistic SVMs wherein the training set is refined by active query from a pool of unlabeled data. Orabona et al.[34] proposed an online algorithm that approximately converges to the standard SVM solution each time new examples are added. This method uses a set of linearly independent observations and tries to project every new observation onto the set obtained so far, thus reducing time and space requirements at a negligible loss of accuracy. Proximal SVM[36] employs a greedy search across the training data to select the basis vectors of the classifier and tunes parameters automatically using the simultaneous perturbation stochastic approximation after incremental additions are made.

Instead of selecting training examples randomly, Chen et al.[9] divided the training set into groups using the k-means clustering algorithm. In active query, a weight is assigned to each example according to its confidence, which is calculated from the error upper bound of the SVM to indicate the closeness of the current hyperplane to the optimal one.

Another key issue in incremental learning is to adapt to the nonstationary underlying data distribution. Cauwenberghs and Poggio[6] developed an incremental and decremental SVM method that divides the training set into three categories: the margin SVs, the error SVs (ones that violate the margin but are not necessarily misclassified), and ignored vectors (ones within the margin). When a new instance is misclassified, the SVM is updated. Bookkeeping is used to categorize examples, the complexity of which is $O(n^3)$ for each incremental example. A later work of Diehl and Cauwenberghs[12] reduced the computational cost by using "leave-one-out" error estimation. Again, the methods assume that the hyperplane does not change significantly.

Klinkenberg and Joachims[23] proposed a method to handle drift in SVMs. The drift represents changes to the underlying distribution of the data collected over an extended period for learning tasks. The method maintains a window to the training data stream and adjusts its size so that the estimated generalization error is minimized. Shilton et al.[37] addressed the sequentially arriving data and parameter variation using a warm-start algorithm. It allows efficient retraining of a SVM after adding a small number of additional examples. Boubacar et al.[5] employed an online clustering algorithm that is developed to learn continuously evolving clusters from nonstationary data. This algorithm uses a fast incremental learning procedure to account for model changes over time. Dedicated to online clustering in multiclass environment, the algorithm is based on an unsupervised learning process with self-adaptive abilities.

6.3 Geometric Incremental Support Vector Machines

Geometric and quadratic optimization views of SVMs were shown to be equivalent.[4,11] A geometric SVM represents each class as a convex hull and finds the minimum distance between the two.[22] To address nonseparable classes, the reduced convex hull (RCH)[11] was developed.[30,31] The method of incremental learning presented here extends the RCH concept and proposes that convex skin represent key examples in training, as well as a means of finding convex skins.

6.3.1 Geometric support vector machines

Let x be a data point in a convex hull C. According to Caratheodory's theorem, x can be represented as a convex combination of a finite number of points in C:

$$x = \sum_{j=1}^{k} \lambda_j x_j, \text{ where } \lambda_j \geq 0, \text{ and} \sum_{j=1}^{k} \lambda_j = 1. \tag{6.1}$$

Given a set of data points X, the convex hull is a linear combination of all the elements in X and can be represented as follows:

$$C(X) = \left\{ \sum_{i=1}^{k} \alpha_j x_i;\ x_i \in X,\ 0 \leq \alpha_j \leq 1,\ \sum_{i=1}^{k} \alpha_i = 1 \right\}. \tag{6.2}$$

Reduced convex hull[4] (also known as soft convex hull[11]) is the set of all convex combinations of elements of X, denoted by $R(X, \mu|\mu < 1)$, as follows:

$$R(X, \mu) = \left\{ \sum_{i=1}^{k} \alpha_j x_i;\ x_i \in X,\ 0 \leq \alpha_i \leq \mu,\ \sum_{i=1}^{k} \alpha_i = 1 \right\}. \tag{6.3}$$

The difference between a convex hull and a RCH is that the weight factor α_i is bounded by μ in a RCH. Using a suitable μ for each class, two overlapping classes can be transformed into a linearly separable case.[4,11,31]

However, the RCH provides no means of finding the extreme points. To overcome this, a compressed convex hull was proposed.[35] It, however, makes explicit assumptions on the kernel, which limits its application.

Geometric SVM represents classes as convex hulls and solves the problem by finding the minimum distance.[22] Given a set of examples $X = \{x_1, x_2, ..., x_n\}$, the function ϕ maps each instance into a features space $\phi(x_i)$. For simplicity, ϕ_i is used here to denote $\phi(x_i)$, and the mapped examples form a feature set $\Phi = \{\phi_1, \phi_2, ..., \phi_n\}$. The convex hull $C(\Phi)$ is rewritten as follows:

$$C(\Phi) = \left\{ \sum_{j=1}^{k} \alpha_i \phi_i | \phi_i \in \Phi, 0 \le \alpha_i \le 1, \sum_{j=1}^{k} \alpha_i = 1 \right\}. \tag{6.4}$$

Similarly, a RCH is the set of convex combinations of instances in Φ with α_i bounded by μ as follows:

$$R(\Phi, \mu) = \left\{ \sum_{i=1}^{k} \alpha_i \phi_i | \ \phi_j \in \Phi,\ 0 \le \alpha_i \le \mu, \sum_{i=1}^{k} \alpha_i = 1 \right\}. \tag{6.5}$$

The decision boundary is then perpendicular to the nearest points between RCHs and can be found following Bennett's method.[4]

6.3.2 Geometric incremental support vector machine (GISVM)

Our method extends the concept of RCH and defines the skin of a convex hull. The idea is that only the examples within the skin are most informative and should be retained for future training, which is similar to Katagiri's idea,[21] but a model for the data distribution is not specified. When additional examples become available, they are used to update the SVM together with the skin of the current convex hull. In such a way, many fewer instances are used in a training process. In addition, with a superset of the possible SVs retained, missing SVs due to significant changes to the data distribution caused by the addition of new examples is avoided.

The skin of a convex hull consists of the outer-most vertices (i.e., examples). Given bounding factors μ_u and μ_l, $0 \le \mu_l < \mu_u \le 1$, the skin $S(\Phi, \mu_l, \mu_u)$ of a convex hull $C(\Phi)$ consists of instances between two RCHs and can be expressed as follows:

$$S(\Phi, \mu_l, \mu_u) = \{\phi_i | \phi_i \in \{R(\Phi, \mu_u) - R(\Phi, \mu_l)\}\}. \tag{6.6}$$

When the data set is dense enough and evenly distributed in the space, the geometric center can be used to find the extreme points of the convex hull. However, this is usually not the case in real-world applications. Due to the lack of knowledge of data distribution, the above procedure could miss less-prominent extreme points. Thus, a recursive method is proposed that finds the vertices (i.e., extreme points) of a convex hull to represent the skin.

It is said that $\phi_j \in \Phi$ is an extreme point of convex hull $C(\Phi)$ if there exists a direction d in terms of two instances, i.e., $d = \phi_b - \phi_a$, and ϕ_a, $\phi_b \in C(\Phi)$, such that

$$\phi_j = \max_{\phi k \in \Phi}(\phi_k - \phi_a, d), \tag{6.7}$$

where $(\phi_k - \phi_a, d)$ is the inner product of the difference vectors with respect to ϕ_a and the direction d.

The extreme points are found in two steps: first, a set of initial extreme points are identified based on the center of gravity; and second, additional extreme points are then found via recursively searching along the direction defined by a pair of extreme points.

For a set of feature vectors Φ, the gravity center $\overline{\Phi}$ is approximated with the arithmetic average, i.e., $\overline{\Phi} = \sum_{i=1}^{n} \frac{1}{n}\phi_i$. The initial set of extreme points is identified by projecting each point $\phi_j \in \Phi$ to the direction $d(\phi_m) = \phi_m - \overline{\Phi}$ and selecting the ones that give the maximum projection magnitude:

$$E_{\text{seed}}(X) = \{\phi_n \mid \arg\max_{\phi n} P(\phi_n,\ d(\phi_m)), \forall\ \phi_m, \phi_n \in \Phi\}, \tag{6.8}$$

where $P(\phi_n,\ d(\phi_m))$ denotes the projection of ϕ_n to $d(\phi_m)$.

The explicit expression of the feature vectors ϕ_i is not needed to compute the extreme points in the above procedure. The projection $P(\phi_n, d(\phi_m))$ in the feature space can be achieved by the kernel operation in the input space as follows. Given two feature vectors ϕ_a and ϕ_b in Φ, the projection of vector ϕ_c is $P(\phi_c, d(\phi_a, \phi_b))$. Thus,

$$\begin{aligned} P(\phi_c, d(\phi_a, \phi_b)) &= \langle \phi_b - \phi_a, \phi_c - \phi_a \rangle \\ &= \langle \phi_b, \phi_c \rangle - \langle \phi_b, \phi_a \rangle - \langle \phi_a, \phi_c \rangle + \langle \phi_a, \phi_a \rangle \\ &= \sum_i b_i\phi_i \cdot \sum_j c_j\phi_j - \sum_i b_i\phi_i \cdot \sum_j a_j\phi_j \\ &\quad - \sum_i a_i\phi_i \cdot \sum_j c_j\phi_j - \sum_i a_i\phi_i \cdot \sum_i a_i\phi_i \\ &= \sum_i \sum_j b_i c_j K(x_b, x_c) - \sum_i \sum_j b_i a_j\ K(x_b, x_a) \\ &\quad - \sum_i \sum_j a_i c_j\ K(x_a, x_c) - \sum_i \sum_i a_i a_i\ K(x_a, x_a), \end{aligned} \tag{6.9}$$

where $\sum_i a_i\phi_i$, $\sum_i b_i\phi_i$, and $\sum_i c_i\phi_i$ are convex representations of feature vectors ϕ_a, ϕ_b, and ϕ_c, respectively. For a vector $\phi_j \in \Phi$, its coefficient vector equals $[0,\ 0,\ \ldots,\ 1,\ \ldots,\ 0,\ 0]'$, within which the index value of the number 1 is j. For a vector $\phi_k \in C\ \Phi$ but $\phi_k \in C(\Phi)$, the values in its coefficient vector are in the range of [0, 1), e.g., the coefficient vector of the gravity center $\overline{\Phi}$ is $\left[\frac{1}{n}, \frac{1}{n}, \ldots, \frac{1}{n}\right]'$.

An example is illustrated in Fig. 6.1(a). The solid squares denote the examples, and the gravity center is marked with a large circle. The projected vectors are marked with solid dots. Using the proposed method, three extreme

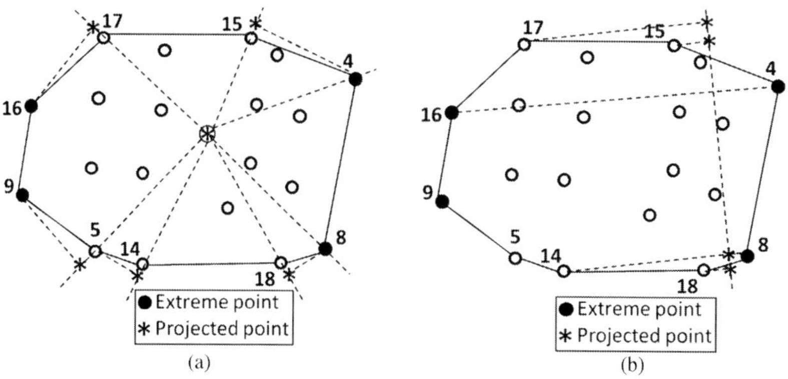

Figure 6.1 Finding extreme (a) data and (b) seed points recursively.

points are identified and highlighted with solid squares. For example, point 16 is identified as an extreme point because it gives the greatest projection to $d(x_{16}, \overline{X})$ [as well as $d(x_{15}, \overline{X})$]. However, instances 14, 15, 17, and 18 are extreme points that are missed by the process.

The primary cause of missing extreme points is the insufficient number of examples, which could be exaggerated in high-dimensional cases. If data points in the feature space are known, classical algorithms such as QuickHull[2] and Gift Wrapping[17] can be used to complete the search. The idea of the proposed algorithm is to recursively search along the perpendicular directions of the convex hull boundaries, which is presented in Algorithms 1 and 2.

Algorithm 1: Search for extreme points.

Require: Φ and E

1. Randomly select ϕ_p, $\phi_q \in E$
2. Randomly select $\phi_m \in \Phi$ and $m \neq p, m \neq q$
3. Identify probing direction d^* using Eq. (6.10)
4. $\Phi_- \leftarrow \{\phi_i | P(\phi_i, d^*) < 0\}$
5. $\Phi_+ \leftarrow \{\phi_i | P(\phi_i, d^*) > 0\}$
6. $E \leftarrow E \cup Probing(\Phi_+, d^*, \phi_p, \phi_q)$
7. $E \leftarrow E \cup Probing(\Phi_-, -d^*, \phi_p, \phi_q)$
8. Return E

This algorithm randomly selects two extreme points $\phi_p, \phi_q \in E$ and another instance $\phi_m \in \Phi$. The searching direction d^* can then be determined as follows:

$$d^* = \phi_m - \phi_p - P\left(\phi_m, d(\phi_p, \phi_q)\right)\frac{\phi_q - \phi_p}{||\phi_q - \phi_p||}. \tag{6.10}$$

A hyperplane through $\phi_q - \phi_p$ and perpendicular to d splits the space into two halves. The projections of instances, i.e., $P(\phi_i, d^*)$, that are on the same sides as ϕ_m are positive, denoted by Φ_+; whereas the projections of the rest instances are negative, denoted by Φ_-. Hence, the further searching for extreme points is divided into two parts, as shown in Algorithm 1.

Searching in each half space is achieved recursively using a pair of identified extreme points ϕ_p and ϕ_q. Let Φ' denote the instances in the half space. With a random instance ϕ_m in Φ', a probing direction d^* can be determined by Eq. (6.10) that points toward the outside of the convex hull; otherwise, change its direction. Hence, an extreme point is identified in Φ' following Eq. (6.7). ϕ_m is paired with ϕ_p and ϕ_q to split the feature space for further probing. The process stops when no additional points exist in Φ'.

Algorithm 2: Recursively probe and search for the extreme points Probing (Φ', d, ϕ_p, ϕ_q).

Require: $\Phi' \subseteq \Phi$, d, ϕ_p, and ϕ_q

1. $F \leftarrow \varnothing$
2. Randomly select $\phi_m \in \Phi'$ and $m \neq p,\ m \neq q$
3. If $\Phi' \neq \varnothing$, then
4. Identify probing direction d^* using Eq. (6.10)
5. If $\langle d^*, d\rangle < 0$, then
6. $d^* \leftarrow \ -d^*$
7. End if
8. $d^* \leftarrow \frac{d^*}{||d^*||}$
9. $F \leftarrow F \cup \{\phi_e | \phi_e = \arg\max_{\phi K \in \Phi'} P(\Phi_k, d^*)\}$
10. For all $\phi_i \in \Phi'$, do
11. If $P(\phi_i,\ d^*) > 0$, then
12. $\Phi'' \leftarrow \Phi'' \cup x_i$
13. End if
14. End for
15. $F \leftarrow F \cup Probing(\Phi'', d, \phi_p, \phi_e)$
16. $F \leftarrow F \cup Probing(\Phi'', d, \phi_q, \phi_e)$
17. End if
18. Return F

Figure 6.1(b) illustrates an example of probing in a half space. The dotted lines depict the projections of the instances. The two extreme points are 17 and 18, which determine the probing direction (d and $-d$ in Algorithm 1). Extreme points 1 and 12 are found.

In the algorithm, the magnitude of vectors $\phi^{(2)} - \phi^{(1)}$ is calculated as follows:

$$\begin{aligned}
||\phi^{(2)} - \phi^{(1)}|| &= \left[\langle\phi^{(1)}, \phi^{(1)}\rangle + \langle\phi^{(2)}, \phi^{(2)}\rangle - 2\langle\phi^{(2)}, \phi^{(1)}\rangle\right]^{\frac{1}{2}} \\
&= \Bigg[\sum_i \sum_j \beta_i^{(1)}\beta_j^{(1)} K(x_i, x_j) + \sum_i \sum_j \beta_i^{(2)}\beta_j^{(2)} K(x_i, x_j) \\
&\quad -2\sum_i \sum_j \beta_i^{(2)}\beta_j^{(2)} K(x_i, x_j)\Bigg]^{\frac{1}{2}}
\end{aligned} \tag{6.11}$$

The range of the projections $R_{X,d}$ of a set of examples X in given directions $d = x^{(1)} - x^{(2)}$, $x^{(j)} = \sum_i x_i \beta_i^{(j)}$ for $j = 1, 2$ would be

$$R_{X,d} = [\min_{x_i \in X} P_{x_i,d}, \max_{x_i \in X} P_{x_i,d}]. \tag{6.12}$$

Given a set $X = \{x_1, \cdots, x_n\}$, the skin segment with an angle θ around $d = x^{(1)} - x^{(2)}$ is

$$SS_{X,d} = \left\{ x_i \,|x_i \in E(X),\ \cos^{-1}\frac{\langle d, x^{(2)} - x_j\rangle}{\|d\| \cdot \left\|x_i^{(2)} - x_j\right\|} \leq \theta \right\}. \tag{6.13}$$

The angle θ_j between vectors $x_j \in X_i$ and center of gravity $x_i^{(g)}$ and $w = x_+^* - x_-^*$ can be found using

$$\left\langle w, x_i^{(g)-x_j} \right\rangle = \|w\| \cdot \left\|x_i^{(g)} - x_j\right\| \cos\theta, \text{ and } \theta = \cos^{-1}\frac{\langle w, x_i^{(g)} - x_j\rangle}{\|w\| \cdot \left\|x_i^{(g)} - x_j\right\|}. \tag{6.14}$$

The decision boundary is perpendicular to $w_2^* - w_1^*$, where w_1^* and w_2^* are the nearest points between the RCHs. Gilbert's algorithms[15] are used to identify the nearest points between convex hulls (as shown in Algorithm 3). Figure 6.2 illustrates an example of GISVM for a linearly nonseparable case. The two classes are enclosed with convex hulls and overlap, as shown in Fig. 6.2(a). The skin of the convex hull is shown in Fig. 6.2(b). Two RCHs are highlighted with solid triangles: the outer RCHs transform the problem into a linearly separable case [see Fig. 6.2(c) for an example], whereas the inner RCHs define the skin thickness. The skin for retention is depicted in Fig. 6.2(d).

Algorithm 3: Gilbert's algorithm for finding the nearest points of two convex hulls.[15]

1. $Z \leftarrow \{\phi_+ - \phi_- | \phi_+ \in \Phi_+, \phi_- \in \Phi_-\}$
2. Randomly select $z^* \in C(Z)$
3. Repeat
4. $z_{\text{old}}^* \leftarrow z^*$
5. $z \leftarrow \arg\min_{z_i \in Z} P(z_i, z*)$
6. $z \leftarrow \arg\min_{z_i \in Z} P(z_i, z*)$
7. Until $\left\|z^* - z_{\text{old}}^*\right\| \approx 0$

6.4 Experimental Results and Discussion

6.4.1 Synthetic and benchmark data preparation

The experiments presented here used synthetic data sets, real-world data sets, and CE videos for evaluation. Two synthetic data sets were created by randomly sampling 2D Gaussian functions and the checkerboard function, namely the XOR data set (see Fig. 6.3 for examples). Ten sets of examples

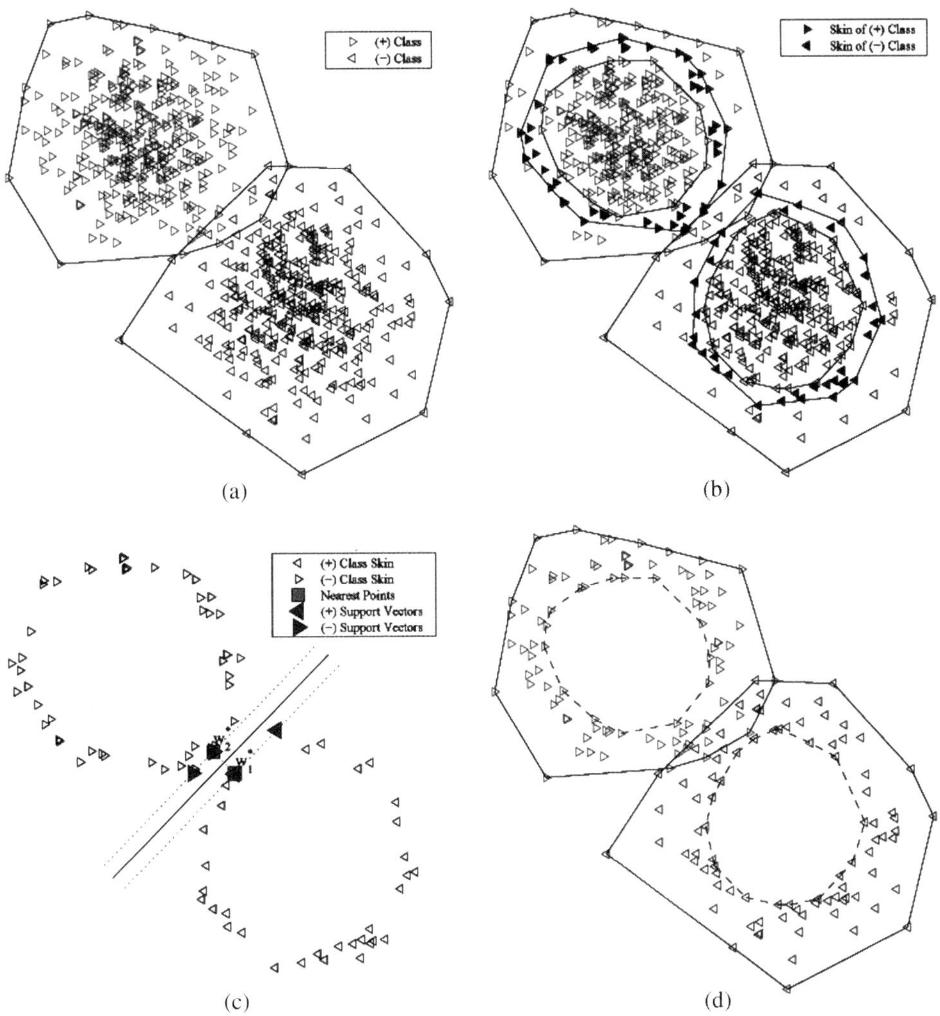

Figure 6.2 Geometric incremental SVM using convex hull skin for linearly nonseparable problems. (a) Two linearly nonseparable classes consisting of 700 examples. (b) The classes become linearly separable using $RCH(\mu = 0.1)$. (c) The decision boundary is found by finding the nearest points between two RCHs. (d) Examples within $S(x_i, 0.1, 1)$ are retained for future model updates.

were randomly generated using each model. The Gaussian data set has 1% overlap, whereas the XOR data set has no overlap. Four real-world data sets were obtained from the UCI machine learning repository.[1] In addition, a mammogram[8] data set was used. Each feature in a data set was normalized to unify its range to between 0 and 1. Table 6.2 lists the properties of the benchmark data sets used in the experiments.

6.4.2 Parameter selection

Figure 6.3 illustrates the decision boundaries of the proposed method applied to synthetic data sets using RBF, polynomial, and linear kernels. The shade in

Table 6.2 Benchmark data sets and their characteristics.

Data Sets	Number of Dimensions	Positive Class	Data Set Size	
			+ Class	– Class
SPECT	22	1	212	55
PIMA	8	1	268	500
YEAST	8	CYT	463	1021
IONOSPHERE	34	b	126	225
MAMMOGRAM	6	2	260	10923

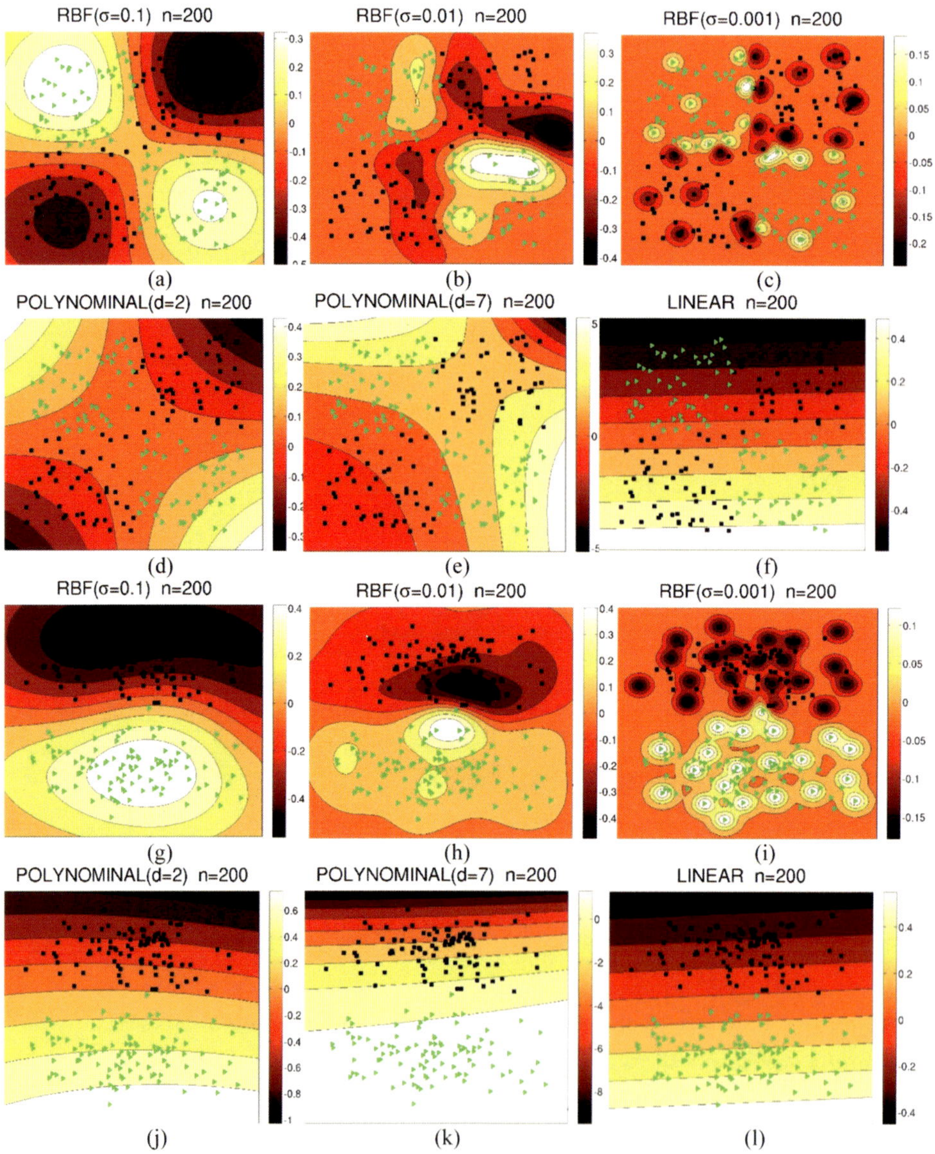

Figure 6.3 Decision boundaries from applying the proposed method to the synthetic data sets using different kernels.

the plots depicts the distance to the decision boundary. The incremental training starts with ten examples, and with each update, five new examples are randomly selected and used. The updates continue until all examples are exhausted. As shown in the figure, when the σ of the RBF kernel is decreased, the final classifier appears overfitted. Among all of the kernels tested, RBF kernels with $\sigma = 0.1$ resulted in better decision boundaries. It is evident that when examples are presented to the GISVM in an incremental fashion, the proposed method achieves superior closeness in modeling the underlying data distribution. The optimal level is reached with RBF kernels of $\sigma = 0.1$.

6.4.3 Efficiency analysis

Figures 6.4(a) and (c) illustrate the maximum number of examples retained (i.e., the examples in the skin of the RCHs) for various σ used in RBF kernels. As σ decreases, the number of examples retained significantly increases. When σ reaches 0.01, the number of examples retained is approximately the total number of examples in the training set. This indicates overfitting of the model, which is consistent with previous experimental results. On the other hand, the number of support vectors (in solid curve) varies slightly. According to these plots, a good choice for σ is 0.1, which provides a close description of the data sets and retains only a small number of examples in the training iterations.

Figures 6.4(b) and (d) show the number of examples retained in the training iterations. With a properly chosen σ, the number of retained examples is small, which implies the stability of the incremental learning

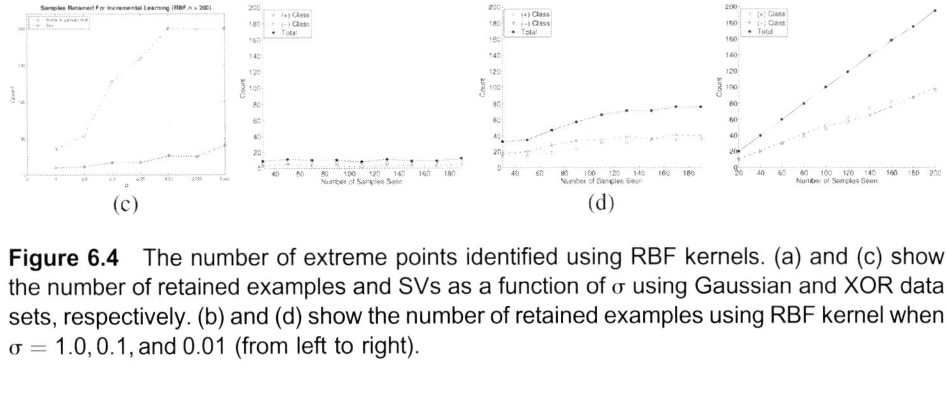

Figure 6.4 The number of extreme points identified using RBF kernels. (a) and (c) show the number of retained examples and SVs as a function of σ using Gaussian and XOR data sets, respectively. (b) and (d) show the number of retained examples using RBF kernel when $\sigma = 1.0, 0.1,$ and 0.01 (from left to right).

process. This also indicates that a much-smaller amount of memory was used to complete the learning.

Table 6.3 lists the time (in seconds) required by the GISVM and batch-learning libSVM to complete the training. Ten repetitions were conducted, and the average time and the standard deviation are reported. Because random examples were used, the training time varies. These experiments assumed that an equal number of examples was used to update a classifier in the incremental learning. The size of examples is referred to as step size Δ.

Limited by the number of examples in the benchmark data sets, two step sizes (i.e., $\Delta = 10$ and $\Delta = 20$) were used in the evaluations. It is clear that the time used by incremental learning is much less than that used by the libSVM. Among all cases, the MAMMOGRAM case consists of the largest number of examples and took significantly more time for training. Although the minimum time to complete training using MAMMOGRAM is in the time range of the GISVM, it can require up to triple the time that the libSVM needs. It is evident that the proposed incremental learning handles data efficiently and can update the classifier in much less time. The average time cost for this method to complete is approximately 13.4% of the time cost for batch-learning SVMs.

It is an interesting observation that a larger step size does not necessarily result in a longer training time. For data sets SPECT, PIMA, and MAMMOGRAM, training of the GISVM took less time using a step size of 20 than a step size of 10. Even in the other cases, the difference is small. This is probably due to the fact that only a small number of examples (i.e., examples within the skin of the RCHs) were carried over to the next round of updates.

Table 6.3 The average training time (in seconds) and standard deviation using batch-learning libSVM and the proposed GISVM. The number of iterations is also reported for the GISVM.

Data Sets	Batch-Learning libSVM	GISVM			
		$\Delta = 10$		$\Delta = 20$	
	Time	Time	Iteration	Time	Iteration
GAUSSIAN	2.9 (0.4)	0.2 (0.04)	19	0.21 (0.06)	9
XOR	5.2 (0.4)	0.34 (0.03)	19	0.54 (0.08)	9
SPECT	2.7 (0.3)	0.35 (0.05)	11	0.25 (0.02)	6
YEAST	97.2 (12.5)	3.59 (0.25)	73	5.16 (0.42)	37
PIMA	71.4 (7.9)	21.14 (0.25)	36	13.39 (1.34)	19
IONOSPHERE	3.8 (0.5)	0.3 (0.04)	15	0.43 (0.05)	8
MAMMOGRAM	4787 (2261)	2859 (541)	445	1540 (338)	223

6.4.4 Accuracy analysis

Figure 6.5 illustrates the classifiers' performance based on accuracy, sensitivity, and specificity during the incremental iterations. Ten repetitions were conducted with randomized initial examples. In each data set, 50% of the data were used for training; the remaining examples were used for testing.

In each case, a SVM classifier was created using all the training data. The best parameters were selected based on their generalization performance with

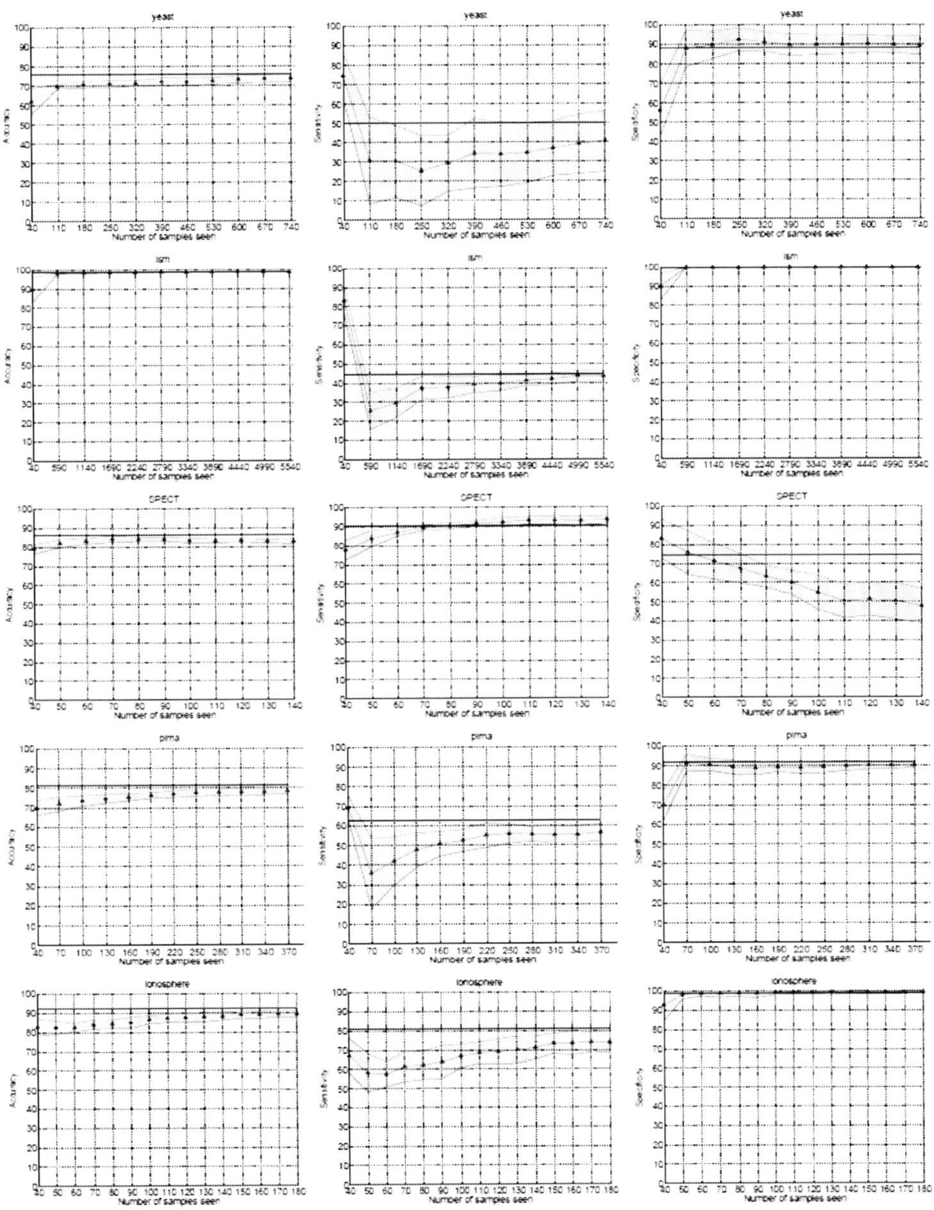

Figure 6.5 Accuracy performance of the GISVM using UCI data sets.

Table 6.4 Parameters used in the proposed GISVM.

Data Sets	Kernel Parameter	Convex Skin	
		μ_u	μ_l
SPECT	$\sigma = 0.05$	0.9	0.6
YEAST	$\sigma = 0.10$	0.9	0.6
PIMA	$\sigma = 0.15$	0.5	0.3
IONOSPHERE	$\sigma = 0.01$	0.5	0.3
MAMMOGRAM	$\sigma = 0.01$	0.8	0.4

the testing data set. Table 6.4 lists the selected kernels and parameters that gave the best performance measures. The results from these classifiers are used as a reference and are depicted as the horizontal lines in Fig. 6.5.

In the proposed incremental learning process, ten examples were randomly selected from each class of the training set, and a SVM was trained. In each incremental step, ten randomly selected examples from the remaining training data set were used to update the classifier. The intermediate classifiers were evaluated with the test data set. For each data set, ten repetitions were conducted, and the average performance is plotted with a solid line in Fig. 6.5. The shaded area depicts the accuracy variation.

With more examples included in the training process, the classifier trained with the proposed method improves its performance; this is evident in the cases of YEAST, SPECT, PIMA, and IONOSPHERE. In the case of MAMMOGRAM (i.e., ISM), the performance is already close to optimal at the beginning, and there is no room for improvement. However, improvement in sensitivity can still be observed in the training, and by the end of iterations, the classifier outperformed the batch learning by a small margin.

Despite a slight drop of specificity of the SPECT data set, the SVMs trained with the proposed method achieved the same performance or even outperformed the batch-learning method. As listed in Table 6.2, the SPECT data set contains more positive examples than negative ones, the ratio of which is approximately 4:1. Hence, the improvement of sensitivity leverages the underperformance of specificity, and the overall accuracy is close to the batch-learning results. It is interesting that in five cases, the intermediate classifier had degradation in early iterations, but the training process was able to recover to the benchmark performance asymptotically as additional examples are included.

6.4.5 Experiments with CE videos

The analysis tool provided by the manufacturer of the Pillcam® capsule endoscope plots the path of the device through the digestive tract based on the wireless signal strength transmitted to the external image downloader carried by the patient. Our experiments on CE videos were performed to automate the classification of the frames in CE videos into digestive organs, namely the esophagus, stomach, small intestine, and colon.

Six CE videos were collected and manually annotated by gastroenterologists. Each video consists of approximately 55,000 frames. Out of the six videos, one was randomly selected to train the classifier, and the other five were used for testing.

In previous experiments with CE videos, the HSV color space was found to have a better classification performance on average.[16] In addition, using the histogram significantly reduces the dimensionality.[*] Hence, the color histogram in the HSV space was adopted as a feature. The color histogram is a very large and sparse matrix: With n bins used in each color component, there are n^3 features using the HSV histogram for every video frame, most of which are zeros or close to zeros. To suppress sparseness and the number of values in features, only the hue and saturation (HS) components were used. As observed in previous experiments,[16,29] using HS components improves control of lighting variations in the GI tract.

The order of classification of multiclass SVMs was determined based on the preliminary evaluation. In the experiments, identification of the esophagus gives the best accuracy followed by the identification of the small intestine. Hence, the order is determined and listed in Table 6.5. The kernels used to train a SVM are also included in this table.

In the learning process, 50 frames were randomly selected from each class of the training video to train a SVM. In each incremental step, 20 frames randomly selected from the remaining training video frames were used to update the classifier. The iteration repeats until the training examples exhaust. Table 6.6 lists the accuracy of the final classifiers. The performance of this method is highly satisfactory. With the majority of frames acquired in the stomach and small intestine, the average accuracies are 86.9% and 94.4%, respectively. Images acquired in the colon are disturbed by the presence of feces.

At the end of incremental training, only 12% of the frames were part of the skins among the four classes for the hierarchical SVMs. Apparently, the

Table 6.5 Order and parameters of hierarchical classification of organs of CE videos.

Order	Dividing Classes	Kernel Parameters
1	Esophagus vs. the rest	RBF ($\sigma = 0.15$)
2	Small intestine vs. stomach and colon	RBF ($\sigma = 0.1$)
3	Stomach vs. colon	RBF ($\sigma = 0.5$)

* Each frame is a 256 × 256 color image. If color is used, the dimensionality of each example is up to 196,608.

Table 6.6 GISVM performance of digestive organs in CE videos.

Video	Esophagus	Stomach	Small Intestine	Colon
1	100%	87.6%	94.2%	85.3%
2	94.4%	85.8%	95.3%	82.2%
3	95.0%	87.2%	94.7%	84.3%
4	100%	86.4%	94.1%	83.7%
5	90.0%	87.7%	93.9%	94.3%

smaller number of examples demands much less memory space for the learning process and thus provides a plausible mechanism for handling a large amount of data. When new examples are added, the classifier is updated efficiently in contrast to the conventional batch-learning methods.

6.5 Conclusion

This chapter presents a GISVM method to learn from large data sets with an emerging trend and dynamic patterns. To overcome high computational demands from a large data set, this method identifies a subset of examples for the training process. It extends the reduced convex hull concept and defines the skin segments of convex hulls. The skin is found by identifying the extreme points of the convex hull. This method is founded on the idea that the examples within the convex hull skin are a superset to the support vectors, including the potential ones in future training. When additional examples are provided, they are used together with the skin of the convex hull constructed from the previous data set.

Using the skin of convex hull in the incremental learning process results in a small number of instances at every incremental step. The set of extreme points are found by recursively searching along the direction defined by a pair of extreme points. Besides the advantages in computational efficiency, the proposed method handles linearly nonseparable cases in multiclass problems.

Experiments were conducted with synthetic, benchmark, and CE data sets. With the synthetic data sets, the proposed method achieves highly satisfactory classifiers that closely model the underlying data distribution with appropriate kernels. The choice of RBF kernel for the synthetic data sets provides a good description of the data sets and retains only a small number of examples in the training iterations.

Using the experiments on benchmark data sets, this chapter demonstrates that the GISVM learning handles data efficiently and updates the classifier in approximately 13.4% of the time needed by the batch-learning SVM. Performance over the incremental steps further verifies the superior stability, improvement, and recoverability of the proposed method. The accuracy over the incremental steps increases steadily. Even in the cases when the performance drops, the classifier is able to recover to previous levels in future iterations because the convex hull skin that contains useful examples is retained. Furthermore, the improvement in the performance measures over the incremental steps (by deleting examples other than the ones on the skin) indicates that retaining the examples on the skin preserves adequate information about the decision boundary of SVMs.

From the experiments on CE videos it was noted that the average performance of classifying a CE video is above 86.9%, which is very competitive. The amount of memory space required in the training process could be one-eighth of what is required by the conventional SVM, which casts new light on processing large data sets with limited resources. Further experiments on CE videos demonstrated that the GISVM can handle data that could not be handled by the libSVM.

References

1. A. Asuncion and D. J. Newman, University of California, Irvine, School of Information and Computer Sciences, UCI Machine Learning Repository, http://www.ics.uci.edu/~mlearn/MLRepository.html (2007).
2. C. B. Barber, D. P. Dobkin, and H. Huhdanpaa, "The quickhull algorithm for convex hulls," *ACM Trans. Math. Software* **22**(4), 469–483 (1996).
3. D. J. C. Barbosa, J. Ramos, and C. S. Lima, "Detection of small bowel tumors in capsule endoscopy frames using texture analysis based on the discrete wavelet transform," *Proc. IEEE Eng. Med. Biol. Soc.*, 3012–3015 (2008).
4. K. P. Bennett and C. Campbell, "Support vector machines: hype or hallelujah?" *SIGKDD Explor. Newsletter* **2**(2), 1–13 (2000).
5. H. A. Boubacar, S. Lecoeuche, and S. Maouche, "SAKM: Self-adaptive kernel machine; a kernel-based algorithm for online clustering," *Neural Networks* **21**(9), 1287–1301 (2008).
6. G. Cauwenberghs and T. Poggio, "Incremental and decremental support vector machine learning," *Proc. Advances in Neural Information Processing Systems*, 409–415 (2000).
7. C. C. Chang and C. J. Lin, "LIBSVM: A library for support vector machines," ACM Transactions on Intelligent Systems and Technology **2**(3), 1–27 (2011).
8. N. V. Chawla, K. W. Bowyer, L. O. Hall, and P. Kegelmeyer, "SMOTE: Synthetic Minority Over-sampling Technique," *J. Artificial Intelligence and Research* **16**, 321–357 (2002).
9. S. Cheng and F. Y. Shih, "An improved incremental training algorithm for support vector machines using active query," *Pattern Recognition* **40**, 964–971 (2007).
10. M. T. Coimbra and J. P. S. Cunha, "Mpeg-7 visual descriptors contributions for automated feature extraction in capsule endoscopy," *IEEE Trans. Circuits and Systems for Video Technology* **16**(5), 628–637 (2006).

11. D. J. Crisp and C. J. C. Burges, "A geometric interpretation of -SVM classifiers," NIPS: Advances in Neural Information Processing Systems, 12 (2000).
12. C. P. Diehl and G. Cauwenberghs, "SVM incremental learning, adaptation and optimization," *Proc. International Joint Conference on Neural Networks* **4**, 2685–2690 (2003).
13. C. Domeniconi and D. Gunopulos, "Incremental support vector machine construction," *Proc. IEEE International Conference on Data Mining*, 589–592 (2001).
14. E. Parrado-Hernandez, I. Mora-Jimnez, J. Arenas-Garca, A. R. Figueiras-Vidal, and A. Navia-Vzquez, "Growing support vector classifiers with controlled complexity," *Pattern Recognition* **36**(7), 1479–1488 (2003).
15. E. G. Gilbert, "An iterative procedure for computing the minimum of a quadratic form on a convex set," *SIAM J. Control* **4**(1), (1966).
16. B. Giritharan, X. Yuan, J. Liu, J. Oh, and S. Tang, "Bleeding detection from capsule endoscopy videos," *Proc. IEEE Eng. Med. Biol. Soc.*, 4780–4783 (2008).
17. R. A. Jarvis, "On the identification of the convex hull of a finite set of points in the plane," *Information Process. Lett.* **2**(1), 18–21 (1973).
18. Y. S. Jung, Y. H. Kim, D. H. Lee, and J. H. Kim, "Active blood detection in a high-resolution capsule endoscopy using color spectrum transformation," *International Conference on Biomedical Engineering and Informatics*, 859–862 (2008).
19. A. Karargyris and N. Bourbakis, "Identification of ulcers in wireless capsule endoscopy videos," *Proc. 6th IEEE International Conference on Symposium on Biomedical Imaging*, Piscataway, NJ (2009).
20. A. Karargyris and N. Bourbakis, "Identification of polyps in wireless capsule endoscopy videos using log gabor filters," *Life Science Systems and Applications Workshop*, 143–147 (2009).
21. S. Katagiri and S. Abe, "Incremental training of support vector machines using hyperspheres," *Pattern Recognition Lett.* **27**(13), 1495–1507 (2006).
22. S. S. Keerthi, S. K. Shevade, C. Bhattacharyya, and K. R. K. Murthy, "A fast iterative nearest point algorithm for support vector machine classifier design," *IEEE Trans. Neural Networks* **11**(1), 124–136 (2000).
23. R. Klinkenberg and T. Joachims, "Detecting concept drift with support vector machines," *Proc. 17th International Conference on Machine Learning*, 487–494 (2000).
24. V. Kodogiannis, "Computer-aided diagnosis in clinical endoscopy using neuro-fuzzy systems," *IEEE International Conference on Fuzzy Systems* **3**, 1425–1429 (2004).

25. V. S. Kodogiannis and M. Boulougoura, "An adaptive neurofuzzy approach for the diagnosis in wireless capsule endoscopy imaging," *International J. Information Technology* **13**(1), 46–54 (2007).
26. P. Y. Lau and P. L. Correia, "Detection of bleeding patterns in WCE video using multiple features," *Proc. IEEE Eng. Med. Biol. Soc.*, 5601–5604 (2007).
27. B. Li and M. Q.-H. Meng, "Analysis of wireless capsule endoscopy images using chromaticity moments," *IEEE International Conference on Robotics and Biomimetics*, 87–92 (2007).
28. B. Li and M. Q.-H. Meng, "Computer-aided detection of bleeding regions for capsule endoscopy images," *IEEE Trans. Biomed. Eng.* **56**(4), 1032–1039 (2009).
29. J. Liu and X. Yuan, "Obscure bleeding detection in endoscopy images using support vector machines," *Optimization and Engineering* **10**(2), 289–299 (2009).
30. M. E. Mavroforakis and S. Theodoridis, "A geometric approach to support vector machine classification," *IEEE Trans. Neural Networks* **17**(3), 671–682 (2006).
31. M. E. Mavroforakis, M. Sdralis, and S. Theodoridis, "A geometric nearest point algorithm for the efficient solution of the SVM classification task," *IEEE Trans. Neural Networks* **18**(5), 1545–1549 (2007).
32. P. Mitra, C. A. Murthy, and S. K. Pal, "Data condensation in large databases by incremental learning with support vector machines," *Proc. 15th International Conference on Pattern Recognition* **2**, 708–711 (2000).
33. P. Mitra, C. A. Murthy, and S. K. Pal, "A probabilistic active support vector learning algorithm," *IEEE Trans. Pattern Analysis and Machine Intelligence* **26**(3), 413–418 (2004).
34. F. Orabona, C. Castellini, B. Caputo, L. Jie, and G. Sandini, "On-line independent support vector machines," *Pattern Recognition* **43**(4), 1402–1412 (2010).
35. X. Peng and Y. Wang, "CCH-based geometric algorithms for SVM and applications," *Appl. Mathematics and Mechanics* **30**(1), 89–100 (2009).
36. C. Renjifo, D. Barsic, C. Carmen, K. Norman, and G. S. Peacock, "Improving radial basis function kernel classification through incremental learning and automatic parameter selection," *Neurocomput.* **72**(1–3), 3–14 (2008).
37. A. Shilton, M. Palaniswami, D. Ralph, and A. C. Tsoi, "Incremental training of support vector machines," *IEEE Trans. Neural Networks* **16**, 114–131 (2005).
38. N. A. Syed, H. Liu, and K. Kay, "Incremental learning with support vector machines," *Proc. Workshop on Support Vector Machines at the International Joint Conference on Arterial Intelligence* (1999).

39. F. Vilarino, L. Kuncheva, and P. Radeva, "ROC curves and video analysis optimization in intestinal capsule endoscopy," *Pattern Recognition Lett.* **27**, 875–881 (2006).

40. S. Agarwal, V. V. Saradhi, and H. Karnick, "Kernel-based online machine learning and support vector reduction," *Neurocomput.* 71, 1230–1237 (2008).

Xiaohui Yuan received his B.S. degree in Electrical Engineering from Hefei University of Technology, China in 1996 and his Ph.D. degree in Computer Science from Tulane University in 2004. After graduating, he worked at the National Institutes of Health as a postdoctoral research scientist from 2005 to 2006 and then joined the University of North Texas in 2006, where he is currently serving as an Associate Professor in the Department of Computer Science and Engineering. He is a recipient of the Ralph E. Powe Junior Faculty Enhancement Award in 2008 and the Air Force Summer Faculty Fellowship in 2011 and 2012. He is a member of IEEE and SPIE. His research interests include computer vision, pattern recognition, data mining, and artificial intelligence.

Balathasan Giritharan received his B.S. degree in Computer Science from the University of Peradeniya, Sri Lanka in 2002, and his M.S. in Computer Science (2007) and Ph.D (2012) in Computer Science and Engineering from the University of North Texas. From 2002 to 2005, he worked as a lecturer at the University of Peradeniya. He interned at Abbott Laboratories in 2011 and then joined CityGrid Media as a Data Scientist in 2012. His research interests include machine learning, image processing, and business intelligence.

Mohamed Abouelenien received his B.S. and M.S. degrees in Electronics and Communication Engineering in 2005 and 2008 from the Arab Academy for Science and Technology in Alexandria, Egypt. He is currently a Ph.D. candidate in the Department of Computer Science and Engineering at the University of North Texas, where he has been appointed as a teaching and research assistant since 2009. His research interests include machine learning, ensemble classification, image processing, pattern recognition, facial recognition, computer vision, and dimensionality reduction.

Jianguo Liu received his B.S. and M.S. degrees in Computational Mathematics from Xiangtan University, China, and his Ph.D. in Applied Mathematics from Cornell University. He currently teaches in the Department of Mathematics at the University of North Texas. His research interests include optimization, numerical analysis, and machine learning.

Xiaojing Yuan received her B.S. degree in Electrical Engineering from Hefei University of Technology, China in 1994, her M.S. degree in Computer Engineering from the University of Technology and Science of China in 1997, and her Ph.D. in Mechanical Engineering from Tulane University in 2003. After working as a postdoctoral researcher with the Image Understanding Lab at Tulane, she joined the University of Houston as a Visiting Assistant Professor in 2004. She is an Associate Professor in the Engineering Technology Department at the University of Houston. Dr. Yuan is a senior member of IEEE and a member of ASEE and ISA. She has over 70 publications in peer-reviewed journals, book chapters, and conferences. Her research and teaching interests include pattern recognition, computational intelligence, image processing, data mining, and information processing in wireless sensor networks.

Chapter 7
Automated Melanoma Screening and Early Detection*

Xiaojing Yuan
Intelligent Sensor Grid and Informatics Lab, University of Houston Engineering Technology Department, University of Houston, TX, USA

Ning Situ
Microsoft Inc., Bellevue, WA, USA

Xiaohui Yuan
Computer Science and Engineering Department, University of North Texas, Denton, TX, USA

George Zouridakis
Engineering Technology, Electrical and Computer Engineering, Computer Science, University of Houston, TX

Melanoma is a malignant tumor of melanocytes. Though less common than other types of skin cancer, it is considered the deadliest form not only because it causes the majority (75%) of deaths related to skin cancer,[1] but also because it can metastasize to other organs in the body. The lifetime risk of developing a melanoma has been skyrocketing in the United States, growing from 1:1,500 people in 1953 to nearly 1:100 in 1996.[2] In fact, the American Cancer Society (ACS) reported that in 2010 an estimated 68,130 new cases of melanoma were diagnosed, resulting in a total of 8,700 deaths.[3] The National Cancer Institute (NCI) at the National Institutes of Health (NIH) estimates that in 2012 the new cases of confirmed diagnosis and resulting deaths from melanoma are

* This work was supported by the University of Houston, Texas Heart Institute, National Science Foundation, and National Institutes of Health.

76,250 and 9,180, respectively.[4] On the other hand, the survival rate of melanoma is above 85% if detected no later than stage I/II,[5] and the cost for an early-stage melanoma treatment is as low as about $1,800 per patient.[6] Advances in imaging technologies enabled non-invasive skin cancer screening and early detection using pigmented skin lesion (i.e., moles) images. Furthermore, such non-invasive and low cost imaging systems (such as dermoscopy†) made it possible for mole screening in the public-health setting by general doctors. However, even the most experienced dermatologist is challenged to perform an assessment and give the correct diagnosis using these images. Because moles vary in size, color, shape, and texture pattern, and the diagnosis severity ranges from completely benign to aggressive and lethal (not to mention the sheer number of lesions that need to be evaluated), such screening is still nonexistent in the current healthcare system even though reports recognize the paramount importance of screening and early detection. An automated melanoma screening and early detection system uses computer-generated quantitative features from dermoscopic skin lesion images to detect the feature models (i.e., signature) of various types of skin cancer.[7]

This chapter begins with an overview of skin-lesion-imaging-based diagnosis systems as well as the available dermoscopy skin lesion datasets in Section 7.1. We present the architecture of an automated melanoma screening and early detection system, AutoScan, and discuss the function of each component in Section 7.2. Section 7.3 describes typical computer-generated feature sets used in melanoma detection systems before presenting our effort to incorporate dermatologists' domain knowledge as a high-level feature. Section 7.4 presents our method to integrate feature selection and decision making by selecting the optimum feature set to build an "on-the-fly" feature model based on their holistic predictive performance. Section 7.5 concludes the chapter and presents future research directions.

7.1 Overview of Automated Melanoma Screening and Early Detection Systems

This section presents common imaging modalities for acquiring pigmented skin lesion images, key functions in a computer-aided skin cancer diagnostic system, skin cancer detection research projects that have successfully transitioned to commercial products, and available skin lesion image datasets.

7.1.1 Optical imaging modalities for pigmented skin lesion image acquisition

Skin lesion imaging technologies are classified based on their resolution and the depth of penetration.[8] In general, optical imaging modality uses a light

† Dermoscopy, an imaging technology, allows visualizing structures inside pigmented skin lesions beyond the capacity of the naked eye.

energy source that shines on the skin and an optical sensor receiver that measures the absorption, reflection, and scattering of the light spectra. The resolution of the skin lesion images the optical imaging modality produces decreases with the depth of penetration. Based on the physiological scale (from subcellular and cellular to nevocellular and surface lesions), skin lesion imaging modalities can be classified as (1) *Primary*: subcellular structures (melanosomes); (2) *Secondary*: cellular level (melanocytes); (3) *Tertiary*: nevocellular nests, collagen, blood vessels; and (4) *Quaternary*: surface, clinical lesions. Six popular optical imaging modalities for acquiring pigmented skin lesion images used by researchers and practitioners are reviewed here and categorized as such.

- **Color video camera**, a quaternary modality, uses a standard color video camera, illumination source, and frame grabber. It was used in the earliest skin lesion imaging system with corresponding image analysis software.
- **Epiluminescence microscopy (ELM)**, also known as dermatoscopy, dermoscopy, or magnified surface microscopy) is a tertiary modality that uses optical magnification to view the oil "immersed" skin lesion. ELM devices use a light ring for illumination, a magnification lens, and a CCD camera to capture and store digitized images. Figure 7.1 shows examples of dermatoscopes and polarized-light dermatoscopes. The magnifying lens of a microscope is typically placed directly on the surface of the skin, with the polarized light ring illuminating subsurface structures. Because of their digital imaging capability, ELM devices are the most-common modality used in automated melanoma screening and early detection systems.
- **Reflectance spectrophotometry**, also a tertiary modality, uses a spectrometer to measure a reflected light spectrum that provides an objective measure of

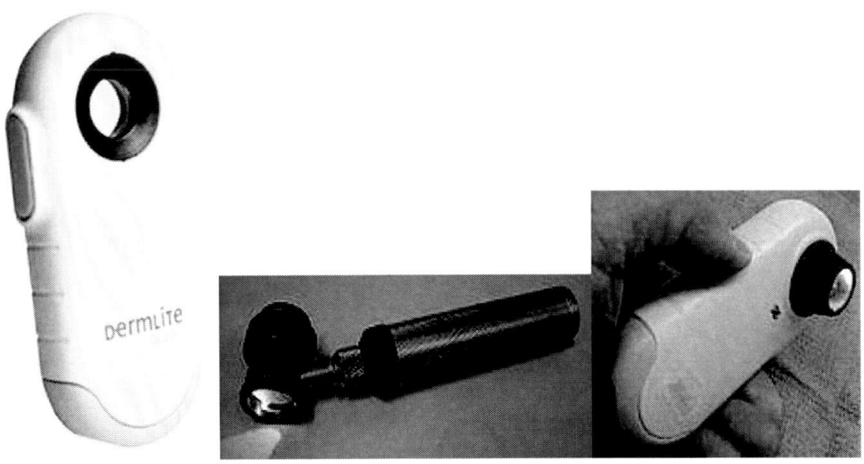

Figure 7.1 Examples of (polarized light) dermatoscopes.[16]

skin color. Because different light wavelengths penetrate different levels of skin, multispectral spectrophotometry is used to measure reflectance at different wavelengths and correlates them to specific subsurface structures that may indicate melanoma.

- **Optical coherence tomography (OCT)**, a tertiary modality, analogous to ultrasound, captures micrometer-resolution, 3D images from within optical scattering media (e.g., biological tissue). Using near-infrared low-coherence light directed at a point on the skin, OCT acquires a vertical slice of a high-resolution skin lesion image by taking a sequence of measurements of light reflectance based on the Michelson interferometry principle.
- **Confocal scanning laser microscopy (CSLM)** is a secondary modality that uses a low-power laser beam to tightly focus on a specific point on the skin. The CSLM sensor detects the light reflected from the focal point. By scanning across the skin surface, CSLM collects a 2D grid that forms a horizontal slice beneath the skin surface.
- **Fluorescence imaging** is also a secondary modality that uses ultraviolet light to excite endogenous fluorophores. Given that cancerous tissue auto-fluoresces differently than healthy tissue at cellular and subcellular levels, fluorescence imaging provides a measurable indicator of the presence of skin cancer.

7.1.2 Key functions of existing skin lesion classification systems

Figure 7.2 shows a functional flowchart summarizing the research on skin lesion image classification.[9,10] After reading the skin lesion images and going through functions such as preprocessing, segmentation, and feature extraction and selection, the system can output a decision on the probability of the skin lesion being melanoma.

Figure 7.3 shows some examples of the pigmented skin lesion images acquired using a dermatoscope (ELM, shown in Fig. 7.1). It is clear that all the images need to go through image pre-processing such as color space conversion, resizing, masking, background subtraction, and noise smoothing. It is also important to even out the difference of light illumination and its effect on skin color and texture pattern. When the image quality is very bad, e.g., the lesion region is covered by heavy hair, as shown in the last row of the left column of Fig. 7.3, it is better to take a new image. The right column of the last row shows the results after the hair-removing algorithm.[11]

After image preprocessing, segmentation extracts the lesion from the background by quantifying its border. As shown in Fig. 7.3, correctly segmenting the lesion will not only make the subsequent steps more efficient but also prevent the signature of the lesion being drowned out by that of the healthy skin in the background. Once the lesion is correctly identified, the system calculates quantitative features to describe a lesion, such as color, size, border smoothness, asymmetry, and texture. There are a large number of

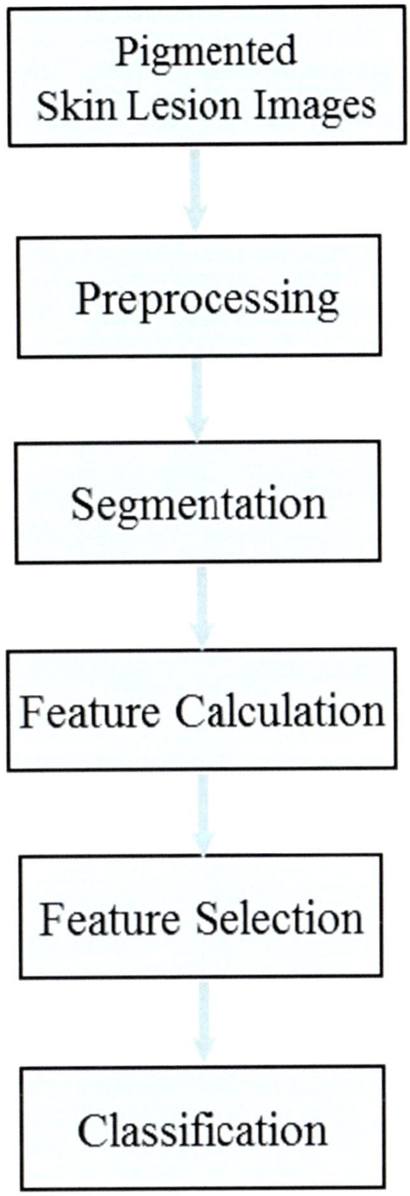

Figure 7.2 Functional flowchart of automated melanoma screening and early detection system.

features that can be extracted from a color skin lesion image. However, not all of them are useful for discrimination. Thus, there is a feature-selection step to choose the set of features that constitutes the signature of a melanoma. The final step is to perform classification based on the feature model built using the identified set of features. A training stage is required to train the feature model, which is then used for making prediction on a new skin lesion image.

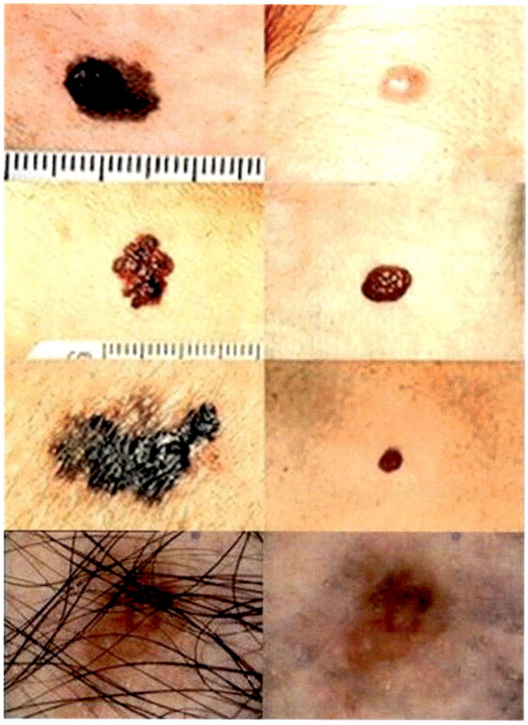

Figure 7.3 Examples of pigmented skin lesion images.

The performance of such a system is the same as the predictive performance of the built feature model. It is evaluated by the receiver operator curve (ROC), which measures sensitivity (maximize true positives), specificity (minimize false positives), and accuracy.

7.1.3 Review of representative skin cancer detection systems

Even though much research has been done to develop new algorithms for each function shown in Fig. 7.2, few commercial systems are available. Table 7.1 summarizes three commercial systems that have successfully transitioned from research projects according to the five functional steps shown in Fig. 7.2 with their overall predictive performance (sensitivity and specificity).

Similar to reflectance spectrophotometry, multispectral ELM can measure reflectance at different wavelengths in the visual and infrared spectra. Melafind®,[12] a FDA approved device for melanoma detection, uses multispectral ELM and captures ten individual pigmented skin lesion images at wavelengths ranging from 430 nm to 950 nm as input. It then performs thresholding-based segmentation on the 430 nm (blue) lesion image, and extracts features from all ten images for classification. After calculating features such as wavelet-based texture pattern, lesion asymmetry, border smoothness, color distribution, and lesion diameter (size), Melafind® uses a

Table 7.1 Comparison of a sample of melanoma detection systems.

Systems	Imaging Modality	Segmentation	Feature Calculation	Feature Selection	Classification
Melafind®, MELA Sciences, USA	Multispectral ELM, 10 spectral bands, 1024 grayscale	Threshold: 420 nm (blue)	10 images for each lesion, 132 ABCD features, 690 wavelet (texture) features	13–15 features; Trained on 246 lesions; Infrared and wavelets	Linear and non-linear classifiers: 95% sensitivity, 70% specificity
SolarScan®, Polartechnics Ltd, Australia	ELM with calibration & artifact removal photomaps	Healthy skin image is used as the baseline, Semi-automated	80 features: color, pattern, and geometry	Trained on 1644 lesions, Multiple models, <12 features	Non-automated research results: 92% sensitivity, 62% specificity
MoleView, Astron-Clinica UK	ELM and spectro-photometry SIAscope (2 mm below skin)	Non-automated, Uses diagnostic protocol	Color (ELM), total melanin, blood, collagen, and dermal melanin	Non-automated, 348 lesions, 2–3 feature combinations	Non-automated protocol: 94% sensitivity, 87% specificity

greedy forward-feature-selection approach to choose thirteen to fifteen features. Features extracted from infrared images and wavelet transformation (i.e., texture) were found to be significant for melanoma detection. Using leave-one-out learning on 246 images with a linear classifier, it can achieve 100% sensitivity and 84% specificity. A recent retrospective study on Melafind®[13] with 1,831 skin lesion images demonstrates that the system has a sensitivity of 98% and an average specificity of 9.9%, which is still superior to that of clinicians (3.7%) ($P = 0.2$). Note that a successful screening system should have high specificity. That is, when the system decides that the lesion is benign, it should not be malignant!

SolarScan® reads pigmented skin lesions acquired from ELM and uses MoleMap to record changes in lesion morphology over time.[14] It uses a semi-automated segmentation procedure and selects a set of features from texture pattern, color, and geometry. Trained on a set of 1,644 skin lesion images and tested on an independent set of 786 images, it achieved 91% sensitivity and 68% specificity.

MoleView[15] uses spectrophotometric intracutaneous analysis (SIA) for melanoma recognition. Its camera, SIAscope, generates eight images of the skin lesion with radiation ranging from 400–1000 nm. Two clinicians identify clinical features (Section 7.1.4) on the acquired images, and logistic regression is used for classification afterwards. The system achieves 80.1% sensitivity and 82.7% specificity.

7.1.4 Overview of skin lesions and commonly used criteria by clinicians

As with many biomedical systems, the main challenge in developing an automated melanoma screening and early detection system lies in the fact that skin lesions vary in size, color, shape, and texture pattern, with diagnostic severity ranging from completely benign to aggressive and lethal melanoma, resulting in numerous possible diagnostic possibilities. The Consensus Netmeeting on Dermoscopy, held in 2001 in Rome (http://www.deroscopyl.org), was the first international meeting to investigate the usage of dermoscopy in preoperative diagnosis of pigmented skin lesions. The Board of the Consensus Netmeeting agreed on a two-step procedure for pigmented skin lesion classification:[16] first identify whether it is melanocytic or nonmelanocytic, and then differentiate benign and malignant melanocytic lesions. Nonmelanocytic lesions are all benign skin growth including basal cell carcinoma, seborrheic keratosis, vascular lesions, and dermatofibroma. The melanocytic skin lesions can be further divided into the following ten categories:[44]

- **Melanoma** is a malignant proliferation of melanocytes that has the potential to metastasize. Though the incidence of melanoma has increased significantly over the last few decades, the prognosis has continued to improve because it was detected at an earlier stage with smaller and thinner, potentially curable, lesions.[17] In fact, the best treatment for melanoma is still early diagnosis and prompt surgical excision of the primary cancer despite progress made in the treatment of metastatic melanoma.[18,19]
- **Clark nevi** are lesions that have different shades of brown coloration. They are regarded as the most-relevant precursor lesions of melanoma. In dermoscopic skin lesion images, Clark nevi can be classified into three types: reticular, globular, and homogeneous. Combinations of these three types are also observed in some lesions. Differentiation of Clark nevi from melanoma *in situ* and early invasive melanoma is the major challenge due to its protean variants.
- **Dermal nevi** include Unna and Miescher nevi,[20] both of which are benign melanocytic nevi. Clinically, both types show distinct features that result in straightforward diagnosis. An Unna nevus is a soft polypoid or sessile, usually papillomatous lesion frequently located on the trunk, arms, and neck. It usually shows a globule or cobblestone pattern with globular structures regularly distributed in the lesion area. Miescher nevi are firm, brownish to nearly skin-colored, dome-shaped papules, with many round and equally sized openings.
- **Reed and Spitz nevi** are well-known simulators of cutaneous melanoma from clinical, dermoscopic, and histopathology perspectives. About 75% of Spitz nevi (SN) demonstrate a symmetric starburst or globular pattern. The remaining cases of SN have an irregular gray-blue pigmentation that is similar to a blue-whitish veil, which is a signature for melanoma. In these cases, differentiation can be made by monitoring the growth of the lesion. If the lesion shows no history of growth, it has a higher chance of being benign.

- **Recurrent nevi** are usually observed after incomplete excision by superficial shaving techniques of dermal or Clark nevi. Recurrent nevi usually appear as asymmetric, bizarrely outlined and poorly circumscribed patches with a dark-brown to black pigmentation resembling a superficial melanoma. As a result, they are commonly misdiagnosed as superficial melanoma or melanoma *in situ*.
- **Blue nevi** are considered as congenital lesions with clearly defined clinical and dermoscopic features that allow clinical diagnosis with a high degree of certainty. They usually demonstrate a homogenous pattern without any local features and have well-defined borders without streaks. However, in rare instances, differentiating blue nevi from nodular melanoma is difficult.
- **Congenital nevi** are benign melanocytic skin neoplasms already present at birth or arising during the first few months of life. They generally appear as flat or elevated light-brown to dark-brown lesions, and are well-known precursor lesions of melanoma. Typical pigmented networks, cobblestone, multi-component patterns, and various-sized dots and globules are all observed in congenial nevi. Small congenital nevi are usually indistinguishable from Clark nevi in dermoscopic images.
- **Combined nevi** are combinations of blue nevi with Clark or Spitz nevi. All these types of nevi could be precursor lesions of melanoma. As a result, when combined nevi are diagnosed, excision is usually recommended unless a very definite diagnosis of benign is presented clinically.
- **Lentigo** refers to a small, brownish macule. In dermoscopic examination of pigmented skin lesions, three types of lentigines are studied: lentigo simplex, reticulated lentigo, and solar lentigo. *Lentigo simplex* is a common benign melanocytic skin lesion that appears as small and sharply demarcated macules with a uniform light-brown or dark-brown color. *Solar lentigo* is a circumscribed brownish macule occurring on chronically sun-damaged skin. It typically has markedly irregular outlines with various shades of coloration ranging from light brown to dark brown. *Reticulated lentigo* has been used to describe a darkly pigmented type of solar lentigo. Dermoscopically, reticulated lentigo is distinctively characterized by a bizarre and asymmetric reticular pattern with a markedly thickened pigment network showing irregular and wide meshes.
- **Labial and genital melanosis** are melanosis of oral and genital mucosae. They are benign melanotic macules characterized by a diffuse pigmentation with a distinct parallel pattern of linear plus curvilinear light-brown to dark-brown streaks.

There are many publications on differential diagnosis using pigmented skin lesion images acquired from dermoscopy. The five most-commonly used algorithms by clinicians are: the ABCD rule,[24,25] the 7-point checklist,[16,26,27]

Table 7.2 Melanocytic detection algorithms.

Algorithms	ABCD Rule	7-Point Checklist	Menzies Method
Criteria	**Asymmetry** Complete symmetry Asymmetry in 1 or 2 axes **Border:** Smooth to rugged **Color:** More color means more severe **Differential structural:** Pigment network Dots Globules Streaks Structureless areas	**Major criteria (2 points):** Atypical pigment network Blue-white veil Atypical vascular pattern **Minor criteria (1 point):** Irregular streaks Irregular pigmentation Irregular dots/globules Regression structures	**Negative features:** Point and axial symmetry Presence of a single color **Positive features:** Blue-white veil Multiple brown dots Pseudopods Radial streaming Scar-like depigmentation Peripheral black dots/ globules Multiple colors (5+) Multiple blue/gray dots Broadened network

the Menzies method,[28–30] pattern analysis,[21–23] and revised pattern analysis.[31] Table 7.2 summarizes the first three algorithms. Many of the patterns listed in the table, such as pigment network, globules, streaks, and blue/gray dots, are also used in the (revised) pattern analysis methods.

As shown in Table 7.2, various global and local features are used in these algorithms for the morphologic diagnosis of pigmented skin lesions. In fact, these global and local features are identified and categorized based on their significance in melanoma diagnosis by clinicians[32] and serve as the backbone for melanoma diagnosis. Representative global features include reticular, globular, cobblestone, homogeneous, starburst, parallel, multicomponent, lacunar, and unspecific patterns. Representative local features include a pigment network, dots and globules, streaks, a blue-whitish veil, vascular structure, pigmentation, and hypopigmentation.

It is clear that differentiating pigmented skin lesions is challenging due to their protean variants. Many published skin-lesion-classification systems focus on separating melanoma and benign lesions because of this.

7.1.5 Pigmented skin lesion datasets

Throughout our study in developing automated melanoma screening and an early detection system, several datasets collected from different imaging modalities were used, partly because the imaging technology to acquire pigmented skin lesion images has advanced tremendously over the past decade.

Set 1: The XLM and TLM skin lesion image datasets were acquired with two Nevoscopes[33,34] that use 5× magnify optical lenses (manufactured by Nikon).[35] An Olympus C2500 digital camera attached to each Nevoscope was used to capture the digitized images. Fifty-one XLM images and sixty TLM images with resolution of 1712 × 1368 were used in our study, with the true region of interest (ROI) manually identified by dermatologists. The remaining lesion images of the 68 XLM–TLM pairs do not show pigmentations and cannot even be segmented by a dermatologist.[36] Thus, they are discarded.

Set 2: One hundred ELM skin lesion images were acquired using an oil immersion technique and studied,[37] of which thirty were melanoma and seventy were benign skin lesions. Three dermatologists performed segmentation on the original image scale, and the average contour was used as the *true* segmentation in our study.

Set 3: The Consensus Netmeeting on Dermoscopy in 2001 resulted in an international collaboration and a collection of numerous pigmented skin lesion images acquired via dermatoscopy at several medical institutions with full annotations for the ABCD rule and 7-point checklists. 1,009 ELM images were obtained from the CD, of which 252 images were benign skin lesions and 757 were melanoma. Among them, 385 images were identified as having low difficulty for differentiating between melanoma and benign by dermatologists.

Set 4: 1,505 ELM skin lesion images from Set 3 and those Menzies[29] used were selected, of which 1,098 were benign lesions and 407 were melanoma. The image resolution ranged from 712 × 454 to 1,024 × 768 pixels, and the lesion size ranged from 7,662 to 804,527 pixels.

Set 5: Based on Set 4, images from Set 2 were included to create a dataset with 1,797 skin lesion images that vary in resolution, size, and quality. This is the most-challenging dataset.

In summary, Sets 3 and 4 are publically available datasets that provide a benchmark for future research and development efforts in melanoma screening and early detection.

7.2 AutoScan: Automated Melanoma Screening and Early Detection

Figure 7.4 shows the architecture of the proposed automated melanoma screening and early detection system, AutoScan. It is similar to traditional skin lesion classification systems, as shown in Fig. 7.2, except for the feature extraction and selection blocks. In AutoScan, after low-level features are

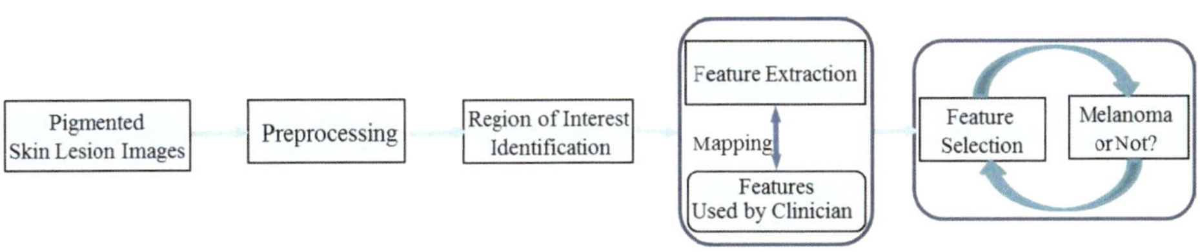

Figure 7.4 The architecture of Autoscan, an automated melanoma screening and early detection system.

extracted from computer algorithms, they are mapped to high-level concepts that the clinician used, as detailed in Section 7.1.3. Afterwards, an optimum set of features (including both low- and high-level features) is selected based on their contribution to the final decision making: melanoma or not. This section briefly reviews functions supported in AutoScan for image preprocessing, ROI identification, and feature extraction before describing in detail how to map the computer-generated low-level features with high-level concepts used by clinicians; the chapter closes with how to select the optimal set of features for the final decision making.

7.2.1 AutoScan preprocessing

In ELM, a halogen light is projected onto an object, rendering its surface translucent.[47] The two modes of ELM in clinical usage are oil immersion[87] and cross-polarization.[14] The cross-polarization mode (XLM) was developed to reduce light reflection from epidermis. In side-transillumination ELM (TLM), a bright ring of light directed at the periphery of a lesion is projected to the lesion center at 45 deg, forming a virtual light source at the focal points approximately 1 cm below the skin surface. Hence, both the surface and subsurface of the skin are translucent.

Common artifacts and distortions in skin lesion dermoscopy images, as shown in Fig. 7.5, include hair, the oil bubble and light ring, and the marks drawn by clinicians indicating the gross lesion location. All of these pose challenges for lesion segmentation, i.e., ROI identification. Thus, the preprocessing steps are used to remove the artifacts and the light ring around the lesion and to remove unnecessary background (e.g., black frames), color space conversion, and noise smoothing to even out the difference of light illumination. As shown in Fig. 7.3, when the image quality is bad, such as when the lesion is covered with heavy hair, it is preferable to obtain a new image.

7.2.2 AutoScan: region-of-interest identification

Extracting the lesion area from the background is an essential step in all computer-aided skin-lesion-classification systems that has continually attracted research efforts. To improve the efficiency of the ROI algorithms,

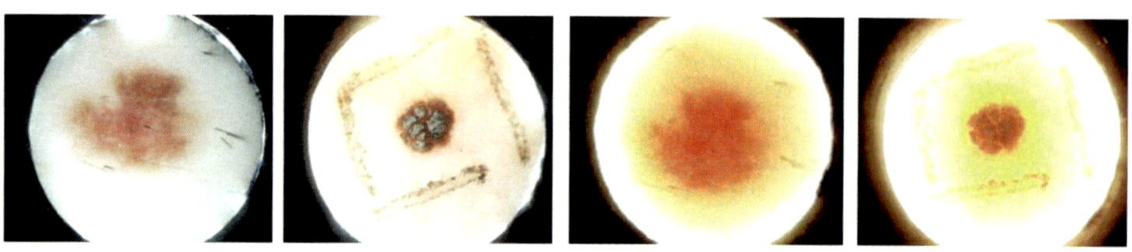

(a) XLM melanoma (b) XLM benign (c) TLM melanoma (d) TLM benign

Figure 7.5 Examples of XLM and TLM skin lesion images.

the ELM skin lesion images are usually downsized, especially if they were taken with a very high resolution.

Thresholding[38–42] and region growing are two simple and widely used image segmentation methods in the literature. They produce satisfactory segmentation when skin lesions have clear boundaries. To account for noise and unclear boundaries, several clustering-based methods were developed that demonstrated improved robustness.[37,42,43] Another popular segmentation method is active contour (or snake),[44–46] in which a curve based on the partial differential equation evolves toward the local optima with respect to an objective function. The objective functions chosen to identify the ROI are mostly edge-based or region-based. Edge-based active contours have been applied[47,48] to segment skin lesion images. Similar to thresholding and region growing, these contours can segment skin lesions with clear boundaries from their background. However, leakage occurs in the presence of weak edges. They are also sensitive to initial conditions. The region-based active contour proposed by Chan and Vese[49] models nonoverlapping homogeneous regions with Gaussian distribution. It performs well for images with two distinctive regions. However, when there are more than two interesting regions in the image, it suffers under-segmentation. In a pigmented skin lesion image, there is only one object of interest. However, many skin lesions are not distributed symmetrically. In fact, the asymmetry and "bubbles" are important structural properties that dermatologists use for diagnostic purposes, as shown in Fig. 7.6.

An active-contour-based region-fusion method was proposed for skin lesion ROI identification[50] to take advantage of region-based active contour while addressing its limitation with a hierarchical approach. In essence, it first segments the lesions into small regions using stricter constraints on homogeneity and a strong edge using a region-based active

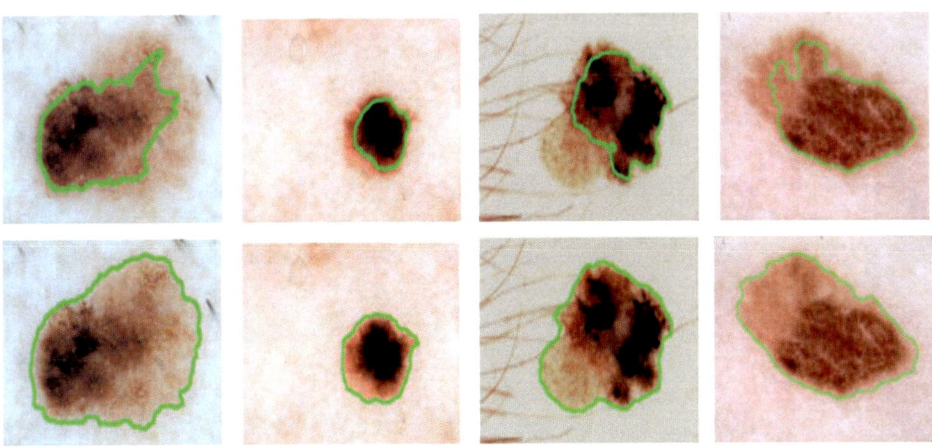

Figure 7.6 Region-of-interest identification results from a region-based active contour (top row) and narrow band graph partitioning (bottom row).

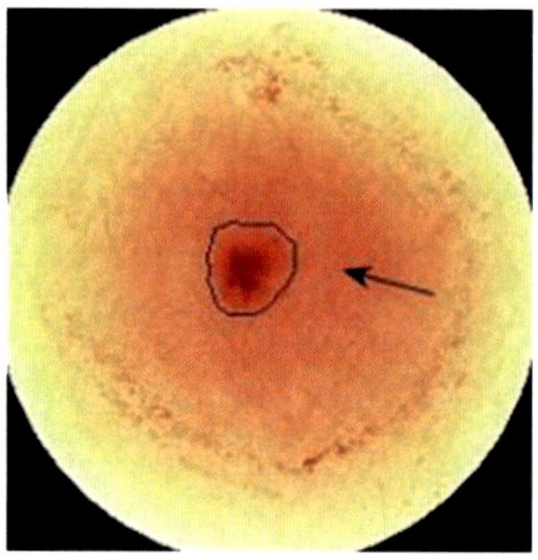

Figure 7.7 A manual segmentation example.

contour before merging them based on a centroid criterion and gradient information. To further improve the robustness and the computational efficiency, a narrow band graph partitioning (NBGP) method[50] was developed with identified ROIs closely matching those manually segmented by clinicians. Figure 7.6 clearly demonstrates the success of NBGP in identifying the ROI, i.e., skin lesion, for those images with a highly asymmetric lesion, weak and/or false edges, and strong hair and bubble artifacts. DullRazor[11] is used to remove hair.

Figure 7.7 illustrates a manually segmented TLM skin lesion image created by a dermatologist. Note that TLM captures both surface pigmentation and subsurface blood flow. It is generally agreed that cancerous cells need more nutrition to grow, resulting in an intense angiogenesis process. However, Fig. 7.7 clearly shows that not all blood flow activity captured by TLM contributes to melanoma from a dermatologist's perspective. The rationale behind the manual tracing is that the blush outside the contour (indicated by arrow in the figure) shows blood volume delivered by underlying larger vessels that does not directly contribute to the lesion. Inside the contour, however, angiogenesis occurs at the capillary level and hence is considered a lesionous body. As shown in the figure, there is no clearly defined boundary between the two. Instead, the red color fades gradually. To address such a challenge, an evolution-strategy-based (ES-based) ROI identification algorithm was developed.[51,52] Evolution strategy[53,54] is a real-number function optimization method that is not affected by uncertainties introduced by quantization errors such as the binary coding for genetic algorithms. The ROI identification problem was formed as an optimization problem with a special objective function designed to deal with skin lesion images based on clustering. Figure 7.8 shows examples of ROI results for three challenging skin lesion images using the ES-based method, manual segmentation by a dermatologist, and the edge-based method. The ES-based method outperforms the edge-based method and closely matches the clinician's manual segmentations.

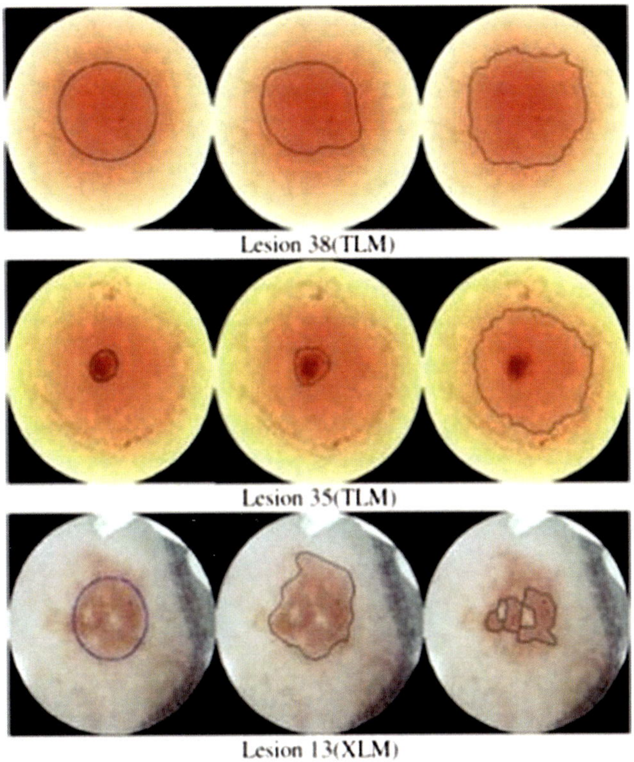

Figure 7.8 Region-of-interest identification results for three challenging skin-lesion images using (column 1) the ES-based method, (column 2) a dermatologist, and (column 3) the edge-based method.

In the next two sections, unless otherwise noted, "skin lesion" refers to the ROI identified and segmented from acquired ELM images of the lesion. Section 7.3 details the feature extraction function and how the extracted low-level features are mapped to high-level concepts clinicians used for melanoma diagnosis. Section 7.4 integrates feature selection with the final decision making in a multitask learning framework.

7.3 Mapping Computer-Generated Features to High-Level Concepts Used by Dermatologists

When dermatologists examine dermoscopy skin lesion images, they make diagnostic decisions based on a set of high-level concepts (Table 7.2). However, it is not easy for a computer to capture such high-level concepts directly. Hence, these high-level concepts are rarely used in automated skin lesion classification systems. This section details efforts to map computer-generated low-level features to these high-level concepts and use them in the final decision making. Because these high-level concepts are usually annotated weakly (i.e., a lesion is known to exist in the image, but its location and other details are unknown), the problem of mapping the computer-generated low-level features to each high-level concept is converted into a multiple-instance learning paradigm. The following section describes a multitask learning framework that integrates feature selection with the final decision making in melanoma detection.

7.3.1 Computer-generated low-level features

Many existing efforts[55–57] to develop a skin lesion classification system have focused on using global features. Computer-generated low-level features are extracted from the skin lesion to represent global features. However, this approach can hardly capture local features, the cumulative effect of which may well determine whether the lesion is benign or malignant. Global features such as the size and border smoothness of the skin lesion are derived directly after the ROI is identified. AutoScan uses a color histogram, color moments, and the color scale invariant feature transform (SIFT) as low-level color features, and the output from a three-level discrete wavelet transform (DWT) as a low-level texture feature.

Low-level intensity features: Instead of using the intensity value of each pixel, the SIFT is used to represent the variation of the intensity pattern within a skin lesion. This transform is a powerful descriptor for natural-scene image classification wherein objects usually have clear edges. Its output is invariant to shift, scale, and rotation,[58] and thus it can capture the intensity pattern. AutoScan uses a 128-dimensional histogram of quantized edge orientation and magnitude of intensity for each lesion.

Low-level color features: One *color histogram* is extracted from each channel of RGB color space for each skin lesion and then quantized into 15 even bins. These histograms are combined into one color histogram vector with 45 elements for each lesion. The RGB color values at each pixel can be viewed as three functions of pixel location. Thus, the moments of the product of these three functions can be calculated and these *color moments*[59] used as descriptors with 27 elements. Applying SIFT in each of the RBG color channel, i.e., *color SIFT*, produces a feature vector with 128×3 elements. To avoid the dimensionality problem, principal component analysis (PCA) is used to reduce its dimension: the first 20 major principal components from the color SIFT are used.

Low-level texture features: Computer-generated low-level texture features, such as the filter response coefficients from the DWT, have been shown to be important when making decision on melanoma or benign lesions.[57,60] AutoScan applies a three-level discrete Haar DWT to capture its global texture pattern. For each of the ten filter channels, energy and standard deviation of the DWT filter response coefficients are used, resulting in a texture feature vector with 20 elements.

7.3.2 Mapping high-level dermoscopic concepts with multiple-instance learning

High-level dermoscopic concepts identified in a dermoscopy skin lesion image are usually categorized into two groups: *global* features and *local* features. Global features generally appear in the whole lesion, whereas local features may appear only in part of the lesion. Their difference is illustrated in Fig. 7.9: The lesion in Fig. 7.9(a) has a global feature—a *globular* pattern—that appears in the whole lesion. The lesion shown in Fig. 7.9(b), however, has a local feature—*dots/globules*—that only appears inside the red boundary. The global feature in this case is denoted as *multicomponent*. Local features are important because many rules used by dermatologists (Table 7.2) rely on them. As described in Section 7.3.1, most of the existing efforts in melanoma detection are based on the global features of skin lesions.

The high-level concept annotation in a skin lesion database is typically *weak* labeled, i.e., we know whether a certain local feature exists but do not know its location or how many times it appears. This is mainly because weak labels are more practical to obtain given that drawing a precise contour around a local feature is not only time consuming, but sometimes difficult as well. To utilize these weak labeled skin lesion images, AutoScan first segments the lesion into small homogeneous regions and considers each of these regions as an instance. Thus the local feature detection problem is converted into a standard multi-instance learning (MIL) problem:[64] a lesion is melanoma (positive) if and only if at least one of its instances is melanoma (positive). Figure 7.10 shows an example of AutoScan using the graph cut method[61–63] to segment the skin lesion into layers of small homogenous regions. For each region, i.e., instance, computer-generated low-level features are extracted. Then standard deviations and entropies are computed based on the five histograms described in Section 7.3.1: intensity pattern, color histogram, moments, SIFT, and wavelet-based texture, assembling a multicomponent descriptor with ten elements.

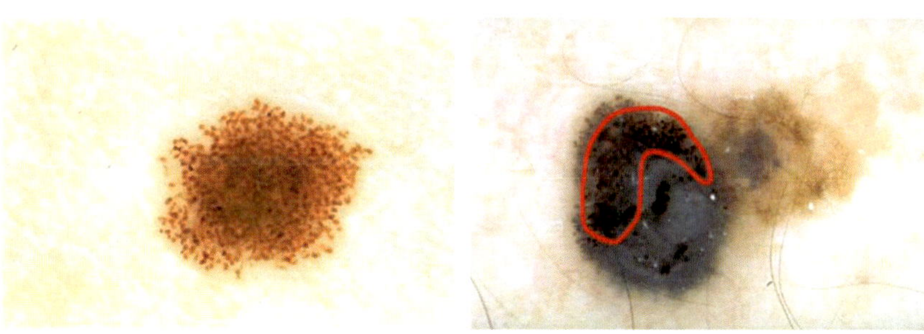

(a) (b)

Figure 7.9 Illustrating the difference between global and local features: (a) a lesion with a global feature (a globular pattern), and (b) a lesion with a local feature inside the blue contour (dots/globules), whose global feature is called multicomponent.

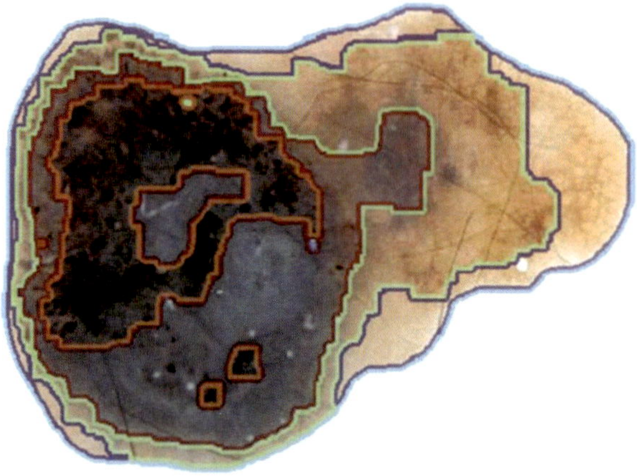

Figure 7.10 Using a graph cut to detect homogeneous regions within a skin lesion.

In the MIL paradigm, if any of the instances is positive, then the whole skin lesion is positive. The major challenge is that during the training phase, even though the label of the whole lesion is known, the label of each instance (region) is not. AutoScan uses a diverse density (DD) function[65] to identify *instance prototypes* (IPs): instances with a high probability of being melanoma (positive). The principle of the DD function is that given any chosen distance function between instances, if an instance is close to at least one instance from positive lesion images and far from all instances from negative lesion images, then its DD function will output a high value. After identifying IPs, MIL can be converted to the traditional (single) instance-based learning problem and solved easily.[66] To improve the performance of MIL, AutoScan extends the traditional DD function with boosting[67] when selecting the IPs. In each round of boosting, the weights of misclassified data are increased, updating the output of the weighted DD function. The following sections describe the details of the enhanced DD function with boosting used in AutoScan.

Suppose there are n^+ positive training lesion images $\{B_1^+, B_2^+, ..., B_{n^+}^+\}$ and n^- negative training lesion images. B_{ijk}^+ is used to represent the kth feature of the jth instance in the ith positive lesion images, and similarly B_{ijk}^- is used to represent the same in negative lesion images.

7.3.2.1 Diverse density function and evidence confidence function

The DD function is developed to evaluate the probability of an instance being positive [68]. Given any instance c (which can be any possible values in the instance space, not necessarily in the training set), the DD function is defined as

$$DD(c) = \prod\nolimits_{i=1}^{n^+} \Pr\left(c|B_i^+\right) \prod\nolimits_{i=1}^{n^-} \Pr\left(c|B_i^-\right), \tag{7.1}$$

where

$$\Pr(c|B_i^+) \propto 1 - \prod_j \left[1 - \Pr(c|B_{ij}^+)\right], \tag{7.2}$$

$$\Pr(c|B_i^-) \propto \prod_j \left[1 - \Pr(c|B_{ij}^-)\right], \tag{7.3}$$

$$\Pr(c|B_{ij}) \propto \exp\left[-\left(\sum_k (c_k - Bi_{jk})^2\right)/\sigma^2\right]. \tag{7.4}$$

Equation (7.4) shows that if c is closer to B_{ij}, then $[1 - \Pr(c|Bij)]$ will be smaller; if c is far from B_{ij}, then $[1 - \Pr(c|Bij)]$ will be larger. Substituting Eq. (7.4) into (7.2), the probability that the instance c is positive becomes higher if it is closer to at least one instance of B_i^+. Similarly, substituting Eq. (7.4) into Eq. (7.3), the probability that the instance c is negative decreases if it is far from all instances of B_i^-.

By maximizing Eq. (7.1) for all possible instances c, we can obtain all instance prototypes (IPs) needed to solve the MIL problem. However, since it is not a convex function with respect to c in instance space, numerical optimization is difficult. To address such difficulty, an *evidence confidence* (EC) function,[67] a modified DD function, is adopted in AutoScan by considering only instances that appear in positive lesion images when optimizing c. To further improve the optimization efficiency, for each of the positively labeled lesion images (i.e., identified melanoma image), only a fixed number of instances with highest evidence confidence are selected as instance prototypes.

7.3.2.2 Boosting enhanced-instance prototype selection

In boosting,[67] the weights of each incorrectly classified example are increased, and the weights of those correctly classified examples are decreased, so that the new model focuses more on the examples that have evaded correct classification. For melanoma detection, let the weights for the training lesion images be $w_1^+, w_2^+, \ldots, w_{n^+}^+$ and $w_1^-, w_2^-, \ldots w_{n^-}^-$; the weighted diverse density (W_DD) function can then be defined as

$$W\_DD(c) = \prod_{i=1}^{n^+} \Pr(c|B_i^+)^{w_i^+} \prod_{i=1}^{n^-} \Pr(c|B_i^-)^{w_i^-}. \tag{7.5}$$

When c is restricted to include only instances from positive training images, Eq. (7.5) becomes the weighted evidence confidence (W_EC) function.

Algorithm 1 shows the boosting enhanced instance prototype selection algorithm using a support vector machine (SVM). The algorithm adopts the weight updating scheme used in Adaboost[67] and the SVM as the base learner. The algorithm boosts both the IP selection and the classifier (SVM). Specifically, in each round of boosting, it changes the set of IPs selected by the W_DD function according to the updated weights from the previous iteration.

Algorithm 1 Adaboost Enhanced Instance Prototype Selection

Initialize the weights $w_1^+, w_2^+, ...w_{n^+}^+$ and $w_1^-, w_2^-, ..., w_{n^-}^-$ and the number of iterations T;

\\ Boosting loop;
for iter = 1 **to** T **do**
  Identify IPs with Eq. (7.5) by maximizing either DD or EC;
  Transform each training lesion into a single descriptor vector with the current set of IPs;
  Train with a SVM using the current set of weights;
  Add the new support vectors to the classifier, as in Adaboost;
  Update the weights, as in Adaboost;
end for

7.2.3 Transforming a lesion image into a descriptor vector

Suppose the selected IPs are $\{c_1, c_2, \ldots, c_n\}$; we can then represent a skin lesion image B_i with many instances using a single descriptor vector[71] based on the distances between each instance and the current set of IPs: $\varphi(B_i) = [d(B_i, c_1), d(B_i, c_2), \ldots, d(B_i, c_n)]$, in which $d(B_i, c) = \min_j(\|B_{ij} - c\|^2)$ is the closest distance between any instances in the lesion image B_i to c. As shown in Algorithm 1, after transforming each training lesion image to a single descriptor vector, any well-developed classifier (such as SVMs) can be used to solve the MIL problem and identify the model to map the computer-generated low-level features with high-level concepts used by clinicians.

7.2.4 Experiment setups and results

360 ELM skin lesion images were chosen from Set 3 (Section 7.1.5), of which 270 were benign and 90 were melanoma, and all of which were annotated by the dermatologists with weak labels of the local features used by clinicians. Among these 360 skin lesion images, 120 were *multicomponent*, whereas the remaining 240 generally had a single dominant dermoscopic global feature. Two typical lesion images are shown in Fig. 7.11. The typical resolution of the images was 500 × 700. Ten local dermoscopic features from the database were considered, as listed in Table 7.3. Only the true lesion area was used to ensure that performance was not influenced by incorrect automated ROI identification.

Five-fold cross-validation is used on all three methods applied to the dataset, and their performance on detecting the ten local features is compared. Given that the dataset is imbalanced, the accuracy alone cannot serve as an effective performance measure. Thus, the area under the receiver operating characteristic curve (AUC) was chosen as the performance metric.

Baseline: The baseline method treats the local dermoscopic features the same as those global features. It trains a SVM with a Gaussian kernel in

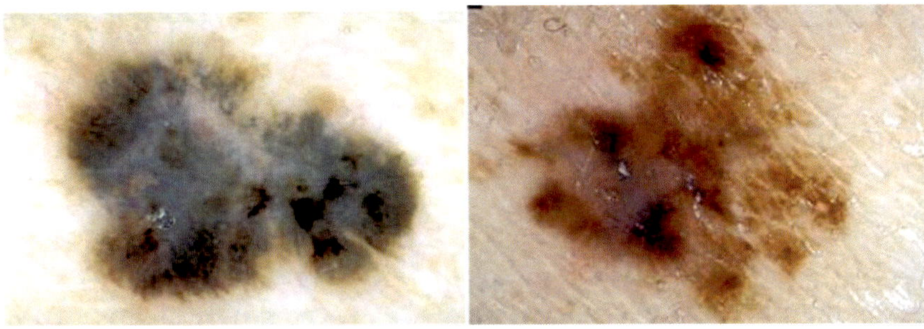

Figure 7.11 Two examples of skin lesion images from 360 lesion images used to map low-level features with high-level concepts.

Table 7.3 Ten local features, the number of lesion images with weak labels, and the performance (AUC %) of three methods. The bold values are the best results among the three methods for each local feature. An asterisk indicates that the performance difference between the method and the best performance is not significant by a paired t-test ($\alpha = 0.05$), i.e., the method produces results comparable to that of the best result.

High-Level Local Features	Number of Images with Weak Label	Baseline	MIL_EC	BIPEC
Pigment Network	178	**84.93**	75.06	78.10
Irregular Pigment Network	84	70.93	70.55	**72.60**
Dots/Globules	287	**79.16**	75.80	78.67*
Irregular Dots/Globules	169	64.18	65.84*	**66.07**
Streaks	178	**77.31**	63.82	66.78
Blue-Whitish Veil	93	80.37	82.67*	**82.99**
Pigmentation	145	78.58*	**78.91**	78.54*
Hypopigmentation	65	**69.38**	62.66	64.99
Regression Structure	86	65.21	67.06*	**68.70**
Vascular Structure	39	**82.08**	72.83	72.83

the form of $\exp(-\gamma\|x - y\|^2)$ (in which x and y are two vectors, and γ is the kernel parameter) with the computer-generated low-level features extracted from the whole skin lesion. The "C" parameter of SVMs and the Gaussian kernel parameter γ are chosen from the set $\{10^{-2}, 10^{-1}, 1, 10, 10^2, 10^3, 10^4, 10^5\}$. The performance from the best setting is reported in Table 7.3.

MIL_EC: This method treats local dermoscopic feature detection as an MIL problem and uses the EC function to select the IPs. The number of IPs selected from one positive image is set to be 1. It uses the same underlying classifier: a SVM with a Gaussian kernel. The "C" parameter of SVM, the Gaussian kernel parameter γ, and the EC parameter σ [Eq. (7.4)] are chosen from the same set used in the baseline method.

BIPEC: This method uses Algorithm 1 to boost the IP selection and the base learner (SVM). The number of boosting rounds is set to five based on empirical observation of the performance. During each boosting round, different training samples used to train the SVM can have different weights. Quasi-random weighted sampling (QWS) is used to sample the training set for each round according to the weights obtained by Adaboost.

Table 7.3 clearly shows that the boosted multi-instance learning (BIPEC) method achieves either the best or comparable-to-best results in six out of ten local dermoscopic features (shaded boxes in Table 7.3): irregular pigment network, irregular dots/globules, blue-whitish veil, regression structure, regular dots/globules, and pigmentation. The first four features are used in the 7-point checklist as important indicators of melanoma, in which the irregular pigment network and blue-whitish veil are two of the three major criteria. The blue-whitish veil is also considered a positive feature for melanoma in the Menzies method. The ability to identify the first four local features is critical in an automated melanoma screening and early detection system. In addition, BIPEC performs better than MIL_EC in eight out of ten local features. This is expected because boosting has been shown to be effective when used to enhance the performance of a weak learner, in this case, MIL_EC.

Even though the baseline method produces the best results in five out of ten local features (with its performance on regular dots/globules comparable to that of BIPEC), these features indicate a benign lesion: regular pigment network, regular dots/globules, streaks, hypopigmentation, vascular structure. In addition, note that the majority of the lesion image dataset (240 out of 360) used in the experiment exhibits one dominant global dermoscopic feature.

7.4 Integrating Feature Selection with Feature-Model Learning

After computer-generated low-level features are mapped to the high-level concepts that clinicians use to make diagnostic decisions regarding melanoma, the natural question is which features should be used in the automated system? One obvious solution is to combine all of the heterogeneous features into one big feature vector. However, such an approach increases the dimension of the feature vector, thus increasing the computational complexity unnecessarily and possibly inducing the potential problem of the "curse of dimensionality," with which most learning methods are not designed to cope. As a result, feature selection (FS) has become a critical step in many learning applications.[64,69–71] Although a NP-hard problem, the benefits of FS are undeniable: in addition to faster and more cost-effective model building, it helps avoid the overfitting pitfall and improve the predictive performance (i.e., accuracy, sensitivity, and specificity) of the learned model.

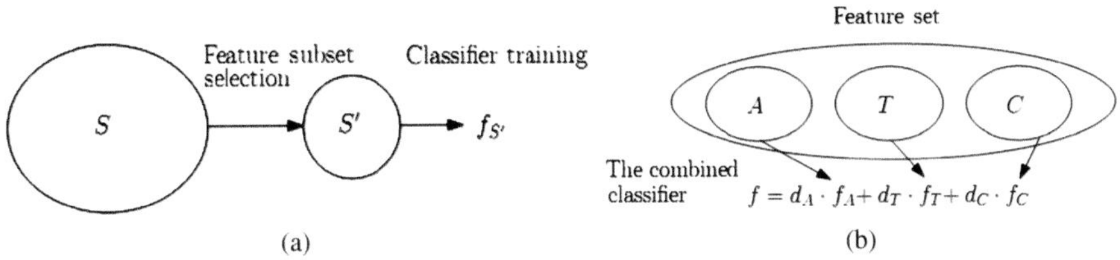

Figure 7.12 Feature selection and combination mechanisms: (a) A typical feature-selection mechanism, where S is a set of features, $S' \subseteq S$, and $f_{S'}$ is the classifier built based on the feature subsets S'. (b) A feature combination mechanism featuring a final classifer f with feature set S as a mixture of the three classifiers learned from three feature subsets A, T, and C: $f = d_A * f_A + d_T * f_T + d_C * f_C$.

7.4.1 Feature selection and combination

Figure 7.12(a) illustrates the basic procedure of feature selection. Given the original feature set S, a FS algorithm chooses a subset $S' \subseteq S$ with which a model is trained. In an automated melanoma screening and early detection system, the computer-generated low-level features can be categorized based on the distinct aspect they represent, such as color, texture, and asymmetry.[75] A feature-combination mechanism, as shown in Fig. 7.12(b), can leverage such "natural splitting" in the feature space. The figure shows that three models are learned from three "natural" views of a lesion: asymmetry (A) f_A, texture (T) f_T, and color (C) f_C. The final model f is a linear combination of these individual models:

$$f = d_A * f_A + d_T * f_T + d_C * f_C.$$

The high-level concepts that clinicians rely on can also be used as another categorization method to split (or group) computer-generated low-level features, with models learned for each of them before they are combined linearly.

There are two major types of feature-selection approaches that have been employed to learn decision models:[110] (1) *wrapper methods*[73–76] that select features based on the predictive performance of the learned model (efficient searching strategies have been developed[69,77–80] to reduce the computational cost), and (2) *filter methods*[81] that use ranking criteria to select features without considering the underlying distribution of the data. Though computationally efficient, the resulting feature subset from the filter method could be suboptimal. RELIEF[82] and FOCUS[83] are two well-known filter methods. In summary, an embedded approach[84] (such as filter methods) incorporates feature selection into the training stage, thus achieving similar performance much faster than that of the wrapper methods. Research has shown that there is often not a single universally optimal feature-selection technique,[85] and sometimes multiple subsets of features may discriminate the data equally well.[86] In fact, the optimal model learned from data with a full feature set is usually different than the model learned using the optimal subset of features.[30]

Our motivation to investigate a feature combination mechanism for automated melanoma screening and an early detection system is twofold. First, when human subjects recognize atypical lesions based on texture, the effect of color is minimal;[87] hence, it is reasonable to build an automated system that processes texture and color information separately, and then combines them at a later stage.[88] Second, model combination has long been adopted to improve the robustness and stability of the discriminative model.[89,90] Gehler and Nowozin[94] recently evaluated the effect of combining heterogeneous features using multiple-kernel learning (MKL)[92–94] (a multitask learning method) and linear programming boosting (LPBoost),[95,96] and their superior results were shown over other object-detection methods on several benchmark datasets. The following sections describe our multitask learning framework to integrate feature selection and the final feature model learning.

7.4.2 Multiple auxiliary kernel learning (MAKL): learning from heterogeneous feature spaces

Multitask learning[97] is a machine learning method that learns the model for a "main" task together with those learned for other related tasks, using a shared representation, which often leads to a better model for the "main" task. It usually assumes that all (main and auxiliary) tasks have the same (i.e., homogeneous) input space,[98,99] which is satisfied when the models learned for all tasks are similar to their average.[100–103] In an automated melanoma screening and early detection system, however, such an assumption cannot be satisfied because different high-level anatomical concepts are typically learned from different sets of low-level features [Fig. 7.12(b)], such as texture,[57] geometry,[42] or color.[104] Figure 7.13 shows an example of the main task and

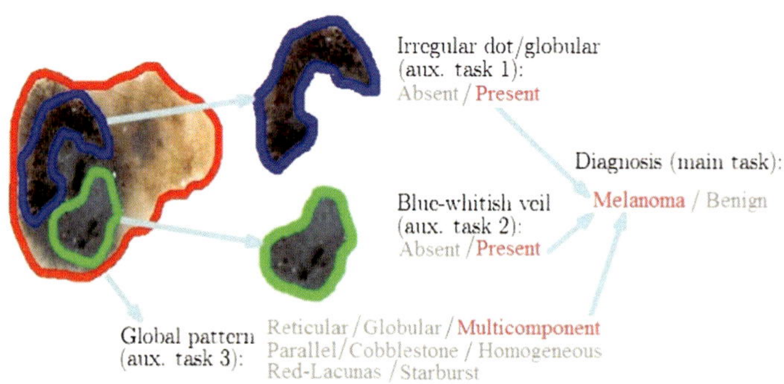

Figure 7.13 An example of the main task and auxiliary tasks in melanoma screening and early detection based on ELM lesions.

auxiliary tasks defined based on high-level concepts. Hence, traditional multitask learning methods, such as MKL, need to be extended to accommodate heterogeneous feature input spaces, in order to benefit skin lesion classification. One possible solution is to map each of the different input spaces into one common space.[101] However, choosing a common feature space is not a straightforward process in general, and it is sometimes infeasible. In addition, transforming between spaces with different dimensions is computationally intensive.

To handle the heterogeneous feature input space, MAKL has been developed to integrate feature combination and feature model building by learning the model of the main task as a weighted average of the output from models learned for auxiliary tasks. It is important to note that we use the weighted average of the output from those auxiliary task models, not averaging their parameters (such as the weight for each feature). In addition, to allow auxiliary task models to adapt to the main task during the learning phase, a small term is added to each auxiliary task model in the weighted sum. Note that each model learned for an individual task (main or auxiliary) could have its own heterogeneous feature set.

Assume that there is a set of T auxiliary tasks $P = \{P^1, P^2, \ldots P^T\}$, and the main task is represented as P_{T+1}. Each task can be represented by a set of features $K^l = \{k_1^l, k_2^l, \ldots, k_m^l\}$, where m is the number of features for task P_l. The number of features m could be different for different tasks and $m + 1$ is the number of features used to learn the main task model. Let $X^l = \{x_1^l, x_2^l, \ldots, x_n^l\}$ be a set of n samples for task P_l, and $x^l = \cup_{t=1}^{m} k_t^l$. The goal is to choose a decision function (i.e., feature model) h from a given hypothesis class such that h achieves the best prediction performance for the *main* task (i.e., melonoma or not?).

Following the generic empirical risk minimization approach in kernel learning, let the model learned for auxiliary task P_l be $h = d^l(w^l + v^l)_{l=1}^T$ with $w^1 \in X^1, v^1 \in X^1, \|w^0\| = 0$, $d^l \geq 0$ for $1 \leq l \leq T$, and $\|\mathbf{d}\|_p^p \leq 1$ for some $p \geq 1$, where $\mathbf{d}$ denotes the vector $(d_0, d_1, \ldots, d_l)$. Note that the relationship between the main task P_{T+1} and a specific auxiliary task P_l is captured by $\|v^l\|$, $1 \leq l \leq T$. A smaller $\|v^l\|$ indicates that the main and auxiliary tasks are more-closely related. In the special case where $\|v^l\| = 0$, $1 \leq l \leq T$, the prediction from the main task model for a sample $(x^l)_{l=1}^T$ becomes $\sum_{l=1}^{T} d^l w_l^l x^l$, which is a weighted sum of the output from models learned for auxiliary tasks.

Without loss of generality, assume that there are n independent and identically distributed (i.i.d.) training samples in the form of $X_i = (x_i^l)_{l=1}^{T+1}, (y_i^l)_{l=1}^{T+1}$, $1 \leq i \leq n$, drawn from a distribution D defined by $X \times \{-1,1\}^0$, a domain that represents both the input feature space and the output hypothesis space for the main task P_{T+1}; and user-defined parameters $B > 0$ and C_l, $1 \leq l \leq T + 1$, the main-task model space H_{T+1} is defined as

$$H_{T+1}(B, X, C_l) = \left\{ (x_l)_{l=1}^{m+1} \rightarrow \sum\nolimits_{l=1}^{m+1} [d_l(w_l + v_l)]^T \right.$$

$$\left. x_l | w_l, v_l \in X_l, 1 \leq l \leq m+1; w_{m+1} = 0 \right\}; \|\mathbf{d}\|_p^p \leq 1, d \geq 0; \tag{7.6a}$$

and

$$\frac{1}{2}\sum\nolimits_{l=1}^{m+1} d_l \|v_l\|^2 \tag{7.6b}$$

$$+\frac{1}{2}\sum\nolimits_{l=1}^{m} C_l d_l \|w_l\|^2 \tag{7.6c}$$

$$+\frac{1}{2}\sum\nolimits_{l=1}^{m}\sum\nolimits_{i=1}^{n} d_l\left(w_l^T x_i^l - y_i^l\right)2 \leq B. \tag{7.6d}$$

The main task model is $(d_l(w_l + v_l))_{l=1}^{T+1}$, and w_l is the lth auxiliary task model with $1 \leq l \leq T$. Equation (7.6a) restricts the p-norm of the weight vector $\|\mathbf{d}\|_p^p$ to be smaller than 1, similar to that of the nonsparse MKL.[110] Equation (7.6b) regularizes the relation between the main task and the specific auxiliary task. Equation (7.6c) regularizes the norm of auxiliary task model using a weighted sum formulation similar to that used in MKL,[106] intuitive MKL,[107] and the group lasso.[108] Equation (7.6d) stipulates that all of the auxiliary task models have a small total quadratic error. If an auxiliary task has a higher impact on the main task (i.e., d_l is high), its error will be penalized more. If $d_l = 0$, there will be no restriction on the error of the lth auxiliary task because it has no effect on the main task model. Using the weighted error term also leads to easier convex optimization. Note that the parameter C_l serves a similar function as the regularization parameter "C" in a SVM. When taking out all of the terms related to w_l in Eqs. (7.6c) and (7.6d) for a fixed l, i.e., a specific learning task,

$$d_l\left[\frac{1}{2}C_l\|w_l\|^2 + \frac{1}{2}\sum\nolimits_{i=1}^{n}(w_l^T x_i^l - y_i^l)^2\right], \tag{7.7}$$

which is exactly the same formulation as that of a SVM with the quadratic loss.

7.4.3 Experiment setup and results

For melanoma screening and early detection, the main task is to learn a model that can make the correct decision about whether the incoming ELM skin lesion image is melanoma or benign. As shown in Fig. 7.13, auxiliary tasks include recognizing global and local dermoscopic features. Instead of learning all ten high-level concepts, three local dermoscopic features are considered as auxiliary tasks: irregular dots/ globules, irregular pigment network, and blue-whitish veil. In addition, all global features shown in Fig. 7.13 are grouped as one auxiliary task: multicomponent or not. The input feature sets for learning auxiliary task models are the same as those detailed in Section 7.3.1.

The same dataset used in Section 7.3 is used here: A total of 360 ELM skin lesion images each with a resolution of 500 × 700. There are 90 melanoma and 270 benign images, with 120 ELM images annotated by clinicians as being multicomponent. Fivefold cross-validation is used on all of the methods, and their performance is compared to the AUC.

Simple: Learn a main task model by concatenating all features of all auxiliary tasks, and train with the main task label using a SVM.

Baseline: A single kernel learned from the feature set of the main task and trained with the main task label using a SVM.

MTK: Use a traditional MTK [XY50] algorithm and concatenate all features of all auxiliary tasks, as in the simple method, to accommodate its requirement for a homogeneous input feature space.

MKL: This method was designed to combine heterogeneous data sources.[109] The nonsparse MKL[105] and 2-norm are used here in the constraint for the weight variables. MKL uses different feature spaces for different auxiliary tasks, but it still uses the main task label to learn.

LPBoost: Linear programming boosting[95] was not designed in a multitask setting. However, it is straightforward to generalize LPB to learn models for multiple tasks. The ν-LPBoost formulation[91] is used here because it has been shown to be more effective than MKL in combining heterogeneous features. In the experiments, v-LPBoost first trains T SVMs for T auxiliary tasks and one for the main task, each with its own feature set. It then builds a linear weighted combination of the $T+1$ models for the main task. This method is the closest to the proposed learning framework. The only difference between the LPBoost and MAKL is that LPBoost performs learning in two steps, whereas MAKL learns all tasks simultaneously.

MAKL: The proposed MAKL method uses a weak label and the individual feature set for auxiliary task model learning. In the experiment, $p = 2$, i.e., the Euclidean norm is used.

To determine the effect of weak annotation (used as auxiliary task labels), all auxiliary task labels are replaced with the main task labels of the lesion in LPBoost and MAKL experiments, while still using their individual feature sets in training. The methods are denoted as LPBoost-main and MAKL-main. In addition, to determine the effect of the weight variable d of the auxiliary tasks, it is set to be the same for MKL and MAKL, resulting in MKL-ave and MAKL-ave.

Exhaustive experiments have been performed to identify the set of optimum parameters used in all these learning methods and the best results

Table 7.4 AUC of different methods minus the AUC of the baseline (diff. AUC).

Methods	Simple	MTK	MKL
Diff. AUC (%) (std %)	−0.39 (3.34)	−0.09 (5.16)	1.73 (3.50)
Methods	**MKL-ave**	**LPBoost**	**LPBoost-main**
Diff. AUC (%) (std %)	3.51 (4.13)	6.17 (6.27)	7.28 (6.70)
Methods	**MAKL**	**MAKL-ave**	**MAKL-main**
Diff. AUC (%) (std %)	**9.42 (5.32)**	**9.39 (5.70)**	6.91 (5.77)

are reported in Table 7.4. Based on a paired t-test at 95% confidence level, the performance difference is significant between all methods and MAKL, except for the MAKL-ave (marked in bold in the table).

Table 7.4 clearly shows that the simple and MTK methods perform significantly weaker than other methods (paired t-test at 90% confidence level), confirming the cognitive research result[87] that the "natural splitting" of the feature set should be taken into consideration when developing automated melanoma screening and early detection systems. In addition, all MAKL-based methods outperform those based on MKL, demonstrating the benefit of accommodating heterogeneous feature space for the main and auxiliary tasks, a norm in real-world applications. The MAKL-based methods also significantly outperform LPBoost-based methods, confirming the benefit of learning all task models simultaneously and allowing auxiliary task models to adapt to that of the main task.

From Table 7.4, it appears that the difference between the weighted scheme MAKL and the average scheme MAKL-ave is not significant. To show the advantage of MAKL over that of its average MAKL-ave, some randomly generated auxiliary tasks in the training phase were added, and the same experimental setting as before was applied. Figure 7.14(a) clearly shows that the performance of MAKL is more robust than that of MAKL-ave when

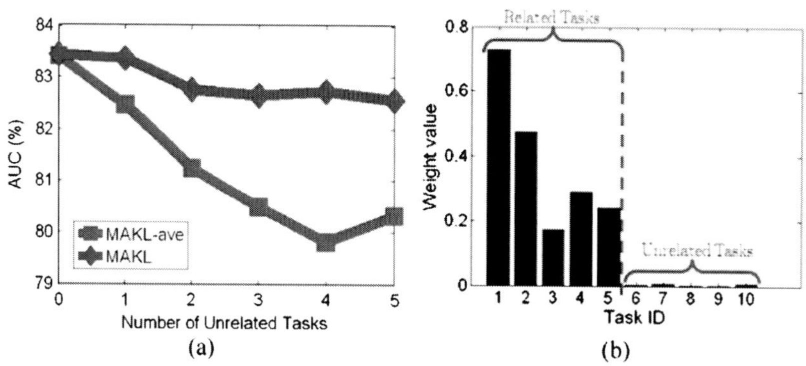

Figure 7.14 Comparison between MAKL and MAKL-ave with the introduction of unrelated auxiliary tasks: (a) AUC vs. the number of unrelated tasks, and (b) the weight learned by MAKL for auxiliary tasks. Tasks 6–10 are randomly generated unrelated tasks.

unrelated auxiliary tasks are present. Figure 7.14(b) shows the weights (i.e., *d*, indicating how much the auxiliary task contributes to the main task) learned by MAKL (averaged over fivefold cross-validation) when five unrelated auxiliary tasks were introduced. MAKL successfully excluded these unrelated auxiliary tasks while learning the model for the main task (melanoma or not?).

7.5 Conclusions and Future Directions

Although AutoScan shows reasonably good results compared with some existing methods, we hope it will serve as the benchmark for future systems. The natural next step for the current AutoScan is to extend the auxiliary tasks from five to include all ten local and global features listed in Table 7.3. Furthermore, given that clinicians are used to utilizing diagnosis criteria such as the ABCD rule, the 7-point checklist, or the Menzies method for melanoma diagnosis, AutoScan can be extended by building a decision support system based on specific criteria selected by dermatologists because it can identify all ten high-level concepts from pigmented skin lesion images already. Perhaps it is possible to develop a generic framework that incorporates and automatically chooses different types of features (intensity, color, texture, and shapes) in information forms (region and edge) while performing ROI identification. Another direction would be to explicitly incorporate the spatial relationship between pixels (either their intensity pattern or texture pattern) in the image, which has been used by clinicians for melanoma diagnosis.

References

1. A. F. Jerant, J. T. Johnson, C. D. Sheridan, and T. J. Caffrey, "Early detection and treatment of skin cancer," *Am. Fam. Physician* **62**(2), 357–368, 375–376, 381–382 (2000).
2. "Managing Pigmented Skin Lesions: Early Detection and Diagnosis of Melanoma," www.thedoctors.com/KnowledgeCenter/PatientSafety/articles/CON_ID_000305 (May 2010).
3. A. Jemal, R. Siegel, J. Xu, and E. Ward, "Cancer statistics, 2010," *CA: Cancer J. Clin.* **60**(5), 277 (2010).
4. "Melanoma," National Cancer Institute at the National Institutes of Health, www.cancer.gov/cancertopics/types/melanoma.
5. C. Balch et al., "Final version of the American Joint Committee on Cancer staging system for cutaneous melanoma," *J. Clinical Oncol.* **19**(16), 3635 (2001).
6. D. Alexandrescu, "Melanoma costs: a dynamic model comparing estimated overall costs of various clinical stages," *Dermatol. Online J.* **15**(1) (2009).

7. H. Ganster, A. Pinz, R. Rohrer, E. Wildling, M. Binder, and H. Kittler, "Automated melanoma recognition," *IEEE Trans. Med. Imaging* **20**(3), 233–239 (2001).
8. M. Moncrieff, "A simple classification of the resolution and depth of imaging systems for pigmented skin lesions," *Melanoma Res.* **12**, 155–159 (2002).
9. A. A. Marghoob et al., "Instruments and new technologies for the *in vivo* diagnosis of melanoma," *J. Am. Acad. Dermatol.* **49**, 777–797 (2003).
10. S. Tomatis, M. Carrara, A. Bono, C. Bartoli, M. Lualdi, G. Tragni, A. Colombo, and R. Marchesini, "Automated melanoma detection with a novel multispectral imaging system: results of a prospective study," *Phys. Med. Biol.* **50**, 1675 (2005).
11. T. Lee, V. Ng, R. Gallagher, A. Coldman, and D. McLean, "Dullrazor®: A software approach to hair removal from images," *Comput. Biol. Med.* **27**(6), 533–543 (1997).
12. M. Elbaum et al., "Automatic differentiation of melanoma from melanocytic nevi with multispectral digital dermoscopy: a feasibility study," *J. Am. Acad. Dermatol.* **44**, 207–218 (2001).
13. G. Monheit et al., "The performance of Melafind: A prospective multicenter study," *Archives Dermatol.* **147**(2), 188–194 (2011).
14. S. W. Menzies et al., "The performance of SolarScan: An automated dermoscopy image analysis instrument for the diagnosis of primary melanoma," *Archives Dermatol.* **141**, 1388–1396 (2005).
15. M. Moncrieff, S. Cotton, E. Claridges, and P. Hall, "Spectrophotometric intracutaneous analyses: a new technique for imaging pigmented skin lesions," *Br. J. Dermatol.* **146**, 448–457 (2002).
16. G. Argenziano et al., "Dermoscopy of pigmented skin lesions: results of a consensus meeting via the Internet," *J. Am. Acad. Dermatol.* **48**, 679–693 (2003).
17. R. Shafir, J. Hiss, H. Tsur, and J. J. Bubis, "The thin malignant melanoma: changing patterns of epidemiology and treatment," *Cancer* **50**, 817–819 (1982).
18. D. S. Rigel and J. A. Carucci, "Malignant melanoma: prevention, early detection, and treatment in the 21st century," *CA: Cancer J. Clin.* **50**, 215–236 (2000).
19. D. S. Rigel, "Malignant melanoma: perspectives on incidence and its effects on awareness, diagnosis, and treatment, *CA: Cancer J. Clin.* **46**, 195–198 (1996).
20. M. Magana-Garcia and A. B. Ackerman, "Naming acquired melanocytic nevi: Unna's, Miescher's, Spitz's, Clark's," *Am. J. Dermatopathol.* **12**, 193–209 (1990).

21. H. Pehamberger, M. Binder, A. Steiner, and K. Wolff, "*In vivo* epiluminescence microscopy: improvement of early diagnosis of melanoma," *J. Invest. Dermatol.* **100**, 356S–362S (1993).
22. H. Pehamberger, A. Steiner, and K. Wolff, "*In vivo* epiluminescence microscopy of pigmented skin lesions. I. Pattern analysis of pigmented skin lesions," *J. Am. Acad. Dermatol.* **17**, 571–583 (1987).
23. A. Steiner, H. Pehamberger, and K. Wolff, "*In vivo* epiluminescence microscopy of pigmented skin lesions. II. Diagnosis of small pigmented skin lesions and early detection of malignant melanoma," *J. Am. Acad. Dermatol.* **17**, 584–591 (1987).
24. F. Nachbar et al., "The ABCD rule of dermatoscopy: high prospective value in the diagnosis of doubtful melanocytic skin lesions," *J. Am. Acad. Dermatol.* **30**, 551–559 (1994).
25. W. Stolz et al., "ABCD rule of dermatoscopy: a new practical method for early recognition of malignant melanoma," *Eur. J. Dermatol.* **4**, 521–527 (1994).
26. G. Argenziano et al., "Dermoscopy: A Tutorial," www.dermoscopy.org/atlas/base.htm, EDRA (2000).
27. G. Argenziano, G. Fabbrocini, P. Carli, V. De Giorgi, E. Sammarco, and M. Delfino, "Epiluminescence microscopy for the diagnosis of doubtful melanocytic skin lesions. Comparison of the ABCD rule of dermatoscopy and a new 7-point checklist based on pattern analysis," *Archives Dermatol.* **134**, 1563–1570 (1998).
28. S. W. Menzies, C. Ingvar, and W. H. McCarthy, "A sensitivity and specificity analysis of the surface microscopy features of invasive melanoma," *Melanoma Res.* **6**, 55–62 (1996).
29. S. W. Menzies, K. A. Crotty, C. Ingvar, and W. H. McCarthy, *An Atlas of Surface Microscopy of Pigmented Skin Lesions*, McGraw-Hill, Sydney, Australia (2003).
30. R. Johr, "Dermoscopy: alternative melanocytic algorithms, the ABCD rule of dermatoscopy, Menzies scoring method, and 7-point checklist," *Clin. Dermatol.* **20**(3), 240–247 (2002).
31. R. P. Braun, H. Rabinovitz, M. Oliviero, A. W. Kopf, and J. H. Saurat, "Pattern analysis: a two-step procedure for the dermoscopic diagnosis of melanoma," *Clin. Dermatol.* **20**, 236–239 (2002).
32. R. P. Braun, H. Rabinovita, M. Oliviero, A. W. Kopf, and J. H. Saurat, "Dermoscopy of pigmented skin lesions," *J. Am Acad. Dermatol.* **52**, 109–121 (2005).
33. Nevoscope website, www.nevoscope.com.
34. P. Dhawan, "Apparatus and method for skin lesion examination," U.S. Patent Number 5,146,923 (1992).

35. V. Terushkin et al., "Transillumination as a means to differentiate melanocytic lesions based on their vascularity," *Archives Dermatol.* **145** (9), 1060 (2009).
36. M. Doshi, "Automated segmentation of skin cancer images," Master's thesis, University of Houston, TX (2004).
37. M. E.Celebi, Y. A.Aslandogan, and P. R.Bergstresser, "Mining biomedical images with density-based clustering," *Proc. ITCC 2005*, pp. 163–168, (2005).
38. E. Maglogiannis, C. Caroni, S. Pavlopoilos, and V. Karioti, "Utilizing artificial intelligence for the characterization of dermatological ages," *NNESMED*, 362–368 (2001).
39. G. Betta, G. Leo, G. Fabbrocini, A. Paolillo, and M. Scalvenzi, "Automatic application of the '7-point checklist' diagnosis method for skin lesions: Estimation of chromatic and shape parameters," *Proc. IMTC05* **3**, 1818–1822, (2005).
40. K. Taouil and N. B. Romdhane, "Automatic segmentation and classification of skin lesion images," *Proc. 2nd International Conferences on Distributed Frameworks for Multimedia Applications*, 1–12 (2006).
41. R. Jailani, H. Hashim, M. N. Taib, and S. Sulaiman, "Border segmentation on digitized psoriasis skin lesion images," *Proc. 2004 TENCON*, 596–599 (Nov. 2004).
42. G. Zouridakis, M. Doshi, and N. Mullani, "Early diagnosis of skin cancer based on segmentation and measurement of vascularization and pigmentation in nevoscope images," *Proc. EMBC 2004*, 1596–1593 (2004).
43. M. E. Celebi, Y. A. Aslandogan, and P. R. Bergstresser, "Unsupervised border detection of skin lesion images," *Proc. ITCC 2005*, 123–128 (2005).
44. V. Caselles, R. Kimmel, and G. Sapiro, "Geodesic active contours," *Proc. ICCV 1995*, 694–699 (1995).
45. M. Kass, A. Witkin, and D. Terzopolous, "Snakes: Active contour models," *Int. J. Comp. Vision* **1**(4), 321–331 (1988).
46. C. Xu and J. L. Prince, "Snakes, shapes, and gradient vector FOW," *IEEE Trans. Image Process.* **7**(3), 359–369 (1998).
47. D. H. Chung and G. Sapiro, "Segmenting skin lesions with partial-differential-equations-based image processing algorithms," *IEEE Trans. Med. Imaging* **19**(7), 763–767 (2000).
48. B. Erkol, R. Moss, R. Stanley, W. Stoecker, and E. Hvatum, "Automatic lesion boundary detection in dermoscopy images using gradient vector flow snakes," *Skin Research Technol.* **11**, 17–26 (2005).

49. T. F. Chan and L. A. Vese, "Active contours without edges," *IEEE Trans. Image Process.* **10**(2), 266–277 (2001).
50. X. Yuan, N. Situ, and G. Zouridakis, "A narrow band graph partitioning method for skin lesion segmentation," *Pattern Recognition* **42**(6), 1017–1028 (2009).
51. X. Yuan, J. Zhang, X. Yuan, and B. P. Buckles, "Multi-scale feature identification using evolution strategies," *Image and Vision Computing Journal* **23**(6), 555–563 (2005).
52. X. Yuan, N. Situ, and G. Zouridakis, "Automatic segmentation of skin lesion images using evolution strategies," *Biomed. Signal Process. Control* **3**(3), 220–228 (2008).
53. N. Hansen, *The CMA Evolution Strategy: A Tutorial*, www.lri.fr/~hansen/cmatutorial.pdf (2005).
54. H. P. Schwefel, *Evolution and Optimum Seeking*, Wiley Interscience, New York (1995).
55. C. Serrano and B. Acha, "Pattern analysis of dermoscopic images based on Markov random fields," *Pattern Recognition* **42**(6), 1052–1057 (2009).
56. T. Tanaka, S. Torii, I. Kabuta, K. Shimizu, M. Tanaka, and H. Oka, "Pattern classification of nevus with texture analysis," *EMBC 2004*, San Francisco, CA (2004).
57. N. Situ, X. Yuan, J. Chen, and G. Zouridakis, "Malignant melanoma detection by bag-of-features," *Proc. EMBC08*, pp. 3110–3113 (2008).
58. D. Lowe, "Distinctive image features from scale-invariant keypoints," *Int. J. Comp. Vision* **60**(2), 91–110 (2004).
59. F. Mindru, T. Tuytelaars, L. Gool, and T. Moons, "Moment invariants for recognition under changing viewpoint and illumination," *Comp. Vision Image Understanding* **94**(1), 3–27 (2004).
60. X. Yuan, Z. Yang, G. Zouridakis, and N. Mullani, "SVM-based texture classification and application to early melanoma detection," *Proc. EMBC 2006*, pp. 4775–4778 (2006).
61. Y. Boykov and V. Kolmogorov, "An experimental comparison of min-cut/max-flow algorithms for energy minimization in vision," *IEEE Trans. Pattern Analysis Machine Intelligence* **26**(9), 1124–1137 (2004).
62. Y. Boykov, O. Veksler, and R. Zabih, "Efficient approximate energy minimization via graph cuts," *IEEE Trans. Pattern Analysis Machine Intelligence*, **20**(12), 1222–1239 (2001).
63. V. Kolmogorov and R. Zabih, "What energy functions can be minimized via graph cuts?" *IEEE Trans. Pattern Analysis Machine Intelligence* **26**(2), 147–159 (2004).

64. T. G. Dietterich, R. Lathrop, and T. Lozano-Perez, "Solving the multiple instance problem with axis-parallel rectangles," *Artificial Intelligence Journal* **89**(1–2), 31–71 (1997).

65. O. Maron and T. Lozano-Perez, "A framework for multiple-instance learning," *Proc. Advances in Neural Information Processing Systems 1998*, 570–576 (1998).

66. W. Li and D. Yeung, "Localized content-based image retrieval through evidence region identification," *IEEE Comp. Society Conf. Computer Vision and Pattern Recognition*, Miami Beach, FL (2009).

67. Y. Freund and R. Schapire, "A decision-theoretic generalization of on-line learning and an application to boosting," *Proc. COLT 1995*, 23–37 (1995).

68. Y. Chen and J. Z. Wang, "Image categorization by learning and reasoning with regions," *J. Machine Learning Res.* **5**, 939 (2004).

69. D. Ververidis and C. Kotropoulos, "Fast sequential floating forward selection applied to emotional speech features estimated on DES and SUSAS data collections," *Proc. EUSIPCO 2006*, Florence, Italy (Sept. 2006).

70. A. Demiriz, K. Bennett, and J. Shawe-Taylor, "Linear programming boosting via column generation," *Machine Learning* **46**(1), 225–254 (2002).

71. S. Andrews and T. Hofmann, "Multiple instance learning via disjunctive programming boosting," *Proc. NIPS 2003*, Vancouver, B.C., Canada (2003).

72. M. Celebi et al., "A methodological approach to the classification of dermoscopy images," *Comp. Med. Imaging Graphics* **31**(6), 362–373 (2007).

73. A. Elisseeff and I. Guyon, "An introduction to variable and feature selection," *J. Machine Learning Res.* **3**(Mar.), 1157–1182 (2003).

74. G. H. John, R. Kohavi, and K. Pfleger, "Irrelevant features and the subset selection problem," *Proc. ICML 1994*, 121–129 (1994).

75. R. Caruana and D. Freitag, "Greedy attribute selection," *Proc. ICML 1994*, 28–36 (1994).

76. P. Langley and S. Sage, "Induction of selective Bayesian classifiers," *Proc. UAI 1994*, 399–406, (1994).

77. P. Somol, J. Novovicova, and P. Pudil, "Flexible-hybrid sequential floating search in statistical feature selection," *Lecture Notes in Computer Science* **4109**, 632–639, Springer, Berlin (2006).

78. A. V. Anghelescu and I. B. Muchnik, "Combinatorial PCA and SVM methods for feature selection in learning classifications (applications to text categorization)," *Proc. KIMAS 2005*, 491–496 (2003).

79. M. W. Mahoney, C. Boutsidis, and P. Drineas, "Unsupervised feature selection for principal components analysis," *Proc. SIGKDD 2005*, 61–69 (2005).

80. B. Sankur and H. K. Ekenel, "Feature selection in the independent component subspace for face recognition," *Pattern Recognition Lett.* **25**, 1377–1388 (2004).

81. H. Yoon, K. Yang, and C. Shahabi, "Feature subset selection and feature ranking for multivariate time series," *IEEE Trans. Knowledge Data Eng.* **17**(9), 1186–1198 (2005).

82. K. Kira and L. Rendel, "The feature selection problem: traditional methods and a new algorithm," *Proc. AAAI-92*, 129–134 (1992).

83. T. Fountain, H. Almuallim, and T. G. Dietterich, "Learning with many irrelevant features," *Proc. AAAI-91*, 547–552 (1991).

84. S. K. Singhi and H. Liu, "Feature subset selection bias for classification learning," *Proc. ICML2006*, 491–502 (2006).

85. Y. Yang, Y. Xiao, and M. Segal, "Identifying differentially expressed genes from microarray experiments via statistic synthesis," *Bioinformatics* **21**(7), 1084–1093 (2005).

86. K. Yeung, R. Bumgarner, and A. Raftery, "Bayesian model averaging: development of an improved multi-class, gene selection and classification tool for microarray data," *Bioinformatics* **21**(10), 2394–2402 (2005).

87. P. Wighton, T. Lee, D. McLean, H. Lui, and M. Stella, "Existence and perception of textural information predictive of atypical nevi-preliminary insights," *Proc. SPIE* **6917**, 69170J (2008) [doi: 10.1117/12.770793].

88. T. Maenpaa and M. Pietikainen, "Classification with color and texture: jointly or separately?" *Pattern Recognition* **37**(8), 1629–1640 (2004).

89. K. P. Bennett, M. Momma, and M. J. Embrechts, "MARK: a boosting algorithm for heterogeneous kernel models," *Proc. KDD-2002*, 24–31 (2002).

90. J. Bi, T. Zhang, and K. P. Bennett, "Column-generation boosting methods for mixture of kernels," *Proc. KDD-2004*, 521–526 (2004).

91. P. Gehler and S. Nowozin, "On feature combination for multiclass object classification," *Proc. ICCV 2009*, 221–228 (2009).

92. F. R. Bach, G. R. G. Lanckriet, and M. I. Jordan, "Multiple kernel learning, conic duality, and the SMO algorithm," *Proc. ICML 2004*, Baniff, AB, Canada (2004).

93. O. Chapelle, V. Vapnik, O. Bousquet, and S. Mukherjee, "Choosing multiple parameters for support vector machines," *Machine Learning* **46** (1–3), 131–159 (2002).

94. I. Grandvalet and S. Canu, "Adaptive scaling for feature selection in SVMs," *Proc. NIPS*, 553–560 (2002).

95. A. Demiriz, K. Bennett, and J. Shawe-Taylor, "Linear programming boosting via column generation," *Machine Learning* **46**(1), 225–254 (2002).
96. J. Leskovec and J. Shawe-Taylor, "Linear programming boosting for uneven datasets," *Proc. ICML 2003*, 456–463 (2003).
97. R. Caruana, "Multitask learning," *Machine Learning* **28**(1), 41–75 (1997).
98. A. Argyriou, T. Evgeniou, and M. Pontil, "Convex multi-task feature learning," *Machine Learning* **73**(3), 243–272 (2008).
99. J. Zhang, Z. Ghahramani, and Y. Yang, "Flexible latent variable models for multi-task learning," *Machine Learning* **73**(3), 221–242 (2008).
100. H. Daume, "Frustratingly easy domain adaptation," *Annual Meeting Association for Computational Linguistics* **45**, p. 256 (2007).
101. T. Evgeniou, C. Micchelli, and M. Pontil, "Learning multiple tasks with kernel methods," *J. Machine Learning Res.* **6**(1), 615 (2006).
102. T. Evgeniou and M. Pontil, "Regularized multi-task learning," *Proc. SIGKDD 2004*, 109–117 (2004).
103. J. Finkel and C. Manning, "Hierarchical Bayesian domain adaptation," *Proc. Human Language Technologies: The 2009 Annual Conference of the North American Chapter of the ACL*, 602–610 (2009).
104. R. Stanley, W. Stoecker, and R. Moss, "A relative color approach to color discrimination for malignant melanoma detection in dermoscopy images," *Skin Res. Technol.* **13**(1), 62–72 (2007).
105. M. Kloft, U. Brefeld, S. Sonnenburg, P. Laskov, K.-R. Müller, and A. Zien, "Efficient and accurate lp-norm multiple kernel learning," *Advances in Neural Information Processing Systems* **22**, 997–1005 (2009).
106. S. Sonnenburg, G. Rätsch, C. Schäfer, and B. Schäolkopf, "Large scale multiple kernel learning," *J. Machine Learning Res.* **7**, 1565 (2006).
107. A. Zien and C. Ong, "Multiclass multiple kernel learning," *Proc. 24th International Conference on Machine Learning*, p. 1198 (2007).
108. M. Yuan and Y. Lin, "Model selection and estimation in regression with grouped variables," *J. Royal Stat. Soc. B* **68**(1), 49 (2006).
109. G. Lanckriet, N. Cristianini, P. Bartlett, L. El Ghaoui, and M. Jordan, "Learning the kernel matrix with semi-definite programming," *J. Machine Learning Res.* **5**, 27–72 (2004).
110. Y. Saeys, I. Inza, and P. Larranaga, "A review of feature selection techniques in bioinformatics," *Bioinformatics* **23**(19), 2507–2517 (2007).

Xiaojing Yuan received her B.S. degree in Electrical Engineering from Hefei University of Technology, China in 1994, her M.S. in Computer Engineering from the University of Science and Technology of China in 1997, her M.S. in Computer Science from Tulane University in 2002, and her Ph.D. in Mechanical Engineering from Tulane in 2003. After finishing her postdoctoral research in the image-understanding lab for one year, she joined the University of Houston in 2004 as a Research Assistant Professor, and started her tenure-track position in 2005. She is an Associate Professor in the Engineering Technology Department and the founding director of the Intelligent Sensor Grid and Informatics (ISGRIN) research lab. She is a senior member of the IEEE and a member of ASEE and ISA. Her research interests include pattern recognition, data mining, biomedical image processing, and information processing in wireless sensor networks.

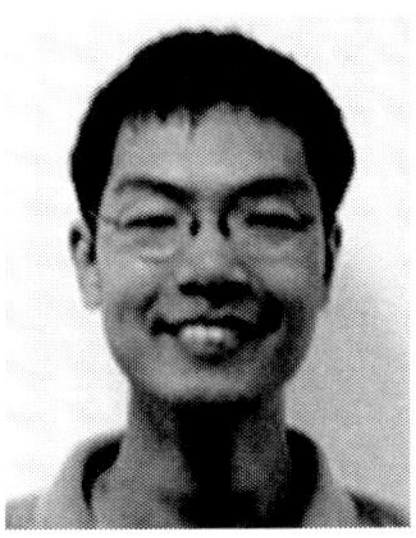

Ning Situ received his B.S. degree in Computer Science from Sun Yat-sen University in 2006. He obtained his Ph.D. at the Biomedical Imaging Laboratory, Department of Computer Science at the University of Houston in 2011. He is currently working in the field of online advertising at Microsoft Corporation.

Xiaohui Yuan received his B.S. degree in Electrical Engineering from Hefei University of Technology, China in 1996 and his Ph.D. in Computer Science from Tulane University in 2004. After graduation, he worked at the National Institutes of Health as a postdoctoral research scientist from 2005–2006, after which he joined the University of North Texas, where he is currently an Associate Professor in the Department of Computer Science and Engineering. He is a recipient of the Ralph E. Powe Junior Faculty Enhancement Award in 2008 and the Air Force Summer Faculty Fellowship in 2011 and 2012. He is a member of the IEEE and SPIE. His research interests include computer vision, pattern recognition, data mining, and artificial intelligence.

George Zouridakis is the founding director of the Biomedical Imaging Lab and a Professor of Engineering Technology, Computer Science, and Electrical and Computer Engineering at the University of Houston. Over the years, he has developed courses, given lectures, and organized and chaired sessions at conferences worldwide on biomedical imaging and brain mapping. His research efforts have been recognized by more than ten research awards and two U.S. patents for handheld imaging devices that can be deployed in the field for rapid automated disease screening. Recent projects have focused on applications in traumatic brain injury, autism, schizophrenia, skin cancer, and Buruli ulcer, and are supported by the National Institutes of Health, the National Science Foundation, and the U.S. Army Medical RAA. His scholarly activities include three books, 13 book chapters, ~160 refereed papers and proceedings, and more than 110 abstracts. He frequently participates in review panels for funding agencies in the U.S. and a number of other countries, and he currently serves as a Faculty Senator, Associate Editor for IEEE-TBME, and Chair of the Houston Chapter of IEEE Engineering in Medicine & Biology Society. His research interests include functional brain mapping and brain connectivity analysis, as well as the development of handheld devices for automated screening of skin diseases.

Chapter 8
A Complex Wavelet-Based Feature Extraction System for Microcalcification Detection in Digital Mammograms*

Ping Zhang
Department of Mathematical and Computer Science, Alcorn State University, Alcorn, MS, USA

Kwabena Agyepong
Department of Advanced Technologies, Alcorn State University, Alcorn, MS, USA

8.1 Introduction

Breast cancer is one of the leading causes of death among women worldwide. In the United States, it is the most-common form of cancer among women. Women in the U. S. have about a one-in-eight lifetime risk of developing invasive breast cancer. Early detection can increase the survival rate and treatment options.

Regular mammographic screening programs for women at certain ages or in high-risk groups are necessary. Mammography is the most common and convenient procedure for detecting nonpalpable cancers. It is an inexpensive

* The Digital Database for Screening Mammography (DDSM) belongs to the University of South Florida, and the authors would like to thank the professors and scientists of USF for providing the data for this research. This project is partially supported by the Research Project of Department of Defense, USA: Development of a Knowledge Base to Support Detection and Diagnosis and Research in Mammography.

practice that is highly effective even when the size of the breast abnormality is minimal. Breast cancer can be divided into three categories: microcalcifications, masses, and architectural distortions. One of the early signs of breast cancer is the presence of microcalcification clusters (MCCs) in the mammogram. Because MCCs have small size, irregular shape, and low contrast, they are often missed or misinterpreted by physicians. Therefore, an automatic and reliable computer-aided diagnosis (CADx) system can be very useful in helping radiologists analyze mammographic lesions that may indicate the presence of cancer. The CADx system can detect and verify those mammogram images where possible cancer evidences have been developed; the images will then be sent to radiologists for final evaluation.

Microcalcifications are tiny deposits of calcium, which appear as small bright spots in the mammogram. Microcalcifications are characterized by clusters, type, and distribution properties. Figure 8.1 shows two images of MCCs (a, b) and two images of a normal mammogram (c, d).

Microcalcification image analysis and detection is an extremely challenging task for the following three reasons: First of all, there is a large variability in the appearance of abnormalities. Likewise, abnormalities are often occluded or hidden in dense breast tissue. Perhaps most importantly, a CADx system for MCC detection is used in serious human disease detection; therefore, a need for near-perfect accuracy is required.

Concerning calcification detection in the regions of interest (ROIs), many methods have been proposed. In a survey paper,[1] mammogram enhancement and segmentation algorithms, and mammographic features, classifiers, and their performances were reviewed and compared. Remaining challenges were also discussed. In another paper,[2] the detection performances of different classifiers, such as support vector machines (SVMs), kernel Fisher discriminant (KFD), relevance vector machine (RVM), and committee machines (ensemble averaging and AdaBoost) were compared, and the test results were reported. Neural networks have been used in many calcification systems. Two automatic microcalcification detection systems were proposed based on the hybrid neural network classifier.[3,4] SVMs were reported in the mammogram detection systems.[5,6] Unsupervised detection of mammogram ROIs was

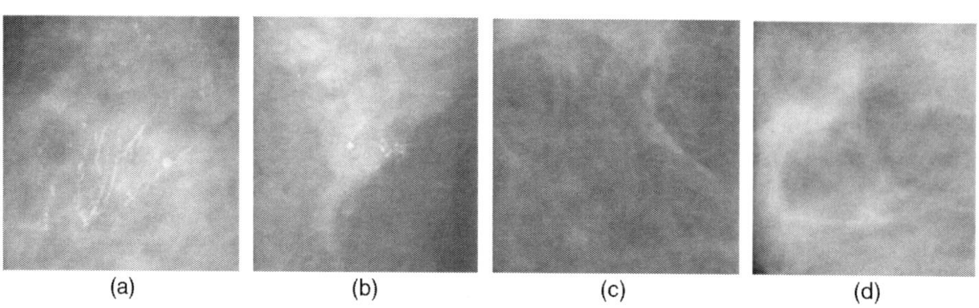

Figure 8.1 Four mammograms: (a, b) calcification images and (c, d) normal images.

introduced.[7] Segmentation of ROIs in mammograms using a topographic approach was introduced in recent literature.[8]

Feature extraction is one of the most important components in mammogram detection. A local feature extraction has been developed.[9] For example, the application of shape analysis to mammographic calcification was introduced.[10] Statistical textural feature for detection of microcalcification was described elsewhere.[11] Wavelets have been widely used in the feature extraction and segmentation in mammogram detection.[12–16] Combining mathematical morphology and neural networks was also proposed in the literature[17] and multifractal analysis has been used in the medical image detection and classification.[18–20]

In order to increase detection rate, a multiple expert system was provided to the detection system.[21] A fuzzy logic was then introduced to detect calcification.[22] A microcalcifications detection algorithm by fitting a model to every location in the mammogram was proposed.[23]

A complex wavelet transform (CWT) is developed in order to keep the attractive attributions of a discrete wavelet transform (DWT) such as approximate half-sample delay property, PR (orthogonal or biorthogonal), finite support (FIR filters), vanishing moments/good stopband, linear-phase filters, etc.[24] In addition, a CWT adds some new merits,[25,26] including approximate shift invariance, good directional selectivity for 2D images, and efficient order-N computation and limited redundancy. The computational complexity of a CWT requires only twice that of a DWT for 1D (2 m times for m-D signal). These good properties have made the CWT successful in many image processing applications recently, including pattern feature extraction and recognition. The dual-tree CWT has shown to be a suitable solution to CWT algorithm implementation.[27–29] However, to our knowledge, the application of Two-Dimensional CWT (2D-CWT) to calcification detection in digital mammograms is a new and challenging research topic.

In this chapter, a microcalcification detection system with a novel hybrid feature extraction scheme is proposed. A system diagram is drawn in Section 8.2. Section 8.3 presents a hybrid feature extraction method, which consists of a texture-based feature and 2D complex wavelet-based multifractal features. Classifier design and system implementation are explained in Section 8.4. The experimental comparisons for calcification detection on four SVM classifiers trained by four feature sets used in this paper are conducted in Section 8.5.

8.2 System Design

The schematic diagram of the proposed system is shown in Fig. 8.2. The system includes five parts: Mammography image input, image preprocessing, hybrid feature extraction, SVM classifier design, and detection output.

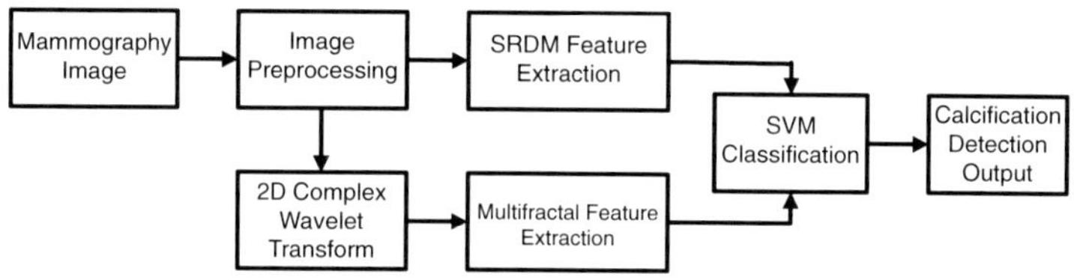

Figure 8.2 Schematic diagram for microcalcification cluster detection.

Image preprocessing includes histogram equalization.[30] Hybrid feature extraction consists of three steps: surrounding region dependence-based method (SRDM) feature extraction, 2D-complex wavelet transform, and multifractal feature extraction. A SVM is employed as a classifier in the proposed system. As the focus of this paper, we will mainly discuss the feature extraction method.

8.3 Hybrid Feature Extraction

Feature extraction is one of the most important steps in the design of a pattern recognition system. Four sets of features are addressed in this section.

8.3.1 Surrounding region dependence-based method

A simple surrounding region dependence-based method[11] is used as one part of first feature set. Firstly, the microcalcification area of a mammogram is divided into different overlapping 128×128 image blocks (the image block size of 128×128 is considered based on mammogram resolution and computation simplicity); then the first feature is extracted from each block. In order to systematically address this feature set, the diagram of the surrounding regions for the current pixel (x, y) is shown in Fig. 8.3. Here, R_1 and R_2 are the inner surrounding region and the outer

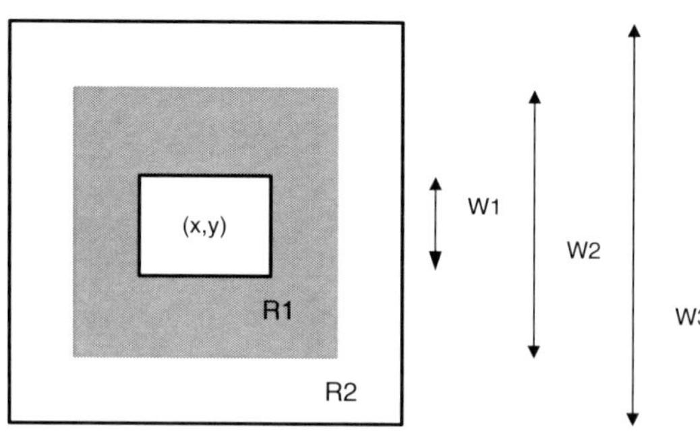

Figure 8.3 Diagram of the surrounding regions.

surrounding region, respectively. w_1, w_2, and w_3 denote the size of the three square windows.

In the following calculations, n is the maximal pixel number of R_1 (inner region), and m is the maximal pixel number of R_2 (outer region):

$$\alpha(i,j) = \#\{(x,y)|c_{R_1}(x,y) = i, c_{R_2}(x,y) = j, (x,y) \in L_x \times L_y\},$$

$$c_{R_1}(x,y) = \#\{(k,l)|(k,l) \in R_1, [S(x,y) - S(k,l)] > q\},$$

$$c_{R_2}(x,y) = \#\{(k,l)|(k,l) \in R_2, [S(x,y) - S(k,l)] > q\},$$

where $S(x, y)$ is the image intensity of the current pixel (x, y), parameter q is a constant, and # equals the number of pixels. $r(i,j)$ is the reciprocal of the element, which is calculated as follows:

$$r(i,j) = \begin{cases} \dfrac{1}{\alpha(i,j)} & \text{if } \alpha(i,j) > 0 \\ 0 & \text{otherwise.} \end{cases} \tag{8.1}$$

The following features are extracted:

Horizontal-weighted sum (HWS)

$$HWS = \frac{1}{N}\sum_{i=0}^{m}\sum_{j=0}^{n} j^2 r(i,j). \tag{8.2}$$

Vertical-weighted sum (VWS)

$$VWS = \frac{1}{N}\sum_{i=0}^{m}\sum_{j=0}^{n} i^2 r(i,j). \tag{8.3}$$

Diagonal-weighted sum (DWS)

$$DWS = \frac{1}{N}\sum_{k=0}^{m+n} k^2 \sum_{i=0_{i+j=k}}^{m}\sum_{j=0}^{n} r(i,j). \tag{8.4}$$

Grid-weighted sum (GWS)

$$GWS = \frac{1}{N}\sum_{i=0}^{m}\sum_{j=0}^{n} ij r(i,j), \tag{8.5}$$

where N is the total sum of elements in the surrounding region, which is defined as

$$N = \sum_{i=0}^{m}\sum_{j=0}^{n} \alpha(i,j). \tag{8.6}$$

Except for these four features, an additional four new features are added in our system:[31]

Variance sum (VS)

$$VS = \sum_{i=0}^{N_x-1}\sum_{j=0}^{N_y-1}(S(i,j) - \overline{S})^2. \tag{8.7}$$

Discrete Laplacian sum (DLS)

$$DLS = \sum_{i=0}^{N_x-1}\sum_{j=0}^{N_y-1}\frac{1}{4}\left(\frac{\partial^2 S(i,j)}{\partial i^2} + \frac{\partial^2 S(i,j)}{\partial j^2}\right). \tag{8.8}$$

Entropy (E)

$$E = -\sum_{i=0}^{N_x-1}\sum_{j=0}^{N_y-1}P(i,j)\log_2 P(i,j). \tag{8.9}$$

Mean grey level (MGL)

$$MGL = \frac{1}{N_x * N_y}\sum_{i=0}^{N_x-1}\sum_{j=0}^{N_y-1}S(i,j), \tag{8.10}$$

where N_x, N_y is the size of selected mammogram. In total, eight features are extracted as texture features.

8.3.2 Wavelet transform

The definition of a CWT is as follows: for a continuous function $f(x)$, it is projected at each step j on the subset V_j, $(\ldots\ldots \subset V_{-1} \subset V_0 \subset V_1 \subset V_2 \subset \ldots)$. The scalar project $c_{j,k}$ is defined by the dot product of $f(x)$ with the scaling function $\phi(x)$, which is dilated and translated:

$$\begin{aligned} c_{j,k} &= \langle f(x), \phi_{j,k}(x)\rangle, \\ \phi_{j,k}(x) &= 2^{j/2}\phi(2^j x - k). \end{aligned} \tag{8.11}$$

The difference between $c_{j+1,k}$ and $c_{j,k}$ is contained in the detailed component belonging to the space W_j, which is orthogonal to V_j. The following equations exist:

$$\begin{aligned} W \oplus V_j &= V_{j+1}, \\ V_j \cap W_j &= \{0\}, j \in Z. \end{aligned} \tag{8.12}$$

Suppose $\psi(x)$ is a wavelet function; the wavelet coefficients can be obtained by

$$w_{j,k} = \left\langle f(x), 2^{j/2}\psi(2^j x - k)\right\rangle. \tag{8.13}$$

Some relationships between $\phi(x)$ and $\psi(x)$ are as follows:

$$\begin{aligned} \frac{1}{2}\phi\left(\frac{x}{2}\right) &= \sum\nolimits_n h(n)\phi(x-n), \\ \frac{1}{2}\psi\left(\frac{x}{2}\right) &= \sum\nolimits_n g(n)\phi(x-n), \end{aligned} \tag{8.14}$$

where $h(n)$ and $g(n)$ represent unit impulse functions of low-pass and high-pass filters respectively, which are related to the scaling function $\phi(x)$ and the wavelet function $\psi(x)$.

According to wavelet theory, a conventional two-dimensional wavelet discrete transform (2D-DWT) can be regarded as equivalent to filtering the input image with a bank of filters whose impulse responses are all approximately given by scaled versions of a mother wavelet. The output of each level consists of four subimages—*LL*, *LH*, *HL*, and *HH*—with 2:1 downsampling.

Mathematically, this recursive algorithm can be expressed in the following equation:

$$\begin{aligned} \psi(x,y)_{LL} &= \phi(x)\phi(y), \\ \psi(x,y)_{LH} &= \phi(x)\psi(y), \\ \psi(x,y)_{HL} &= \psi(x)\phi(y), \\ \psi(x,y)_{HH} &= \psi(x)\psi(y). \end{aligned} \tag{8.15}$$

For example, the *LL* wavelet is the product of the low-pass function $\phi(x)$ along both the first and second dimension. The *LH* wavelet is the product of the low-pass function $\phi(x)$ along the first dimension and the high-pass function $\psi(y)$ along the second dimension.

If the wavelet filters are real, then Mallat's dyadic wavelet decomposition fast algorithm can be used.[24] However, 2D-DWT has two drawbacks: lack of shift invariance and poor directional selectivity.

8.3.3 Complex wavelet transform for feature extraction

2D-CWT provides true directional selectivity and pixel-shifting insensitivity. The six subband images of 2D-CWT can be represented by the following wavelet core functions:[25]

$$\begin{aligned} \psi_1(x,y) &= \psi(x)\psi(y), \\ \psi_2(x,y) &= \psi(x)\overline{\psi(y)}, \\ \psi_3(x,y) &= \phi(x)\psi(y), \\ \psi_4(x,y) &= \psi(x)\phi(y), \\ \psi_5(x,y) &= \phi(x)\overline{\psi(y)}, \\ \psi_6(x,y) &= \psi(x)\overline{\phi(y)}, \end{aligned} \tag{8.16}$$

where $\phi(x) = \phi_h(x) + j\phi_g(x)$, and $\psi(x) = \psi_h + j\psi_g(x)$; both are complex functions. $\psi_i(x,y)$, ($i = 1, \ldots, 6$) are six subbands of complex coefficients at each level, which are oriented at angles of ±75 deg, ±45 deg, and ±15 deg.

2D-CWT can be implemented using a dual-tree structure.[26] For each tree, its structure is similar to 2D-DWT. It has two decomposition operations on each level, namely row decomposition and column decomposition, except that the different filters are applied for perfect reconstruction, and the output of the subband images is congregated into complex wavelet coefficients. Figure 8.4 shows our proposed 2D-CWT feature extraction scheme.

The dual-tree complex wavelet decomposition consists of two trees, A and B, both of which have the same structure. In order to realize perfect reconstruction from decomposed subimages, a low-pass filter and a high-pass filter at the first level need to be specially designed and denoted as h00, g00 for tree A and h10, g10 for tree B, which are called prefilters. The other complex filters in the higher levels are set to h01 and g01 for tree A, and h11 and g11 for tree B.

A calcification image of size $N \times N$ is decomposed into four subband images—*LL*, *LH*, *HL*, and *HH*—at the first level of each tree. Each of the subband images has a size of $N/2 \times N/2$. At the higher levels, the decompositions are based on an *LL* subband image from the previous level. For example, if a 128 × 128 block of ROI submammograms is decomposed into the fourth level, the final size of each subband image is 16 × 16. The complex wavelet coefficients will be used as features.

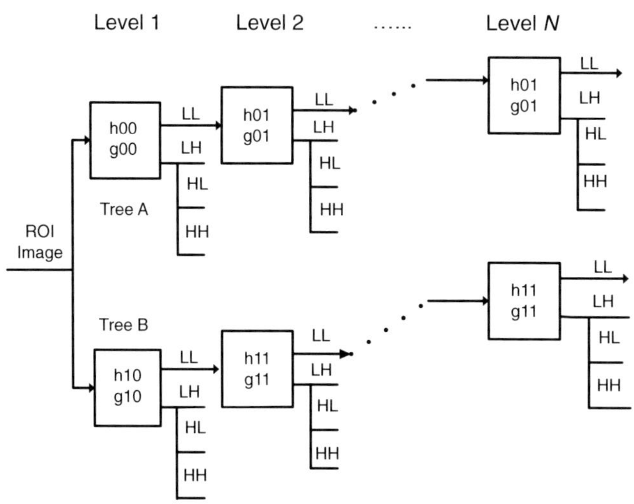

Figure 8.4 The schematic diagram of 2D-CWT for feature extraction.

8.3.4 2D-CWT multifractal feature

Wavelet coefficients have been used for features in many pattern recognition systems because wavelet transformation explores the detailed information on different decomposition layers and different orientations. Due to the fact that the shape and size of the small light spots in the calcification ROI image are changeable case-by-case, improvements to a 2D-complex-wavelet-based feature are needed in order to keep more stable features.

The multifractal application to image classification and recognition has received a lot of attention. The term "multifractal" was coined by Mandelbrot in 1975 to describe the irregular structure of many natural objects and phenomena. Central to multifractal geometry is the concept of self-similarity. Considering a bounded set R in Euclidean n-space, the set is said to be self-similar when R is the union of N_r distinct (non-overlapping) copies of itself, each of which has been scaled down by a ratio $r < 1$ in all coordinates. The similarity dimension Ds is given by

$$N_r \cdot r^{Ds} = 1, \quad \text{where } Ds = \log(N_r)/\log(1/r). \tag{8.17}$$

The ranges in the value of Ds characterize the type of multifractal.

A few methods to compute the multifractal dimension have been proposed in the literature, such as the walking-divider, box counting, prism counting, epsilon-blanket, perimeter-area relationship, fractional Brownian motion, power spectrum, and hybrid methods.[32]

A novel 2D-CWT-based multifractal feature extraction scheme is proposed in this chapter. Four subimages of *LL*, *LH*, *HL*, and *HH* in the fourth layer of 2D-CWT for each tree of the ROI (128 × 128) are used to calculate multifractal number as follows:

$$\begin{aligned} d(k) = {} & \sum_{x=0}^{M-1}\sum_{y=0}^{N-k-1} \sqrt{(I_{re}(x,y) - I_{re}(x,y+k))^2 + (I_{img}(x,y) - I_{img}(x,y+k))^2}/(M*(N-k)) \\ & + \sum_{y=0}^{N-1}\sum_{x=0}^{N-k-1} \sqrt{(I_{re}(x,y) - I_{re}(x+k,y))^2 + (I_{img}(x,y) - I_{img}(x+k,y))^2}/(N*(M-k)), \end{aligned} \tag{8.18}$$

$$F(k) = \log(d(k+1)) - \log(d(1)), \tag{8.19}$$

where $k = 1, 2, \ldots, l$, and l is number of features extracted. A vector of $[F(1), F(2), \ldots, F(l)]$ will be fed into a SVM classifier for training and testing. As 2D-CWT is implemented using a dual-tree structure (shown in Fig. 8.4), there are two sets of *LL*, *LH*, *HL*, *HH* subimages: $I_{re}(x,y)$ is the real part of 2D-CWT intensity value at the fourth layer, whereas $I_{img}(x,y)$ is the imagery part of the 2D-CWT intensity at the same layer with $I_{re}(x,y)$. In Eq. (8.19), if only six features are extracted from each subimage at the fourth layer of each tree, then there are in total 6 × 8 = 48 features extracted.

8.4 Classifier Design

Support vector machines[34] rely on representing original features to a higher dimension by an appropriate nonlinear mapping $\phi(\;)$. Data from two categories can be separated by a hyperplane.

In this chapter, the kernel function is a Gaussian radial basis kernel:

$$K(x,z) = \exp\left(-\|x-z\|^2/\sigma^2\right). \tag{8.20}$$

Training a SVM for a pattern recognition problem leads to the following quadratic optimization problem:

$$\text{Minimize: } W(\alpha) = -\sum_{i=1}^{l}\alpha_i + \frac{1}{2}\sum_{i=1}^{l}\sum_{j=1}^{l} y_i y_j \alpha_i \alpha_j k(x_i, x_j); \tag{8.21}$$

$$\text{Subject to: } \begin{array}{l} \sum_{i=1}^{l} y_i \alpha_i = 0, \\ \forall i : 0 \le \alpha_i \le C, \end{array} \tag{8.22}$$

where the number of training examples is denoted by l, and α is a vector of l variables. Each component α_i corresponds to a training example (x_i, y_i). x_i is the feature vector of the ith training sample, whereas y_i is its label. The solution is the vector α, for which Eq. (8.21) is minimized and the constraints in Eq. (8.22) are fulfilled. A SVM classifier is especially used for two-class problems, the layout of which is as follows:

Number of nodes in the input layer:	Number of features
Value in the output layer:	[–1, 1] (normal case and calcification cancer)

8.5 Experiment Results

In order to test the microcalcification detection performance of the proposed feature extraction algorithm and system, 72 cancer mammograms and 100 normal mammograms from the well-known mammogram DDSM database[33] are used in the experiments, from which 22 cancer mammograms and 30 normal mammograms are chosen for testing. Other mammograms are used for training. In the DDSM database, the boundary of clustered MCs is given in the overlay information, which is based on the radiologists' knowledge and used to verify the correctness of the proposed algorithm and system.

As mentioned in Section 8.3.1, the image of a mammogram is divided into different overlapping 128 × 128 subimages. In total, 1,000 subimages of microcalcification ROIs and another 1,000 subimages of noncancerous areas are created for training. 400 microcalcification subareas (128 × 128) and 400 normal subareas are selected from the testing set.

A comparison of the calcification-detection performance of four different feature sets has been conducted. Four sets of features introduced in this

Table 8.1 Four feature sets and dimensions.

Feature Set	Feature Dimension
SRDM feature set	8
SRDM + multifractal feature set	38
SRDM + 2D-CWT feature set	168
SRDM + 2D-CWT-based multifractal feature set	56

chapter are employed to train four SVM classifiers with the same configuration except for the number of input features. Table 8.1 lists the number of features in each of the four feature sets.

In the first feature set, eight features were introduced in Section 8.3.1. In the second feature set, the multifractal feature set is directly extracted from original mammography images; therefore, Eq. (8.18) is adopted as follows:

$$d_{re}(k) = \sum_{x=0}^{M-1}\sum_{y=0}^{N-k-1} \sqrt{(I_{org}(x,y) - I_{org}(x,y+k))^2/(M*(N-k))} + \sum_{y=0}^{N-1}\sum_{x=0}^{N-k-1} \sqrt{(I_{org}(x,y) - I_{org}(x+k,y))^2/(N*(M-k))}, \quad (8.23)$$

where $I_{org}(x, y)$ is the image density at pixel (x, y) of the original mammogram. 30 features are extracted based on Eq. (8.23). Therefore, the number of features in the SRDM + multifractal feature set is 38.

The third feature set consists of eight SRDM features and a 2D-CWT feature set. The 2D-CWT feature set is extracted based on original mammography images and by the following means: The amplitude coefficients for three high-frequency components and both the amplitude and phase information for the low-frequency component at the fifth layer (with subimage size of 4×4) are extracted as features. The number of features $= 4 \times 4$ (for each subband image) $*$ 3 (high-frequency subband images for each tree) $*$ 2 (trees) $+ 4 \times 4$ (for each subband image) $*$ 2 (trees) $*$ 2 (parts: real and imaginary) $= 160$. In total, 168 features are extracted.

In order to suppress the large number of dimensionality in the third feature set, the fourth feature set is formed as a SRDM + 2D-CWT-based multifractal feature set. As shown in Section 8.3.4, the 2D-CWT-based multifractal feature set has 48 components. By adding eight SRMD texture-based features, 56 features are extracted.

Experiments show that four classifiers trained by the four feature sets have different detection rates. The SVM classifier trained by the fourth feature set (hybrid feature set: SRDM + 2D-CWT-based multifractal feature set) can achieve over 83% correct detection rate on the testing set. The detection performances are listed in Table 8.2.

The receiver operating characteristic (ROC) curves are plotted in Fig. 8.5 as the ratio of the true positive fraction against the false positive fraction.

Table 8.2 Four feature sets and their detection performances on the test set. Note that a true positive rate (TP) is a percentage made by the classifier system that correctly corresponds to the lesion test images, whereas a true negative rate (TN) is a percentage made by the classifier system that correctly corresponds to the nonlesion testing images.

Feature Set	TP	TN	Detection Rate
SRDM feature set	76%	77%	76.50%
SRDM + multifractal feature set	79%	80%	79.50%
SRDM + 2D-CWT feature set	80%	82%	80%
SRDM + 2D-CWT-based multifractal feature set	82%	84%	83%

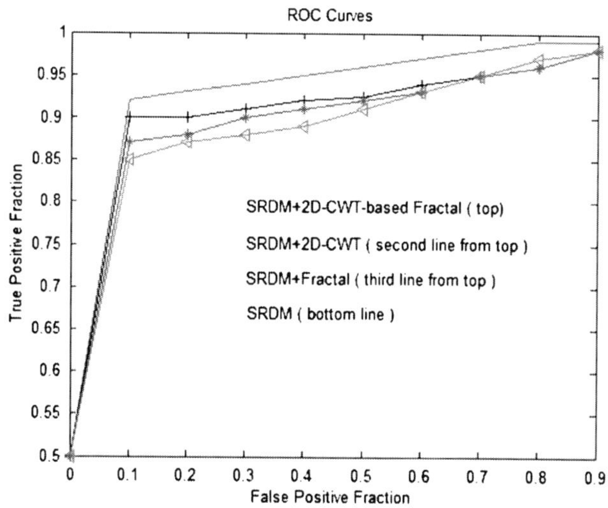

Figure 8.5 ROC curves with different feature sets.

From the comparison of the ROC curves, it can be concluded that our proposed hybrid feature set (the fourth feature set in Table 8.2) has the best discriminant ability among the four feature sets.

Among the four feature sets, the SRDM (texture-based feature) has a minimum of 8 features; however, it has the lowest detection rate and the lowest ROC performance. The second feature set combines the SRDM with multifractal feature. Both detection rate and the ROC performance have been increased. The third feature set consists of the SRDM and the 2D-CWT feature set. The detection performance was improved slightly; however, the number of the feature set is too large (168). According to a rule of thumb, if the third feature set is used to train an SVM classifier, much more training samples are needed. In the fourth feature set, the merit of the 2D-CWT is kept. Furthermore, the multifractal feature set keeps a steady feature attribution for the irregular shape of calcification spots in the mammograms. As a result, both detection rate and the ROC performance are significantly improved.

Figure 8.6 shows some examples of calcification areas that have been correctly detected by some classifiers trained with the four feature sets. The proposed MCC algorithm has been applied to whole mammogram images. Figure 8.7 shows three calcification areas that have been correctly detected in digital mammograms.

The proposed hybrid feature extraction method (SRDM + 2D-CWT-based multifractal feature set) is a new scheme based on SRDM. In this newly proposed feature set, the textual information, wavelet-oriented information, and multifractal information of the mammogram are congregated to form a new hybrid feature. From Table 8.2 and Fig. 8.6, it can easily be concluded that the proposed hybrid feature method has achieved a better detection rate and a better ROC curve than the SRDM-based feature set, which has been published in the literature.[11,31]

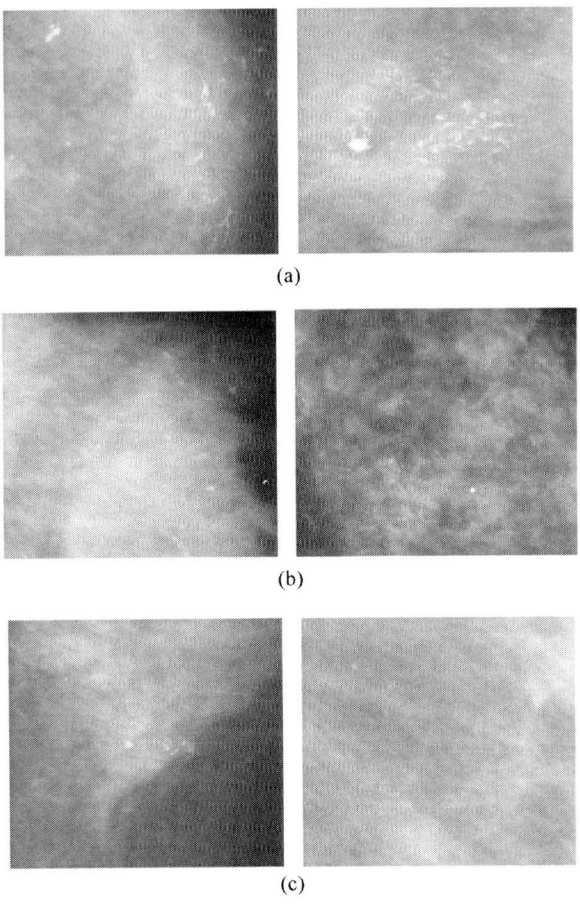

Figure 8.6 Detected MCC areas by classifiers with different feature sets listed in Table 8.2: (a) all four classifiers trained by the four feature sets, (b) only the last three classifiers trained by the last three feature sets, and (c) only the classifier trained by the fourth feature set.

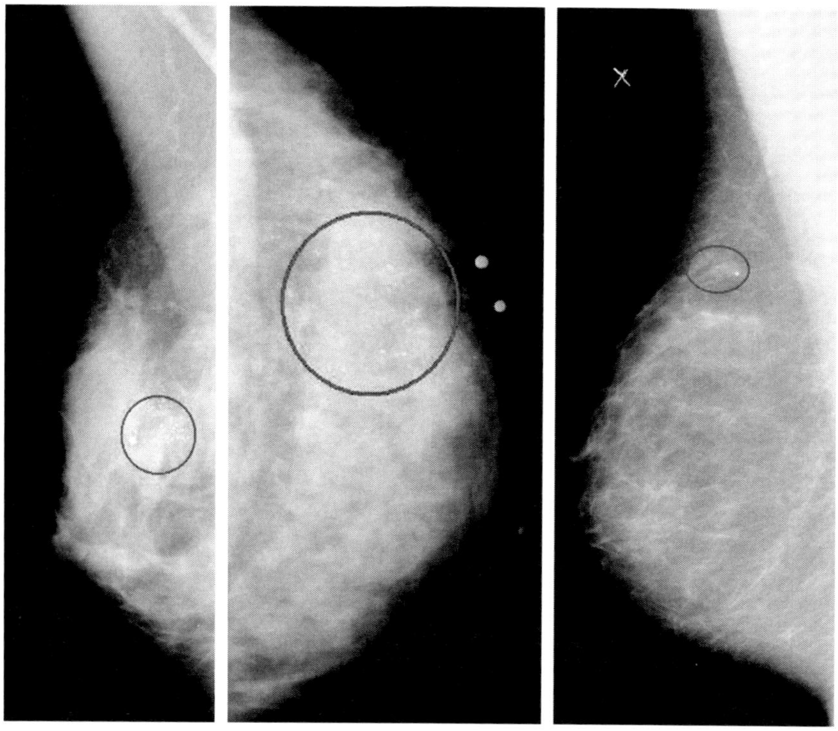

Figure 8.7 Examples of MCC areas in three mammograms detected using the proposed feature and system.

8.6 Conclusion

A novel hybrid feature method is proposed and successfully applied to microcalcification detection in digital mammograms. The hybrid features combine texture-based features and 2D complex wavelet-based multifractal features. 2D-CWT can decomposite a ROI area into different subimages with different frequency bands and directional orientations, which provides true directional selectivity and pixel-shifting insensitivity. Furthermore, the 2D-CWT-based multifractal feature set can suppress the drawback of the 2D-CWT feature's sensitivity to the irregularity of microcalcification in the mammogram. Experiments demonstrated that the proposed scheme has improved micro-calcification detection performance compared to other feature extraction methods previously published. Future work will focus on constructing a classification system in the ensemble classifiers with hybrid features in order to increase the system's reliability and detection rate at the same time in one system.

References

1. H. D. Cheng, X. Cai, X. Chen, L. Hu, and X. Lou, "Computer-aided detection and classification of microcalcifications in mammograms: a survey," *Pattern Recogn.* **36**(12), 2967–2991 (2003).

2. L. Wei, Y. Yang, R. M. Nishikawa, and Y. Jiang, "A study on several machine-learning methods for classification of malignant and benign clustered microcalcifications," *IEEE Trans. Med. Imaging* **24**(3), 371–380 (2005).
3. A. Papadopoulos, D. I. Fotiadis, and A. Likas, "An automatic microcalcification detection system based on a hybrid neural network classifier," *Artificial Intel. Med.* **25**, 149–167 (2002).
4. R. Panchal and B. Verma, "Neural-association of microcalcification patterns for their reliable classification in digital mammography," *Int. J. Pattern Recogn. Artificial Intel.* **20**(7), 971–983 (2006).
5. I. El-Naqa, Y. Yang, M. N. Wernick, N. P. Galatsanos, and R. M. Nishikawa, "A support vector machine approach for detection of microcalcifications," *IEEE Trans. Med. Imaging* **21**(12), 1552–1563 (2002).
6. J. Ye, S. Zheng, and C. C. Yang, "SVM-based microcalcification detection in digital mammograms," *Proc. Int. Conf. Comput. Sci. Software Eng.* **6**, 89–92 (2008).
7. M. Haindl, S. Mikes, and G. Scarpa, "Unsupervised detection of mammogram regions of interest," *Lecture Notes in Computer Science, Knowledge-based Intelligent Information and Engineering Systems* **4694**, 33–40, Springer, Berlin (2010).
8. B. W. Hong and B. S. Sohn, "Segmentation of regions of interest in mammogram in a topographic approach," *IEEE Trans. Inform. Tech. Biomed.* **14**, 129–139 (2009).
9. D. J. Marchette, R. A. Lorey, and C. E. Priebe, "An analysis of local feature extraction in digital mammography," *Pattern Recogn.* **30**(9), 1547–1554 (1997).
10. L. Shen, R. M. Rangayyan, and J. E. L. Desautels, "Application of shape analysis to mammographic calcifications," *IEEE Trans. Med. Imaging* **13**(2), 263–274 (1994).
11. K. J. Kim and H. W. Park, "Statistical textural features for detection of microcalcifications in digital mammograms," *IEEE Trans. Med. Imaging* **18**(3), 231–238 (1999).
12. S. N. Yu, K. Y. Li, and Y. K. Huang, "Detection of microcalcifications in digital mammograms using wavelet filter and markov random field model," *Comput. Med. Imaging and Graphics* **30**(3), 163–171 (2006).
13. R. N. Strickland and H. Hahn II, "Wavelet transforms for detecting microcalcification in mammograms," *IEEE Trans. Med. Imaging* **15**(2), 218–229 (1995).
14. T. C. Wang and N. B. Karayiannis, "Detection of microcalcifications in digital mammograms using wavelets," *IEEE Trans. Med. Imaging* **17**(4), 498–509 (1998).

15. C. B. R. Ferreira and D. L. Borges, "Analysis of mammogram classification using a wavelet transform decomposition," *Pattern Recog. Lett.* **24**(7), 973–982 (2003).
16. G. Boccignone, A. Chianese, and A. Picariello, "Computer aided detection of microcalcifications in digital mammograms," *Comput. Biol. Med.* **30**(5), 267–286 (2000).
17. S. Halkiotis, T. Botsis, and M. Rangoussi, "Automatic detection of clustered microcalcifications in digital mammograms using mathematical morphology and neural networks," *Signal Process.* **87**(7), 1559–1568 (2007).
18. T. Stojic, I. Reljin, and B. Reljin, "Adaptation of multifractal analysis to segmentation of microcalcifications in digital mammograms," *Physica A* **357**, 494–508 (2006).
19. D. R. Chen, R. F. Chang, C. J. Chen, M. F. Ho, S. J. Kuo, S. T. Chen, S. J. Hung, and W. K. Moon, "Classification of breast ultrasound images using multifractal feature," *J. Clinical Imaging* **29**, 235–245 (2005).
20. L. Bocchi, G. Coppini, J. Nori, and G. Valli, "Detection of single and clustered microclfications in mammograms using multifractals models and neural networks," *Med. Eng. Phys.* **26**, 303–312 (2004).
21. M. D. Santo, M. Molinara, F. Tortorella, and M. Vento, "Automatic classification of clustered microcalcifications by A Multiple Expert System," *Pattern Recogn.* **36**(7), 1467–1477 (2003).
22. H. D. Cheng, J. L. Wang, and X. J. Shi, "Microcalcification detection using fuzzy logic and scale space approaches," *Pattern Recogn.* **37**(2), 363–375 (2004).
23. M. J. Bottema and J. P. Slavotinek, "Detection and classification of lobular and DCIS (small cell) microcalcifications in digital mammograms," *Pattern Recogn. Lett.* **21**(13–14), 1209–1214 (2000).
24. C. K. Chui, *Wavelets: A Mathematical Tool for Signal Analysis*, SIAM, Philadelphia (1997).
25. I. W. Selesnick, R. G. Baraniuk, and N. G. Kingsbury, "The dual-tree complex wavelet transform," *IEEE Signal Process. Mag.* **22**, 123–151 (2005).
26. N. G. Kingsbury, "Image processing with complex wavelets," *Phil. Trans. R. Soc. Lond. A* **357**, 2543–2560 (1999).
27. C. C. Liu and D. Q. Dai, "Face recognition using dual-tree complex wavelet features," *IEEE Trans. Image Process.* **18**, 2593–2599 (2009).
28. G. Y. Chen and W. F. Xie, "Pattern recognition with SVM and duel-tree complex wavelets," *Image Vision Comput.* **25**, 960–966 (2007).
29. T. Celik, H. Qzkaramanh, and H. Demirel, "Facial feature extraction using complex dual-tree wavelet transform," *Comput. Vision Image Understanding* **111**, 229–246 (2008).

30. R. C. Gonzalez and R. E. Woods, *Digital Image Processing*, 3rd ed., Pearson Prentice Hall (2008).
31. J. Xu and J. S. Tang, "Detection of clustered microcalcification using an improved textual approach for computer aided breast cancer diagnosis system," *CSI Comm.* **1**, 17–19 (2008).
32. M. J. Turner, J. M. Blackledge, and P. R. Andrews, *Fractal Geometry in Digital Imaging*, Academic Press, San Diego, CA (1998).
33. M. Heath, K. Bowyer, D. Kopans, R. Moore, and W. P. Kegelmeyer, "The digital database for screening mammography," *Proc. Fifth Int. Workshop on Digital Mammography*, 212–218 (2001).
34. N. Cristianini, J. Shawe-Taylor, and R. Holloway, An Introduction to Support Vector Machines and Other Kernel-Based Learning Methods, Cambridge University Press, Cambridge, UK (2000).

Ping Zhang is an associate professor with the Department of Mathematics and Computer Science, Alcorn State University (ASU), Alcorn, MS. He received his B.S. degree in Electronic Engineering and his M.S. degree in Biomedical Image Processing from Chongqing University; he received his Ph.D. in Computer Science at the Center for Pattern Recognition and Machine Intelligence (CENPARMI) at Concordia University, Montreal, Canada in 2006. Before joining ASU, he worked at EPSON Canada as a senior scientist, Nanyang Technological University, Singapore as a research fellow, and Chongqing University as an Associate Professor. Dr. Zhang has more than 15 years research experience in pattern recognition, image recognition, computer vision, and biomedical image processing. He has authored more than 40 journal and international conference papers. He is a senior member of the IEEE.

Kwabena Agyepong is an associate professor with the Department of Advanced Technologies at Alcorn State University, Alcorn, MS. He received his B.S. in Electrical Engineering from the University of Science and Technology, Kumasi, Ghana. He obtained his M.S. and Ph.D. from the University of Cincinnati. As a principal investigator, he has managed and implemented several research projects and has published many research papers in journals and international conferences. His research interests include information processing, database management, and medical image detection.

Chapter 9
Computer-Aided Prostate Cancer Diagnosis: Principles, Recent Advances, and Future Prospective*

Clara Mosquera-Lopez and Sos S. Agaian
Department of Electrical and Computer Engineering, University of Texas at San Antonio, Texas, USA

Alejandro Velez-Hoyos
Department of Pathology, Pablo Tobon Uribe Hospital, Medellin, Colombia

Ian Thompson
University of Texas Health Science Center at San Antonio, Texas, USA

9.1 Introduction

Over the next few years, a dramatic increase in the number of people developing different types of cancer is expected. It is estimated from global statistics that 10 million new cancer patients are diagnosed each year, and this number will double by the year 2020. Although several billions of dollars are spent on cancer research, there is still no definitive cure for this disease.[3] Meanwhile, various techniques have been developed to be used in all phases of cancer diagnosis and management with the aim of helping doctors define an appropriate treatment plan for patients and monitor treatment efficacy. Imaging forms an essential part of cancer clinical protocols and can provide morphological, structural,

* This work has been partially supported by NSF Grant (RD-0932339) and by CTRC P30 Cancer Center Support Grant (CA054174) from the National Cancer Institute.

metabolic, and functional information. As a result, cancer diagnosis is becoming increasingly image-based. There are three main objectives of using imaging with respect to cancer management: (1) the earliest possible detection of benign or malignant lesions or tumors, which is probably the strongest factor in reducing mortality for certain cancers; (2) correlation of imaging results with other clinical parameters; and (3) accurate staging and follow-up after treatment. In order to aid radiologists and pathologists in cancer diagnosis, various computer-based methods have been developed to detect, classify, and grade the malignancy of tumors by using certain visual criteria.[4]

Due to the importance of imaging in cancer diagnosis and treatment, computer-aided diagnosis (CAD) has become one of the major research subjects in medical imaging and diagnostic radiology and pathology. The goal of CAD is to increase the productivity of radiologists/pathologists by improving the accuracy and consistency of diagnoses, reducing image reading time, and providing computer-based tools for image visualization and annotation. The general approach for CAD development is to find the location of a lesion (detection) and also to estimate the probability of the presence of disease (differential diagnosis). Ultimately, a CAD system may become an integrated tool in all areas of medical imaging. An important benefit of using computers to diagnose disease is the reproducibility and consistency of the diagnostic methods because the performance of computers is not affected by fatigue, perceptual errors, or variability in classification criteria. The use of computers to assist clinical diagnosis is well established in radiology.[2] In 1998, the U. S. Food and Drug Administration (FDA) approved the clinical use of the first CAD system for mammography developed for detection of breast cancer.[5] Since then, other systems for disease detection have been used. For instance, several CAD systems have been employed in breast screening[6–10] or have been prospectively analyzed in order to be used in clinical practice for breast cancer detection.[11,12] Support tools for skin cancer[13,14] and leukemia[15] have been also presented. In addition, several CAD systems for prostate cancer diagnosis have been developed with the aim of providing diagnostic information obtained from quantitative image analysis. The information provided by accurate CAD systems regarding cancer localization and grade may be used by doctors as a "second opinion" in order to make final diagnosis decisions.

Prostate cancer is one of the most diagnosed cancers in the United States.[16] Cells that have the appearance of prostate cancer can be found in the prostate gland of nearly 50% of men over the age of 50.[17] Prostate cancer is considered to be the second significant cause of cancer behind skin cancer, and the second leading cause of cancer death behind lung cancer among American men.[16,18] Statistics for 2013 from the American Cancer Society show that about 238,590 new cases will be diagnosed this year, and 29,720 men will die of prostate cancer.[16] The lifetime risk of a man being diagnosed with clinically apparent prostate cancer is around 11%, and the lifetime risk of dying of prostate cancer is 3.6%.[17] However, strategies based on early and accurate diagnosis that distinguishes the harmless (indolent) cancers from the potentially lethal ones plays a key role in therapy selection and consequently in patient survival.[17] Today, microscopic analysis of needle biopsy tissue sections remains the gold-standard method for cancer detection and grading.

Currently, pathologists visually evaluate histopathology images and grade them according to the Gleason grading system, which was endorsed as a standard for prostate carcinoma in 2003 by the World Health Organization (WHO),[19] and it is widely used by pathologists around the world in clinical practice. The system proposed by Dr. Donald Gleason assigns a grade from 1 to 5 depending on the architectural pattern of the glands of the prostate tumor,[20,21] as shown in Table 9.1. Gleason grade 1 refers to well-differentiated

Table 9.1 Characteristics of Gleason patterns and examples.

Grades 1 and 2: These patterns consist of well-differentiated, uniform, single glands that grow close together. The main difference between pattern 1 and pattern 2 is that the latter presents more-variable single glands slightly spaced apart, and boundaries of the tumor are less well-circumscribed.[31] Gleason pattern-1 and pattern-2 glands tend to be larger than intermediate-grade carcinomas.[28] These two grades are rare and closely resemble normal prostatic tissue.

Grade 3: This pattern is composed of moderately differentiated glands that show marked variations in size and shape. The neoplastic gland size is usually smaller than that seen in low-grade tumors.[28] Grade-3 tumors infiltrate non-neoplastic prostate acini and may show papillary or cribriform patterns. In 2005, it was decided by the International Society of Urological Pathology (ISUP)[32] that cribriform glands would be considered pattern 3 only if they are well circumscribed, ovoid to round, and of similar size to normal glands.[31]

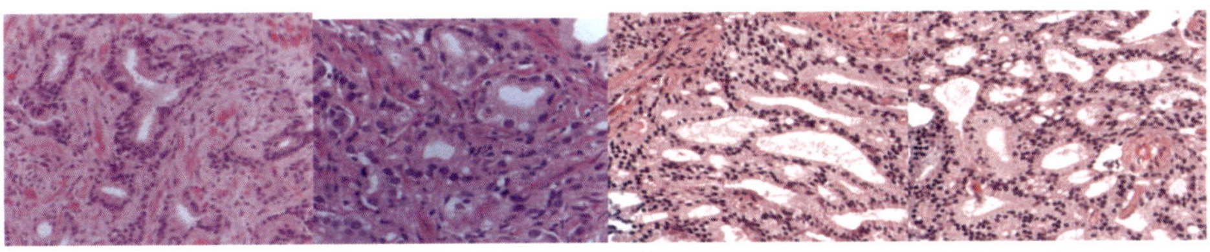

Grade 4: This pattern consists of large, irregular cribriform glands or fused glands with poorly formed glandular lumina. In grade-4 tumors, there are no individual or separated gland units, as seen in grades 1–3.[28] Cribriform glands larger than benign glands are also included in this pattern.[31]

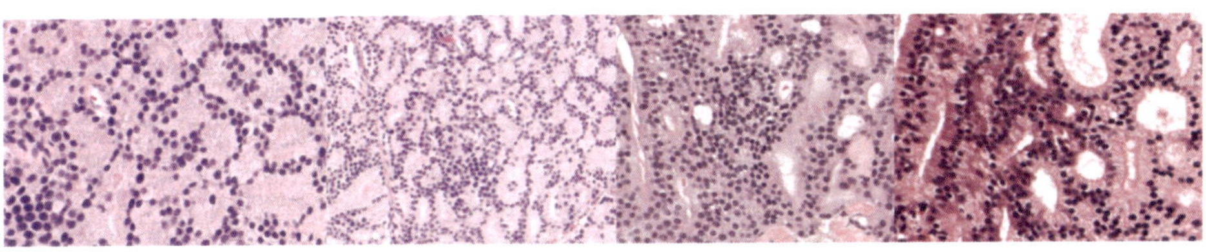

Grade 5: This pattern includes anaplastic carcinoma with minimal glandular differentiation and diffusely infiltrating prostatic stroma. Tumors showing comedocarcinoma and solid sheets, cords, and individual cells without acinar formation were included in this group by the ISUP.[31,32]

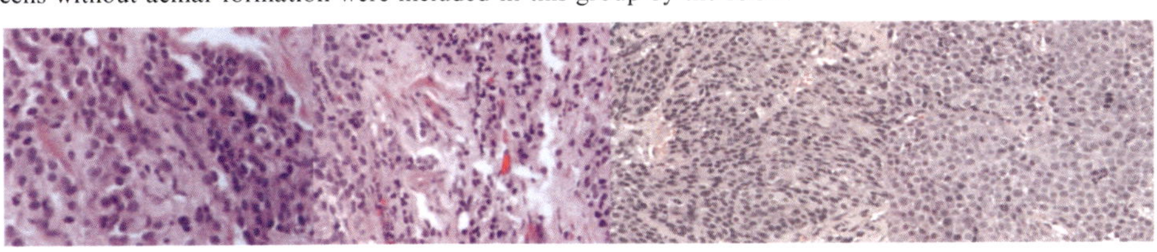

glands that resemble normal tissue, whereas Gleason grade 5 represents poorly or nondifferentiated glands. Therefore, high grades (grade 4 and 5) are closely related to a more-aggressive disease, whereas low grades correlate with a more-favorable patient outcome. In contemporary clinical practice, prostate-tumor grading starts with pattern 3.[22–24] Gleason patterns 1 and 2 are often incidental findings in the peripheral zone. Recent studies have argued that the inclusion of grades 1 and 2 could produce diagnostic confusion and potential errors (i.e., reporting these grades usually reflects undergrading and does not correlate with radical prostatectomy).[25–27] In prostate cancer staging, special importance is given to the recognition of Gleason grades 3 and 4 because there is a significant prognostic difference between them; thus, cases with these grades must be properly diagnosed. Table 9.1 summarizes the important characteristics of each Gleason pattern and provides sample images. Moreover, the sum of the primary (predominant) and secondary (second-most prevalent) Gleason grades found in a histopathology image is reported as a Gleason score. When a tumor has only one histologic pattern, the primary and secondary pattern are given the same number. The Gleason score on a biopsy is a powerful indicator for prostate cancer prognosis that correlates with all of the important pathologic parameters at radical prostatectomy, prognosis after radical prostatectomy, patient outcome radiotherapy, and many molecular markers.[28]

Today, the characterization of prostatic tissue specimens by pathologists is a decisive step in the diagnosis of prostate cancer.[1] However, the grading process is time consuming and subjective. The results of microscopic evaluation of tissue are influenced by pathologist experience and fatigue, variability in the interpretation and application of the grading criteria, and complexity of tissue samples. The combination of subjectivity and level of expertise of pathologists leads to high intra- and interobserver variations, affecting the accuracy of pathology reports. It has been reported in recent studies that intra- and interobserver reproducibility of the Gleason grading system ranges from 60–90%, and comparisons between the assigned grade in the needle biopsy and the grade of the matched whole prostate gland reflects undergrading of the needle biopsy specimen in 42% of the cases and overgrading in 15% of the cases.[29] Undergrading is a serious problem in circumstances where the treatment of a low-Gleason-grade tumor (containing grades 1 and 2) would vary from a treatment of a Gleason score 5 or 6.[30] These issues create the need for a reproducible grading system based on quantitative characterization of histopathology images using appropriate features that would likely result in significantly improved accuracy in cancer diagnosis.

Recently, several prostate cancer CAD systems using different approaches have been proposed to classify/grade prostate cancer histopathology images.

Practically, prostate-cancer computerized grading systems can be divided into two main classes: *tissue-structure-based systems* and *textural-based systems*.

Tissue-structure-based systems employ the characteristics of glands and their components (lumen, epithelial cytoplasm, and epithelial nuclei) along with the nuclei cells' spatial arrangement and density to detect prostatic tumors and to identify the Gleason grade of biopsy specimens.[33–39]

Textural-based systems use algorithms for image analysis in spatial and transform domains to characterize the pattern of Gleason grades. Spatial domain algorithms include co-occurrence matrix analysis, fractal dimension, and run-length features, among others, whereas transform domain features include wavelet energy of detail coefficients and low-resolution images.[40–50]

Each approach has advantages and limitations. Several studies have shown that CAD systems utilizing morphometric features correlate well with cancer prognosis.[51–56] However, in high-grade prostatic tumors characterized by the lack of individual or separated gland units, some basic elements of the tissue, such as lumen, are absent or can be occluded by mucin or cytoplasm;[57] for these cases glandular features might be not as effective as features resulting from textural analysis. Several combinations of morphometric and textural parameters have been also proposed to distinguish among Gleason patterns.

This chapter reviews recent developments in the field of detection and grading of cancerous tissue from digitized microscopic histopathology images, emphasizing both prostate cancer CAD technologies. Also discussed are current developments in pattern recognition techniques to classify whole slide images or smaller ROIs within prostatic tissue samples. This chapter focuses on specific issues related to the developed recognition systems that encompass image conditioning algorithms, segmentation methods, feature vectors (used to characterize tumors), classification methods, and system performance analysis. The following sections describe the general components of CAD systems for prostate cancer detection and diagnosis. Section 9.2 presents a brief overview of the current clinical approach for detecting and grading prostate cancer and the procedure for digitized histopathology image acquisition when a biopsy is performed as part of the prostate cancer diagnosis protocol. Section 9.3 describes the steps followed by computer-assisted prostate cancer diagnosis systems. This section includes image processing algorithms (used for noise filtering, color normalization, image segmentation, and others), feature vectors used for image classification, an overview on learning algorithms, and methods for system performance evaluation. Finally, concluding remarks and future directions in prostate cancer CAD systems are discussed in Section 9.4.

9.2 Clinical Approach for Prostate Cancer Detection and Grading

The gold standard for prostate cancer diagnosis is histopathological analysis of biopsy tissue samples.[58–60] Preliminary screening methods such as prostate specific antigen (PSA) and digital rectal exam (DRE) are used to identify patients who need a biopsy, as shown in Fig. 9.1. Elevated PSA levels and abnormal results of DRE tests are indicators of prostate cancer risk, but they are not definitive methods for diagnosis. On one hand, a PSA test yields low specificity and sensitivity—for example, serum PSA levels over 4 ng/mL are normally associated with high risk of PCa—however, studies have shown that certain populations of men with lower levels of PSA share the same risk level.[61,62] On the other hand, detection by palpation through DRE is also limited to relatively large and superficial lesions.[61]

9.2.1 Core needle biopsy

If prostate cancer is suspected, a needle biopsy is performed to confirm the presence or absence of disease and to grade the prostatic carcinoma, if detected. A prostate needle biopsy is a surgical procedure usually performed in conjunction with transrectal ultrasound (TRUS guided-needle biopsy), in which small tissue samples are removed from the prostate gland.[63] Because the biopsy operation is a sampling procedure, several shortcomings may arise. One important aspect to take into account is that due to the multifocal nature of PCa, there is always some level of risk for missing pathologically significant lesions resulting in false negatives or repeated biopsies.[61]

The biopsy samples are then processed in order to produce glass microscopic slides. Tissue processing is made up of a series of steps including fixation, embedding, sectioning, staining, and coverslipping. The goal of fixation is to preserve tissue permanently in as lifelike a state as possible. Once the removed samples have been fixated, the tissue is embedded in paraffin for sectioning. Sectioning tissue involves nothing more than cutting it into smaller portions that can be placed on a slide. The next step is tissue staining, which is

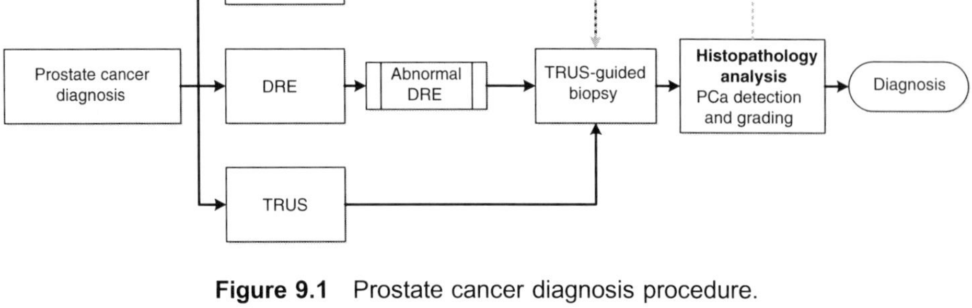

Figure 9.1 Prostate cancer diagnosis procedure.

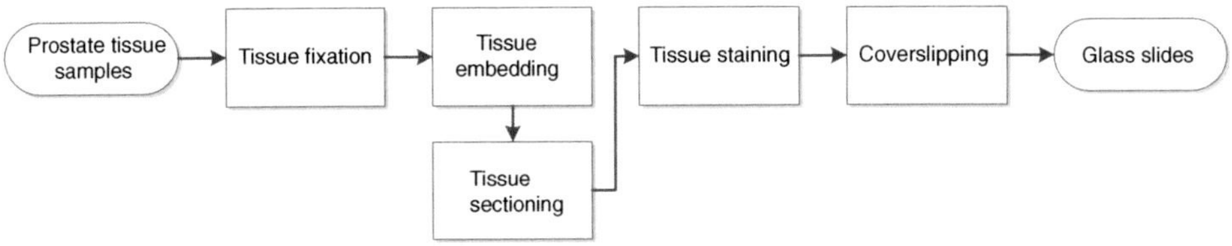

Figure 9.2 A block diagram of tissue processing steps.

applied to stain various cellular components of tissue. It is important to note that before any stain can be done, the slides must be deparaffinized.[64] In general, Hematoxylin and Eosin (H&E) are used for prostatic tissue staining. Hematoxylin has a deep-blue-to-purple color, and Eosin has a pink color. Due to chemical reactions between stains and tissue components, the nuclei are stained blue, whereas the cytoplasm and extracellular matrix have varying degrees of pink staining.[65] Staining is a key step because it highlights histological and textural features that are diagnostically important. Finally, the stained tissue sections are covered with a thin piece of glass to protect the tissue and to provide better optical quality for microscopic examination; this step is called coverslipping. Once the processing steps have been completed, the glass slide is ready for a pathologists' evaluation. Figure 9.2 shows a block diagram of tissue processing steps.

9.2.2 Digital pathology imaging

After the biopsy slides are processed, visualization and assessment of the tissue is performed by a trained pathologist. Today, the evaluation of tissue in clinical practice is performed with a microscope. However, over the last decade, pathology has benefited from the development of digital imaging devices that are able to produce digital slides.[66] These devices have also made possible the development of computer-aided systems for histopathology image analysis. There are two types of apparatus capable of digitizing slides at high magnifications: motorized digital microscopes and whole-slide scanners. Motorized microscopes are those in which the functionality and original components remain the same, but a high-resolution camera is joined with the microscope to produce digital slides.[67] The digital images produced by a microscope are taken from specific ROIs, but the entire tissue present on the slide cannot be digitized at once. A computer connection is available for device control and image visualization and storage. It is important to note that multispectral images can be also obtained from a camera-equipped microscope by using liquid crystal tunable filters (LCTF), which enable the acquisition of images at different wavelengths to form the so-called multispectral cube.[1,68]

Whole-slide scanners, which can produce a digital version of a whole-tissue glass slide, include components similar to those in an automated

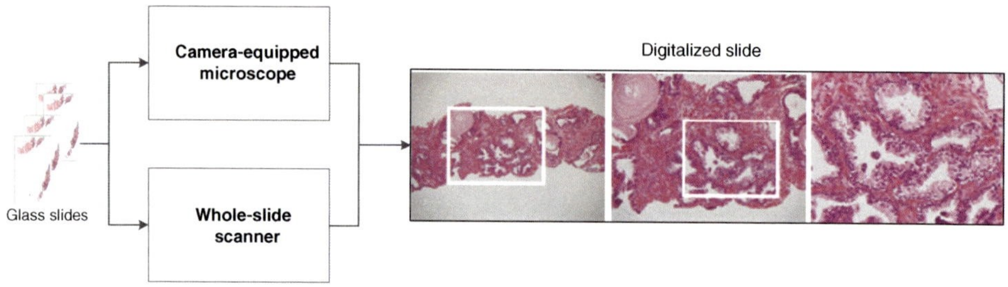

Figure 9.3 Digital pathology imaging.

microscope but with some modifications. For instance, they do not have eyepieces or position and focus control. All whole-slide digitation systems consist of an optical microscope system, an acquisition system, software to control the scanning process, and an interface for computer connection. Modern slide devices utilize the principle of line scanning or square matrix acquisition. In line scanning, the scanner scans the image row by row using an array of sensors, whereas the matrix acquisition method uses square-by-square scanning in order to take a high number of image frames. The slide scanner software then stitches the digital images of these lines or frames together to produce one digital image.[67] The digitalized version of a single core of a prostate biopsy digitized at 40× magnification consists of approximately 225 million pixels. Thus, a prostate biopsy composed by 12–20 samples will contain approximately 2.5–4 billion pixels of data. The large amount of data produced in digital pathology creates challenges for the development of fast and efficient multiresolution algorithms for image analysis.[1] Figure 9.3 shows examples from commercial pathology imaging devices.

9.3 State-of-the-Art in Histopathology-Image-Based, Computer-Aided Prostate Cancer Diagnosis

Systematic research related to the development of CAD schemes for prostate cancer diagnosis from digitized histopathology images began in the late 1990s. Since then, many investigators have been developing systems (whose descriptions have been published in a great number of research papers and patents) essentially based on an understanding of the processes that would be involved in image readings by pathologists. This strategy seemed quite logical because the knowledge of expert pathologists (who have been carrying out the very complex and difficult tasks of image reading) is an important clue to design feature vectors that correctly characterize images of different cancer patterns. CAD systems have not been created to replace the pathologist but rather to be a support tool to make the work faster and more reproducible. A great challenge for CAD systems is to be accurate enough to be used in routine clinical work in many hospitals around the world.

The primary goal of CAD systems for prostate cancer diagnosis is to assist pathologists in disease detection and grading. Like any pattern recognition system, the design of a system for prostate cancer classification requires the definition of all necessary preprocessing algorithms for image conditioning, definition of pattern classes, and design of quantitative representation of each class through feature vectors, cluster analysis, classification, and system validation. A recognition system can be operated in two modes: training and classification.[33] Training mode refers to the use of the information in a dataset to train classifiers, which will then assign labels to unknown test patterns according to the learned classification rules when the system operates in classification mode. A flowchart for a generic CAD system for prostate cancer diagnosis and grading is shown in Fig. 9.4.

A CAD system for PCa detection and grading generally consists of the following key steps:[69]

1. Image preprocessing steps, which remove the irrelevant background and noise, enhance or highlight important details of the images, and normalize image size and color.
2. Segmentation, which is used to label important objects within the image.
3. Feature extraction, which is used to represent data by deriving relevant properties at an image or object level. Features may include size, shape, color, texture, statistics, etc.
4. Classification of the images, which consists of separating objects into different classes according to a classification rule.
5. Visualization of images and data, which allows the user to see the results of the classification methods. For example, images of cancer localization or Gleason grade maps may be displayed as a final result.
6. Cross-validation, which concerns the evaluation of CAD performance using statistical indicators.

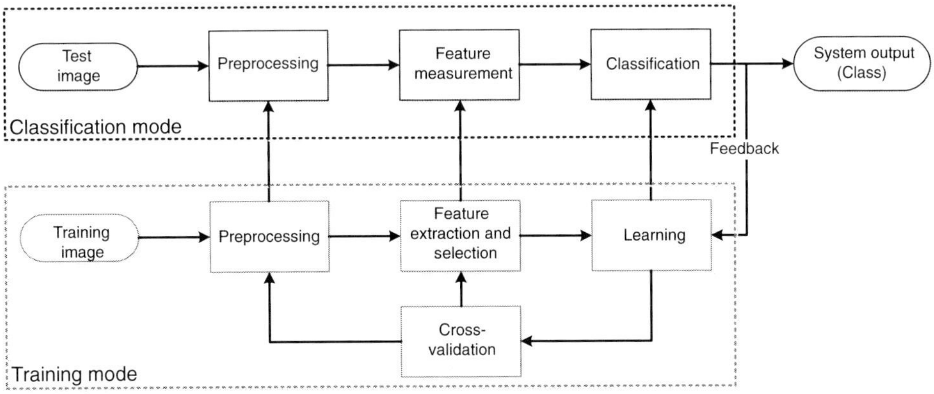

Figure 9.4 Generic architectures of CAD for prostate cancer diagnosis and grading (core tasks).

9.3.1 Image preprocessing

The role of preprocessing is to separate the pattern of interest to be classified from the background,[70] remove noise,[71–76] enhance details of the pattern,[77–101] detect edges or especial structures in the image,[102–110] normalize the size, magnification, orientation, and color of the image, and any other operation that will contribute in defining a numerical representation of a given pattern.[70] Generic and specific algorithms for noise removal, image enhancement, and edge detection may be applied to histopathology images. Those algorithms have proven to have a positive incidence in cancer recognition. This section describes several methods for image color normalization that have been applied to histopathology images to deal with issues related to color variability. In addition, we present a summary on histology image segmentation often used as a preparation step before feature extraction in tissue-structure-based classification systems.

9.3.1.1 Color normalization

Color is one of the most informative features of an image, and it plays a central role in the development of image processing algorithms and computer vision systems. However, the color of histopathology images may be affected by several factors. Such factors include variations in staining and scanning conditions due to image acquisition protocols, capturing-device properties, and lighting conditions. Consequently, the performance of color-based classification systems is degraded when the color of the images is not standardized. Thus, color standardization is practically a necessary condition for developing accurate color-based automatic recognition systems. Standardization mainly consists of controlling the settings of a set of images such that their original device-dependent color space is transformed into some device-independent color space. Several studies have been carried out to reduce color variations in histopathology images coming from different sources. One approach[111] uses standard color filters selected for histology H&E stained slides (i.e., color chart) in order to calibrate and profile imaging devices. The color produced by a particular scanner is mapped to the reference colors by means of a polynomial transformation, in which the color transformation matrix is constructed according to the parameters of a specific scanner or imaging device. After the polynomial transformation is done, the resulting image is gamma corrected in order to produce the final result.

Another approach to color normalization of histopathology images is based on local color transference.[112] The algorithm uses fuzzy c-means to segment the reference and input images in meaningful regions and generates a weighting function based on the pixels' fuzzy membership index, which modulates the color transference operations. Examples of the application of this normalization method on images from different sources are presented in Figure 9.5.

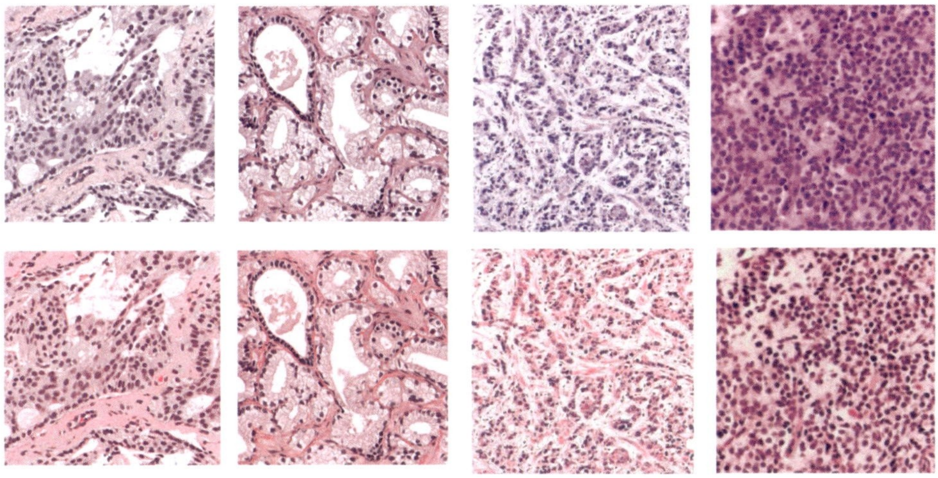

Figure 9.5 Color standardization results: (top) original images and (bottom) color-standardized images.

Another approach maps the color distribution of an over/understained image to that of a well-stained target image. Magee et al.[113] used a multimodal extension to linear normalization in lab color space, which generates several transforms depending on the image composition (one for each segmented tissue structure present in the image), and the distribution of each class is then mapped to each class of the image being processed. The results of this algorithm are illustrated in Figs. 9.6(d)–(f).

Another path followed by several researchers to accomplish color normalization is based on color deconvolution. Specific color deconvolution vectors are estimated[113] to separate the stains composing a color histology image [Figs. 9.6(g)–(i)]. The image corresponding to each stain is processed using filters to eliminate undesirable effects and residual stains or to highlight tissue details, such that the color attributes of the final histology image are improved and normalized.

9.3.1.2 Histopathology image segmentation

Segmentation of certain histological structures is a required step for systems that take morphometric information of the tissue as feature vectors. The presence, extent, shape, size, and other morphological features of cancer nuclei and glands provide important clues in prostate cancer detection and grading. For example, the size of glands in prostatic carcinoma tends to reduce as the Gleason grade of the tumor is higher.[1] Also, the density and distribution of cancer cells in a specific area of the tissue are important indicators used for grading of prostatic tumors. Therefore, automated detection and segmentation of tissue structures are imperative steps in many modern CAD systems. In this section, we discuss several proposed methods for segmentation of prostate glands and other histological structures of the tissue.

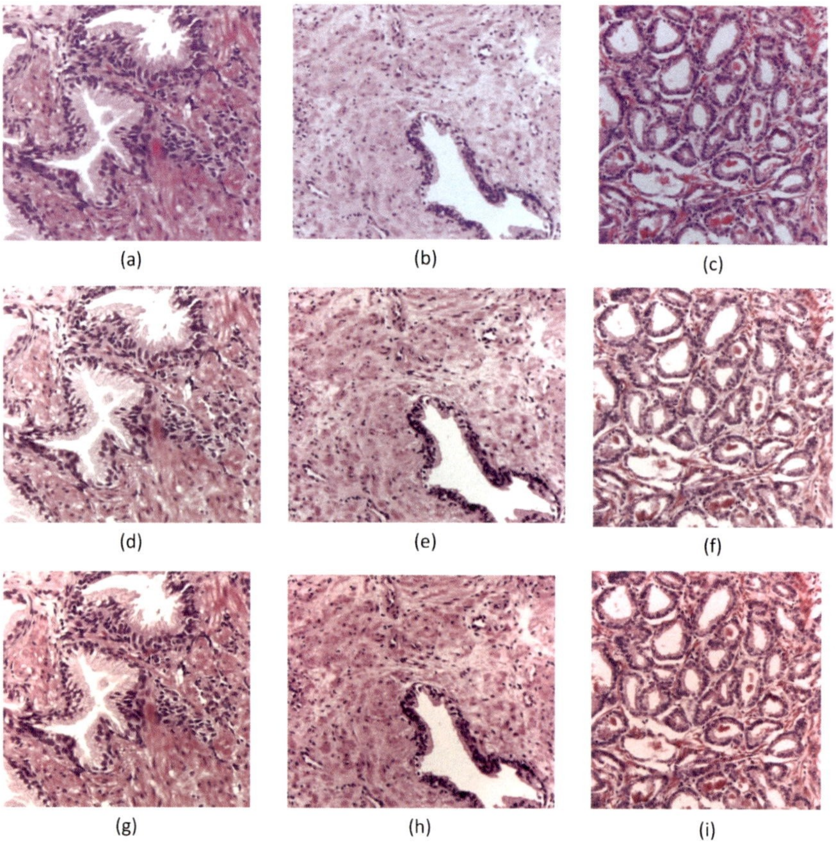

Figure 9.6 Color standardization results: (a)–(c): original images; (d)–(f) color-standardized images using color distribution mapping; and (g)–(i) color-standardized images using the color deconvolution approach.

A. Nuclear segmentation

Several works have been conducted on nuclear segmentation. Clustering algorithms such as thresholding,[68] k-means, and fuzzy c-means[1,114] (as well as Bayesian classification,[37] color region mapping,[115] watershed transform, and novel edge detection algorithms) have been employed to separate nuclear information from other structures in histopathology images. Because nuclei appear with a dark-blue color in histopathology images, they can be segmented using pixel-level analysis. However, a recent study by Nguyen et al.[57] proposes a maximum-object-likelihood binarization algorithm. The nucleus objects are initially separated from the background by thresholding the blue channel of the histology image; the area and circularity of each blob is then used as a feature vector to refine the segmentation. Moreover, knowledge about the appearance of nuclei in cancerous glands (light blue with prominent nucleoli or dark spots) and normal glands (uniform dark or light blue without nucleoli spots) is used to classify the segmented nuclei into normal or cancerous nuclei. This study showed that the accuracy of detecting cancerous regions in whole-slide biopsy images is improved with the introduction of cytological features related to nuclear classification. More details about the features used for cancer detection and the reported accuracy of this study are described in Section 9.3.2.

B. Gland segmentation

Segmenting glands from a microscopic image is a challenging problem, even manually, because of the variety and irregularity in size and shape, and their poor differentiation in high-grade cancerous images.[115] Various segmentation methods have been proposed to achieve accurate gland segmentation. Naik et al.[116] introduced a gland segmentation method based on the fact that each gland has three main components arranged in a particular fashion: the lumen area is surrounded by epithelial cell cytoplasm, with a ring of nuclei defining the outer boundary of the gland region. In order to detect pixels corresponding to the structures of interest, a Bayesian classifier is trained on the image pixel values. After lumen candidate areas are determined, gland size constraints are incorporated as well as structure-based constraints (i.e., a lumen region should be immediately surrounded by cytoplasm, and cytoplasm is bordered by a ring of nuclei). Next, gland boundary segmentation is performed using level sets defined mathematically as follows:[117]

$$\frac{\partial \varphi}{\partial t} = F|\nabla \varphi| = 0, \tag{9.1}$$

where the function F defines the speed of evolution. The curve evolution is controlled by the nuclei likelihood image and the initial contour is initialized using the segmented lumen areas. The curve is evolved outward until the difference between the boundaries of two consecutive iterations is less than a predetermined value. Once the gland contour has been delineated, additional gland size constraints are applied to discard regions that are too large to be considered glands. An example of the gland segmentation results of the algorithm is shown in Fig. 9.7. In this figure, blue contours correspond to

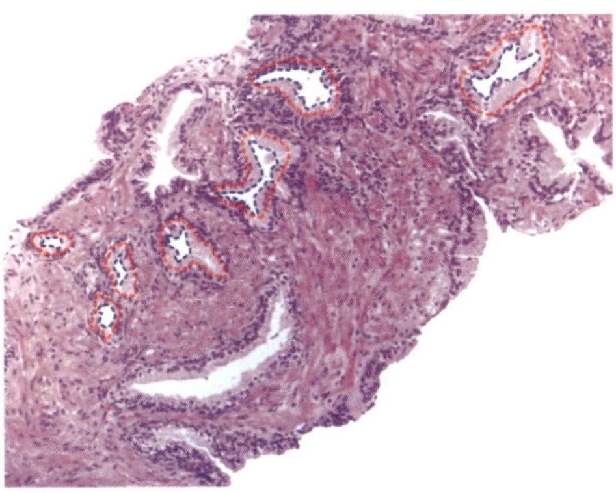

Figure 9.7 Gland segmentation results in a prostate tissue image by using level sets.

lumen regions, and red boundaries define the segmented prostate glands. As can be observed from the figure, the segmented glands consist solely of lumen and internal cytoplasm regions, and in most cases the nucleus boundaries are not included,[118] enlarged true glands are removed probably due to the size constraints applied during the lumen candidates selection and final adjustments.

A high-throughput computer-aided system was developed by Xu et al.[119] The system comprises three modules: a hierarchical-frequency-weighted mean shift normalized cut (HNCut) for initial detection of glands, a geodesic active contour (GAC) model for gland segmentation; and a diffeomorphic-based similarity (DBS) feature extraction for classification of glands as benign or cancerous. The proposed approach can rapidly detect, segment, and classify prostatic glands. However, in several presented cases, the boundaries of glands do not include cytoplasm and nuclei structures as part of the gland. Figure 9.8 shows a representative example of the gland segmentation results.

Another method for complete gland segmentation (including nucleus boundaries) was presented by Nguyen et al.[118] They used a color space transformation from RGB to CIELAB in order to label tissue structures (lumen, blue mucin, cytoplasm, nuclei, and stroma) according to their color composition. Once the pixels have been classified, a two-stage algorithm is used for gland boundary generation. In the first step, nucleus objects are enlarged by combining them with cytoplasm pixels. In the second step, enlarged nuclei are grouped to form gland boundary segments. In order to reconstruct a complete gland, previously labeled lumen structures are unified with boundary segments through a lumen boundary expansion procedure. The lumen expansion algorithm runs until the gland has the estimated gland size or the stroma pixels are reached.

Mosquera Lopez et al.[115] developed another approach for prostate gland reconstruction. In order to segment gland units, the components of glands are

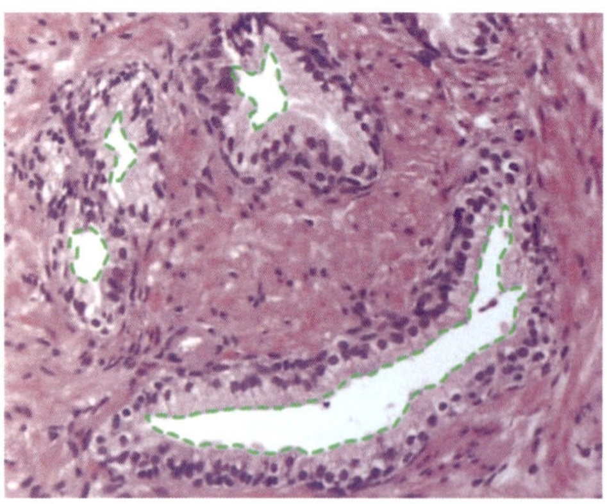

Figure 9.8 Gland segmentation results in a prostate tissue image using HNCut and GAC.

Figure 9.9 Gland segmentation process: (a) unifying lumen and cytoplasm pixels, (b) segmented gland units, and (c) inner boundaries of segmented gland units.

identified by decomposing the histopathology image using color region mapping. After applying lumen size constraints to avoid the presence of noise, selected lumen structures are unified with epithelial cytoplasm pixels to conform to the interior part of each gland. Next, morphological operations are performed on epithelial nuclei and stroma pixels to define the inner boundary of the gland and to separate glands using stroma pixels. The described procedure is illustrated in Fig. 9.9.

9.3.2 Feature extraction

The type of features extracted for image representation defines the classification of the CAD system for prostate cancer detection and grading. As mentioned previously, research on CAD systems for prostate cancer diagnosis can be categorized into tissue structure-based and textural-based systems depending on the analysis methods used to get information from the histology images. Some of the features used in prostate cancer and other types of cancer classification are inspired in the way pathologists evaluate the tissue under the microscope. In this section a summary of the feature vectors used in the major studies on prostate cancer recognition is presented along with the reported accuracy of the systems utilizing those features.

Jafari-Khouzani and Soltanian[43] used a multi-wavelet transform to represent cancerous images. The energy and entropy of multi-wavelet coefficients of each resulting sub-matrix were computed along with textural features extracted from a co-occurrence matrix to classify a data set of 100 images into Gleason grades 2–5. The maximum reported accuracy was 97%. Huang et al.[45] developed a system that analyzes the texture complexity of histological images using two fractal measurements: (1) fractal dimension calculated by using a differential box counting method, and (2) entropy-based fractal dimension. The accuracy of the system using various classifiers (i.e., Bayesian classifier, k-NN and SVM), that is around 95%, was estimated using cross-validation methods. Khurd et al.[47] developed a texture classification system to differentiate between Gleason grade

3 and 4 cancerous images extracting texton features (i.e., basic texture elements), which are clustered using random forests or k-means algorithms. The classification results showed a maximum net accuracy of 94% when k-means is the clustering algorithm and SVM with a Radial Basis Function (RBF) kernel is employed as classifier. Yoon et al.[48] developed a computer-aided classification system where textural features were extracted from a cardinal multiridgelet transform (CMRT) to differentiate images of Gleason 3 from Gleason 4. SVM with a Gaussian kernel was used for the classification task, and the accuracy of the system is 93.75% using the leave-one-out cross-validation method. Almuntashri et al.[49] presented a method for automatic classification of prostate cancer biopsy image by combining features from wavelet transform and wavelet-based fractals analysis in order to capture textural complexity of cancerous images. Experimental results showed average classification accuracy (with one-vs.-all SVM classifiers) of 95% in a set of 45 images of Gleason grades 3, 4, and 5. Mosquera-Lopez and Agaian[50] presented a system for classification of prostatic carcinomas of grades 3–5 by using statistics of the distribution of wavelet energy within cancerous patches to join the probability of wavelet coefficients obtained by wavelet decomposition of the channels of color images and measurements of color fractal dimension. The average accuracy of the system was estimated by cross-validation yielding a correct classification rate of 97%.

Although most of the research in prostate cancer CAD systems has been done on cancer grading of selected regions of interest, there are several textural-based systems whose main objective is cancer detection in whole slides. For example, Doyle et al.[42] developed a boosted Bayesian multi-resolution system for prostate cancer detection using textural features including first-order statistics, co-occurrence features and Gabor filter response in a pixel basis taken a small neighborhood around the pixel under study. They reported classification accuracies of 69%, 70%, and 68% for low-, medium-, and high-resolution, respectively.[57] Color fractals[120] have recently been used along with probabilistic pairwise Markov models to distinguish cancerous regions from benign tissue, yielding an area under the ROC curve of 0.821.[41]

Other classification approaches consist of segmenting tissue structures and extracting features based on individual properties of each one of them. Mainly, nuclear and glandular features have been considered important characteristics in the detection of prostate cancer and its severity. Tissue-structure-based systems exploit the correlation between the size, shape and arrangement of histological structures within the histology image with Gleason grades. For example, Doyle et al.[37] used centroids of segmented nuclear structures to create Voronoi, Delaunay, and minimum spanning tree graphs in order to capture the spatial arrangement of nuclei in pathological images represented by the area and edge length features as well as the nuclear density. They classified H&E stained images into cancer, non-cancer and cancer confounders groups using statistics from the constructed graphs. Wetzel et al.[35] constructed a spanning tree graph to connect segmented cell nuclei over a tumor image and quantify the architectural arrangement of cells. The proposed system was able to correctly match the grade value of pathological image at 80%.

Nguyen, Jain, and Allen[38] used glandular features to classify histology image in benign, grade 3, and grade 4. Features include statistics related to size and shape of luminal and glandular areas as well as nuclei density and gland area covered by blue mucin. The best accuracies achieved by the system are 97.75%, 94%, 87.3%, and 98.58% at differentiating benign vs. grade 3 tissues, benign vs. grade 3 tissues, grade 3 vs. grade 4, and benign vs. carcinoma, respectively. Mosquera-Lopez et al.[115] developed a system to classify cancerous tissues belonging to Gleason grades 3–5 by using a combination of morphology and architectural features of the tissue. The extracted features include 2D color histograms in the HSV color space, size, and shape of glandular structures, nuclei density, and nuclear arrangement from graph characteristics. The average accuracy of the system is 97.63%, 96.57%, and 87.30% when distinguishing Gleason 3 vs. Gleason 4, Gleason 3 vs. Gleason 5, and Gleason 4 vs. Gleason 5, respectively.

Several combinations of feature vectors coming from morphometric and textural image analysis are found in the literature. Roula et al.[68] investigated the accuracy of utilizing Haralick features,[121] gland area and nuclear area extracted from a multi-spectral microscopy image to separate stroma, benign prostatic hyperplasia (BPH), prostatic intraepithelial neoplasia (PIN), and prostatic carcinoma obtaining an average classification error of 5.57%. Diamond et al.[122] used morphometric and Haralick texture features to identify stroma, normal, and cancerous regions in samples of prostatic tissue from whole-mount radical prostatectomy. Classification of noncancerous regions was performed using morphometric characteristics of the histology image under the assumption that normal tissue exhibit larger areas of associated lumen. On the other hand, classification of stroma and cancerous tissue was done by looking at Haralick features. In this study, 79.3% of the sub-regions of interest were correctly classified. Tabesh et al.[123] integrated object- and image-level features describing the color, texture, and morphometric characteristics of histopathology images. Object-level features include statistics of the intensity of the segmented tissue structures, whereas image-level features consider color channel histograms, fractal measurements, and statistics of wavelet coefficients. The developed system achieves an accuracy of 96.7% classifying tumor and non-tumor images and a maximum accuracy of 81% in Gleason grading classification tasks. In addition, a system for cancer detection on whole slides was presented by Nguyen et al.[57] in which cytological features from segmented cancer nuclei are combined with textural features derived from first-order statistics, second-order statistics, and Gabor filter response. The accuracy of the system is reported by using TPR = 78% and FPR = 6%. Table 9.2 further details the most used features in prostate cancer CAD systems.

Table 9.2 Features extracted from digitized images of prostate needle biopsy.

Feature Categories	Features Description	CAD System
Texture features	**Haralick features from co-occurrence matrix**[121] Co-occurrence matrix can be defined as follows:[68] $$Co(i, j, d, \theta) = \alpha$$ α couples of pixels having i and j, respectively, as grey levels and separated by a distance d in a direction angle θ.	[57], [42], [68], [43], [122], [44]
	Co-ocurrence matrix features: Inertia, uniformity or energy, variance, entropy, joint entropy, maximum probability, contrast, correlation, measurements of correlation, sum average, sum variance, sum entropy, difference average, difference variance, difference entropy, inverse difference moment, cluster shade, cluster.	
	Gray level run-length matrix[124,125] $$GLRLM(i, j \mid \theta) = \beta$$ β is the total number of occurrences of runs of length j at gray level i in a given direction θ.	[40], [126]
	Gray level run-length matrix features: Short run emphasis, long run emphasis, gray level nonuniformity, run-length nonuniformity, run percentage.	
	First-order statistics of pixel values: Average, median, variance, standard deviation, range, maximum value, minimum value.	[57], [42], [68],
	Texture moments: Texton histograms.	[47]
	Wavelet features: Energy and entropy of wavelet detail coefficients, statistics of wavelet coefficients (mean, variance, standard deviation), joint probability of wavelet coefficients obtained for each image channel, low resolution images.	[43], [48], [49], [50], [123]
	Fractal features: Fractal dimension, color fractal dimension, fractal code, entropy-based fractal dimension, wavelet-based fractal dimension.	[41], [45], [49], [50], [46], [123]
	Filter responses: Gabor features, Sobel, Kirsch, gradient, derivative.	[42]
Lumen features	Statistics of lumen area, perimeter, circularity, elliptic fit (major axis length, minor axis length, orientation, eccentricity), ratio of average to maximum of lumen area and lumen perimeter.	[35], [38], [36], [34], [123], [115]
Glandular features	Statistics of glandular area, perimeter, circularity/roundness, spatial density of gland distribution, gland area covered by blue mucin, statistics of the distances from the lumen center to the nuclei boundary (gland radius), circularity, elliptic fit (major axis length, minor axis length, orientation, eccentricity).	[35], [68], [37], [38], [36], [123] [122], [115]
Nucleus features	Nucleus density, cancerous nucleus density, percentage of nucleus area in glandular area, area, circularity.	[57], [68], [38], [123], [122], [115]
Color features	Color channel histograms, 2D color histograms, color channel differences, channels ratio, entropy of color histograms.	[115], [123]
Graph features[1]	**Voronoi tessellation:** number of nodes, number of edges, cyclomatic number, number of triangles, number of k-walks, spectral radius, eigenexponent, Randic index, roundness factor, area, area disorder, roundness factor homogeneity, network cycle features (non-triangular cycles, average network cycle length, maximum network cycle length).	[37], [39]
	Delaunay triangulation: number of nodes, number of edges, number of triangles, statistics of edge length and triangles' area, degree, cyclomatic number, number of k-walks, spectral radius, eigenexponent, Wiener index, Randic index, eccentricity, fractal dimension, network cycle features (non-triangular cycles, average network cycle length, maximum network cycle length).	[35], [37], [39], [115]
	Minimum spanning tree: number of nodes, edge length, degree, number of neighbors, Wiener index, eccentricity, Randic index, Balaban index, fractal dimension, network cycle features (non-triangular cycles, average network cycle length, maximum network cycle length).	[35], [37], [39]

9.3.3 Classification

In general, detection and grading of prostate cancer using automated systems is done by supervised pattern recognition. In the context of pattern recognition, a pattern is a vector of features describing the properties of an object or class. Supervised classification approaches require the use of annotated samples (features with their respective class label) to train the classifier to determine the decision boundaries (classes separation) in a given feature space.[127] Before classification, features selection and feature space dimensionality reduction is a key step. The goal of features selection is to choose the most discriminative features (features that contain more information about patterns) to improve the class separation and consequently the classification performance. On the other hand, dimensionality reduction techniques, as indicated by its name, aim to reduce the dimension of the feature space in general by mapping the feature vector onto a lower-dimensional space through some coordinates transformation. Linear methods for dimensionality reduction include principal component analysis (PCA), linear discriminant analysis (LDA),[128] and independent component analysis (ICA).[129] In contrast, an example of nonlinear dimensionality reduction is manifold learning. This technique was employed in a system for prostate cancer grading by Sparks and Madabhushi.[130] The main goal of using manifold regularization is to reduce the dimensionality of the feature space, but preserving a nonlinear relationship between object instances.

Once an appropriate feature set is selected, several classification methods can be used for prostate cancer diagnosis. Such classification methods include *k* nearest neighbors (*k*-NN), Bayesian classifier, Support Vector Machine (SVM), neural networks, Markov random field (MRF) classifier, Gaussian classifier, and classical linear discrimination (CLD), among others. Each classification approach has its own advantages and disadvantages. Several studies[38,45,123] have published comparisons among the various procedures in order to demonstrate which one is better for prostate cancer diagnosis under specific circumstances. A large comparative study on machine learning techniques for prostate cancer diagnosis was conducted by Alexandratou et al.[131] In their work, 16 supervised machine learning algorithms were compared based on their performance. Classification problems regarding cancer detection (tumor vs. non-tumor), low- vs. high-grade recognition; and the multiclass problem Gleason grading were addressed. Thirteen Haralick texture characteristics were calculated based on grey level co-occurrence matrix of microscopic prostate tissue. For the best-performing algorithm in each case the accuracy obtained was 97.9% for cancer detection, 80.8% for low–high grade discrimination, and 77.8% for accomplishing both detection and Gleason grading. Logistic regression and sequential minimal optimization for training a support vector machine were among the top scoring algorithms in each classification problem.

Cascade systems have been also proposed for automatic prostate cancer diagnosis. For instance, Doyle et al.[132] presented a cascaded multi-class pairwise classifier to grade regions of interest of prostate tissue biopsies. The proposed classifier incorporates domain knowledge to partition the multi-class problem into several binary-class tasks, reducing the intra-class heterogeneity that causes errors in one-versus-all multi-class approaches. In their cascaded approach, successive classifications are performed, beginning with the most broad (i.e., cancer detection) and proceeding to increasingly granular separations (pattern 3+4 vs. pattern 5 and epithelium + atrophy vs. stroma, dark gray line), and finally classifying the most similar classes within each group (pattern 3 vs. pattern 4 and epithelium vs. stroma, light gray line). This reduces classification error by ensuring that the separations are performed between dissimilar classes.

Another classification approach that has been used in prostate cancer detection is adaptive boosting or AdaBoost. This algorithm combines various weak classifiers to generate a strong classifier, which in most cases has a better performance than individual classifiers. A boosted multi-resolution classifier was presented by Doyle et al.[42] in which Bayesian weak classifiers (one per each extracted feature at three different magnifications) to divide images' pixels into cancerous and noncancerous groups. The classification procedure starts at the lower resolution and only detected cancerous areas are evaluated in the next magnification level. This approach is based on a combination of boosted and cascade classifiers.

Because several classification methods might be used in a complementary way, exploring cascade classification systems and other multiclassifier combinations may be beneficial for computer-aided prostate cancer diagnosis (detection and grading). In such a case, different decision rules can be used and integrated in a system in order to improve the overall labeling because some of classifiers are better at resolving one aspect of the labeling problem, whereas another method may be superior in a different respect.[127]

9.3.4 System accuracy assessment

Evaluating classification performance is important for several reasons: (1) when building classifiers, the parameters used for classification can be tuned. For example, at this point, several tests should be done in order to choose predictor variables or features, to estimate parameters, to explore data transformations, and so on. (2) When evaluating given classifiers, it can be determined whether they are good enough for the purpose or whether they provide sufficient improvement over an existing method to merit switching.[133] One of the most used methods for estimating classification performance is cross-validation. Cross-validation can be done by applying three different schemes, namely *k*-fold, hold-out, and leave-one-out. The basic form of cross-validation is *k*-fold, and the other schemes are special cases derived from k-fold. Implementation of cross-validation methods is done as follows:[134]

- **k-fold validation:** To implement this method, the data is randomly divided into k equal (or nearly equal) folds. Next, k iterations of training and test are carried out, such that at each iteration a different segment is held out for validation and k–1 folds are used to fit the model.
- **Hold-out validation:** For this scheme, the dataset is split into two non-overlapped segments: one for training and the other for testing. Hold-out avoids some samples to be used for both learning and validation, yielding a better estimation for the generalization performance of the algorithm used for recognition.
- **Leave-one-out validation:** This is a special case of k-fold cross-validation, where k is the number of data. In this scheme only one sample is held out for testing. The results of leave-one-out cross-validation are considered to be almost unbiased, but they have large variances.

In order to obtain reliable performance estimation of a given classifier, it is recommended to have a large number of iterations.

9.3.4.1 Performance indicators

- **Correct classification rate** is the ratio of correct classified cases to the total number of classified cases[135]

$$CCR = \frac{TP + TN}{total\ samples},$$

  where TP and TN are true positive and true negative samples, respectively.
- **Sensitivity** is the proportion of cases belonging to a given class, which are correctly classified as cases of that class. That is, the sensitivity is the true positive rate (TPR)[136]

$$TPR = \frac{TP}{TP + FN},$$

  where FN represents the false negative samples.
- **Specificity** is the proportion of non-cases of a specific class, which are correctly classified as non-cases of that class. That is, the specificity is the true negative rate (TNR)[136]

$$TNR = \frac{TN}{TN + FP},$$

  where FN represents the false positive samples.
- **Positive predictive value (PPV)** is the proportion of those predicted to belong to a class, which really belong to that class[136]

$$PPV = \frac{TP}{TP + FP}.$$

- **Negative predictive value (NPV)** is the proportion of those predicted to be non-cases of a given class, which really do not belong to that class[136]

$$NPV = \frac{TN}{TN + FN}.$$

If sensitivity and specificity (which are both conditional probabilities) are combined, weighted by the marginal probabilities of the classes, the probability of correct classification is obtained.

Another accuracy indicator often used in prostate cancer detection is the area under the ROC curve. A ROC curve is defined as a plot of the false positive rate (FPR), on the vertical axis, against the TPR, on the horizontal axis. A good classification rule is reflected by a ROC curve which lies in the upper left triangle of the square.[137] The area under the ROC curve (AUC) measures classifier discrimination capability; that is, the ability of the classifier to correctly separate classes. An AUC of 1 represents a perfect classifier, and an AUC of 0.5 represents a worthless test.

Finally, with all classification tasks described using some examples, a summary of the developed systems for computer-assisted prostate cancer detection and grading is presented chronologically (sorted by publication year) in Tables 9.3 and 9.4, respectively. The main characteristics of each system are described briefly. The tables include the systems that have been described hereinbefore and other systems found in the literature.

Table 9.3 Computer-aided systems for prostate cancer detection.

Author(s)	Features	Dataset	Classification method	Accuracy	System category
Farjam et al. (2007)[33]	Variance of size of glandular objects, roundness of glands, and cancer index	Two data sets containing 91 and 199 pathological samples at different magnifications and illumination conditions	*k*-NN	95–98%	Tissue-structure-based system
Sun et al. (2009)[40]	Run-length matrix features: low gray-level run emphasis, high gray-level run emphasis, run percentage, mean and standard deviation	Nine subimages of a tissue sample from a radical prostatectomy at 50× magnification	Multilayer perceptron (MLP)	89.5%	Texture-based system
Monaco et al. (2009)[34]	Gland area	20 prostate histological sections	Bayesian and Markov random field (MRF) classifier	AUC: 0.87	Tissue-structure-based system
Yu et al. (2011)[41]	Color fractal dimension and probabilistic pairwise Markov Model (PPMM)	27 radical prostatectomy specimens digitized at 40× magnification	Markov random field (MRF) classifier	AUC: 0.831	Texture-based system

(*Continued*)

Table 9.3 (Continued).

Author(s)	Features	Dataset	Classification method	Accuracy	System category
Bouatmane et al. (2011)[138]	Haralick features, glandular area, and nuclear area	592 textures multi-spectral images of size 128 × 128 examined at 40× magnification	*k*-NN one vs. all binary classifiers plus round-robin (RR) sequential forward-feature selection	99.83%	Multifeature system
Nguyen et al. (2011)[57]	Combination of cytological features and texture features including first-order statistics, second-order statistics and Gabor features	17 whole-slide images digitized at 20× magnification. 6 images were used for training, and 11 for testing	SVM with RBF kernel	TPR: 78% FPR: 6%	Multifeature system
Doyle et al. (2012)[42]	First-order statistics, filter response (Sobel, Kirsch, gradient, derivative), co-occurrence features, and Gabor features	100 whole-slide images at 40× optical magnification	Boosted Bayesian classifier	AUC: 0.84, 0.83, and 0.76 for the lowest, medium, and highest image resolution, respectively	Texture-based system

Table 9.4 Computer-aided systems for prostate cancer grading.

Author(s)	Features	Dataset	Classification method	Accuracy	System category
Wetzel et al. (1999)[35]	Glandular features and nuclear-based spanning tree and Delaunay features	54 prostate cases digitized at 10× magnification	Content-based image retrieval (CBIR) system	80%	Tissue-structure-based system
Roula et al. (2002)[68]	Haralick features, global variance of pixels, glandular area, and nuclear area.	10 multispectral images of whole mount sections from radical prostatectomy	Supervised Classical Linear Discrimination (CLD)	94%	Multifeature system
Jafari-Khouzani et al. (2003)[43]	Energy and entropy features calculated from multiwavelet, and co-occurrence matrix features	100 images at magnification 100× belonging to Gleason patterns 2–5	*k*-NN	97%	Texture-based system
Diamond et al. (2004)[122]	Glandular and nuclear area as well as Haralick features	Images of size 100 × 100 digitized at 40× magnification	Not specified	79.3%	Multifeature system
Tabesh et al. (2007)[123]	Color channel histograms, fractal dimension, fractal code, wavelets along with color, texture, and morphometric properties of the histological objects from the MAGIC[139] system	367 images digitized at 20× magnification for cancer detection, and 268 images of Gleason grades 2–5 for cancer grading	Linear and quadratic Gaussian, *k*-NN, and SVM	96.7% 81%	Multifeature system

(*Continued*)

Table 9.4 (Continued).

Author(s)	Features	Dataset	Classification method	Accuracy	System category
Wittke et al. (2007)[36]	Area fraction, specific line length, specific Euler number, and the quotient of specific line length to area fraction of epithelium, lumina, and epithelium plus lumina	78 grayscale digital images of prostatic adenocarcinoma	Graph-based classification	Training set: 92.31% Test set: 64.10%	Tissue-structure-based system
Naik et al. (2008)[37]	Glandular shape and size features as well as Voronoi, Delaunay, and minimum spanning tree graph features	44 subimages of benign tissue (17), grade 3 carcinoma (16), and grade 4 carcinoma (11)	SVM	91.48%	Tissue-structure-based system
Alexandratou et al. (2008)[44]	Haralick features	50 samples of histopathological data belonging to Gleason grades 2–5	Multiparameter statistical method of multiple logistic discrimination analysis	87%	Texture-based system
Huang et al. (2009)[45]	Differential box counting fractal dimension and entropy-based fractal dimension	205 images	Bayesian, k-NN, and SVM	94.6%	Texture-based system
Nguyen et al. (2010)[38]	Statistics of size and shape of lumen, glands, as well as nuclei density and blue mucin area	78 subimages of size 501 × 52 of benign tissue (30), grade-3 carcinoma (28), and grade-4 carcinoma (20)	SVM, multilayer Perceptron, and k-NN	88.4%	Tissue-structure-based system
Tai et al. (2010)[46]	Wavelet-based fractal dimension. Classical fractal dimension and entropy-based fractal dimension are computed from each wavelet subband	1000 pathological images	SVM	86.3%	Texture-based system
Khurd et al. (2010)[47]	Basic texture elements	75 images at magnification 10× of size 1392 × 1040 of grade-3 carcinoma (25) and grade-4 carcinoma (50)	Random forest with SVM	94%	Texture-based system
Yoon et al. (2011)[48]	Variance and entropy of cardinal multiridgelet transform (CMRT) coefficients	42 images of size 768 × 768 of grade-3 carcinoma (16) and grade-4 carcinoma (26)	Gaussian kernel SVM	93.75%	Texture-based system
Almuntashri et al. (2011)[49]	Haar wavelet energy and wavelet-based fractal dimension	45 images of size 512 × 512 of Gleason grades 3 (15), 4 (15), and 5 (15)	SVM	95%	Texture-based system

(Continued)

Table 9.4 (Continued).

Author(s)	Features	Dataset	Classification method	Accuracy	System category
Khurd et al. (2011)[39]	Network cycle features and graph features	25 images of Gleason grade 3 and 50 images of Gleason grade 4. Images are size 1392 × 1040 and were acquired at 10 × magnification	SVM	AUC: 0.995	Tissue-structure-based system
Mosquera-Lopez et al. (2012)[110]	HSV color features, glandular and nuclear features, and architectural features from Delaunay triangulation	71 images of size 512 × 512 of Gleason grades 3 (30), 4 (30), and 5 (11)	SVM linear kernel	95%	Multifeature system
Mosquera-Lopez et al. (2013)[50]	Wavelet energy distribution, joint probability of wavelet coefficients, and color ratio-based fractal dimension	71 images of size 512 × 512 of Gleason grades 3 (30), 4 (30), and 5 (11)	SVM linear kernel	97%	Texture-based system

9.4 Conclusions, Future Directions, and Potential New Strategies

Computer-aided prostate cancer diagnosis has become an exciting area of research in the past last years. Several efforts have been made in the development of accurate systems for automated detection and grading of prostatic disease from digitized biopsy images. Most of the systems use textural features, spatial arrangement characteristics of tissue structures, and morphometric properties of glands to detect prostatic carcinomas and to grade them using the Gleason grading system. The features used to characterize Gleason patterns are inspired in how pathologists assess tissue samples. However, doctors frequently use a complex set of features that are often difficult to formulate in computational forms. The capabilities of the existing systems can be easily extended to properly assign Gleason scores to each studied slide, and to produce localized cancer maps that show the distribution of all Gleason grades found within each histopathology image.

Researchers in automated prostate-cancer diagnosis still face several challenges. For instance, the large amount of data in pathology requires fast systems for image analysis that can assist pathologists in real clinical applications such as real-time cancer detection and surgery quality control. On the other hand, prostate cancer diagnosis based on 3D tissue evaluation is an investigation field almost unexplored until now. 3D systems will allow pathologists to have more information about prostatic tissue structures as well as the volumetric characteristic of the detected carcinomas. A migration from electronic files of 2D CAD systems to 3D CAD systems may create new opportunities for the future prostate cancer diagnosis and prediction systems.

Moreover, a system evaluation framework should be created in order to assess the accuracy and other statistical parameters of the developed recognition systems. Today, as indicated by Gurcan et al.,[1] it is so difficult to compare computer-assisted prostate cancer diagnosis systems in a systematic and objective manner, because each system is built and tested under different conditions (e.g., different datasets and ground truth annotations). In addition, the reported performance is done using different metrics. In order to compare several CAD systems for prostate cancer diagnosis or for another purpose, it is not enough to have a better indicator measured on a test set, or in a cross-validation or other comparison based on sample data. It is necessary to carry out statistical tests, so that we can be confident that any differences represent genuine underlying differences in performance, and are not mere random sampling effects.[133] One resource that might help in the generation of the mentioned evaluation framework is a standardized annotated database that contains histopathology images of different prostate cancer cases. This dataset will help researchers not only to train their CAD systems, but also will allow comparison of performance among developed systems.

Once the performance of the CAD system can be validated, the quantitative information obtained from the analysis of histopathology images can be integrated into prostate cancer risk assessment methodologies[140] that result in risk stratification using predictive models. This field of study offers the possibility of combining information from prostate specific antigen (PSA) screening, prostate cancer family history, and biopsy pathology reports to analyze a patient's cancer risk using learning algorithms. A combination of several predictors, including those used in risk analysis, might be also used to predict prostate cancer in young men, clinically significant cancers in all men, and aggressive cancers in older men while limiting unnecessary biopsies and overdetection of certain cancers. New predictive models should attempt to diagnose only those cancers that will affect a patient's life using all of the patient's available information.

Another problem related to prostate cancer diagnosis and prediction involves studying the relationship between human prostate and breast cancer. A study of evolution indicates that the prostate and breast appeared at the same time 65 million years ago with the development of mammals[17] and some other studies have shown a high correlation (correlation coefficient 0.81) between the incidence and age-adjusted rates for prostate cancer and breast cancer. Other types of cancers such us endometrium cancer and ovary cancer are also correlated with prostate cancer. This may implicate the study of estrogenic factors, because the breast, endometrium, and ovary are estrogen-responsive tissues, and estrogen exposure has been shown to affect carcinogenesis.[141] These factors can also be included into intelligent CAD systems to refine their diagnosis and prediction capabilities.

On the other hand, progress in computer networking and digital pathology allows for the expansion of several medical services offered using communication networks such as the Internet. Complete information of patients and their clinical history may be available everywhere; thus, new services may be developed, such as online facilities for second opinions and consultation about cancer prediction and risk assessment based on available and shared patient information. For instance, telepathology applications will link smaller centers with expert consultants in order to better diagnose cancer cases. The use of communication networks brings new problems. First, it will be necessary to develop secure network-based CAD systems that combine information of patients from several sources to analyze prostate cancer cases. Second, developed systems should be suitable for deployment as a cloud service. Third, there is a challenge in constructing intelligent interfaces to integrate all available resources. The designed interfaces should satisfy several interrelated criteria: They should be integrated, expressive, secure, capable of sharing information quickly, client-oriented, cooperative, easy to use, easy to access, and customizable. Finally, it would be important to design strategies for secure retrieval, delivery, and storage of complex, digital prostate-imaging data.

Because there are many different devices with networking capabilities and acceptable computation resources (less expensive devices) these days, such as smartphones, tablets, and portable computers, the design of CAD systems should be optimized for all of these devices. A key limitation of existing CAD systems is that most of them require workstations or computers due to the computational complexity of their analysis algorithms. This fact makes it expensive to use those systems and creates the need to develop prostate-cancer diagnosis and prediction systems that are compatible with less-expensive devices.

Lastly, it is important to point out that all developments in computer-aided prostate cancer diagnosis should be done in constant collaboration between computer vision investigators and clinical and research pathologists to ensure that the resulting work has the potential to be applied to real medical problems. It is a pathologist who can provide objective feedback about the performance of the developed systems and give insights to improve the CAD results.[1] Opinions from oncologists and radiologists during the research process are also important because they are potential users of the systems. All of the tools discussed in this review may be expensive to implement, but a well-planned deployment will decrease cost and improve the quality of medical care.

References

1. M. N. Gurcan, L. E. Boucheron, A. Can, A. Madabhushi, N. Rajpoot, and B. Yener, "Histopathological image analysis: a review," *IEEE Rev. Biomed. Eng.* **2**, 147–171 (2009).
2. R. A. Castellino, "Computer aided detection (CAD): an overview," *Cancer Imaging* **5**(1), 17–19 (2005).
3. K. Sikora, "Developing a global strategy for cancer," *Eur. J. Cancer* **35**(14), 1870–1877 (1999).
4. S. Haque, D. Mital, and S. Srinivasan, "Advances in biomedical informatics for the management of cancer," *Annals of the New York Academy of Sciences - Techniques in Bioinformatics and Medical Informatics* **980**, 287–297 (2002).
5. U. S. Food and Drug Administration, "Summary of Safety and Effectiveness Data: R2 Technologies (P970058)" (1998).
6. T. W. Freer and M. J. Ulissey, "Screening Mammography with Computer-aided Detection: Prospective Study of 12,860 Patients in a Community Breast Center Radiology," *Radiology* **220**(3), 781–786 (2001).
7. D. Gur, J. H. Sumkin, H. E. Rockette, M. Ganott, C. Hakim, L. Hardesty, W. R. Poller, S. R. Wallace, and L. Wallace, "Changes in breast cancer detection and mammography recall rates after the introduction of a computer-aided detection system," *J. Natl. Cancer Inst.* **96**(3), 185–190 (2004).
8. R. L. Birdwell, P. Bandodkar, and D. M. Ikeda, "Computer-aided detection with screening mammography in a university hospital setting," *Radiology* **236**(2), 451–457 (2005).
9. T. Cupples, J. E. Cunningham, and J. C. Reynolds, "Impact of computer-aided detection in a regional screening mammography program," *AJR Am. J. Roentgenol* **185**(4), 944–950 (2005).
10. S. V. Destounis, P. DiNitto, W. Logan-Young, E. Bonaccio, M. L. Zuley, and K. M. Willison, "Can computer-aided detection with double reading of screening mammograms help decrease the false-negative rate? Initial experience," *Radiology* **232**(2), 578–584 (2004).
11. M. Morton, D. H. B. K. R. Whaley, and K. K. Amrami, "Screening mammograms: interpretation with computer-aided detection–prospective evaluation," *Radiology* **239**(2), 375–383 (2006).
12. J. C. Dean and C. C. Ilvento, "Improved cancer detection using computer-aided detection with diagnostic and screening mammography: prospective study of 104 cancers," *AJR Am. J. Roentgenol.* **187**(1), 20–28 (2006).

13. I. Sanchez and S. Agaian, "A new system of computer-aided diagnosis of skin lesions," *Proc. SPIE* **8295**, 82951A (2012) [doi: 10.1117/12.906796].
14. I. Sanchez and S. Agaian, "Computer aided diagnosis of lesions extracted from large skin surfaces," *Proc. IEEE Systems, Man, and Cybernetics (SMC)*, Seoul (2012).
15. M. Madhukar, S. Agaian, and A. T. Chronopoulos, "New decision support tool for acute lymphoblastic leukemia classification," *Proc. SPIE* **8295**, 829518 (2012) [doi: 10.1117/12.905969].
16. American Cancer Society, "What are the key statistics about prostate cancer?," www.cancer.org/cancer/prostatecancer/detailedguide/prostate-cancer-key-statistics (17 January 2013).
17. D. Tindall and P. J. Scardino, "State of research for prostate cancer: Excerpt from the report of the Prostate Cancer Progress Review Group," *J. Urology* **57**(4), 28–30 (2001).
18. O. W. Brawley, D. P. Ankerst, and I. M. Thompson, "Screening for prostate cancer," *CA Cancer J. Clin.* **59**(4), 264–273 (2009).
19. D. J. Luthinger and M. Gross, "Gleason grade migration: Changes in prostate cancer grade in the contemporary era," *PCRI Insights* **9**, 2–3 (2006).
20. G. T. Mellinger, D. Gleason, and J. Bailar, "The histology and prognosis of prostatic cancer," *J. Urology* **97**(2), 331–337 (1967).
21. D. Gleason and G. T. Mellinger, "Prediction of prognosis for prostatic adenocarcinoma by combined histological grading and clinical staging," *J. Urology* **111**(1), 58–64 (1974).
22. R. B. Shah, "Current perspectives on the Gleason grading of prostate cancer," *Arch. Pathol. Lab. Med.* **133**, 1810–1816 (2009).
23. J. I. Epstein, "An update of the Gleason grading system," *J. Urology* **183**(2), 433–440 (2010).
24. J. I. Epstein, "Gleason score 2-4 adenocarcinoma of the prostate on needle biopsy: a diagnosis that should not be made," *Am. J. Surg. Pathol.* **24**, 477–478 (2000).
25. D. M. Berney, "Low Gleason score prostatic adenocarcinomas are no longer viable entities," *Histopathology* **50**(6), 683–690 (2007).
26. PathologyOutlines.com, Inc., "Prostate carcinoma grading (Gleason)," www.pathologyoutlines.com/topic/prostategrading.html (7 August 2012).
27. A. Velez-Hoyos, A. Mendoza-Luna, and J. F. Uribe-Arcila, "Reporte actual de las biopsias: consenso de la Sociedad Internacional de Patología - ISPU," *Urología Colombiana* **16**, 99–106 (2007).
28. D. E. Baydar and J. Epstein, "Gleason grading system, modifications and additions to the original scheme," *Turkish J. Pathol.* **25**(3), 59–70 (2009).

29. P. A. Humphrey, "Gleason grading and prognostic factors in carcinoma of the prostate," *Modern Pathol.* **17**, 292–306 (2004).

30. J. Epstein, C. Pound, and P. A. W. y P. C. Walsh, "Disease progression following radical prostatectomy in men with Gleason score 7 tumor," *J. Urology* **160**(1), 97–100 (1998).

31. B. Delahunt, R. J. Miller, J. R. Srigley, A. J. Evans, and H. Samaratunga, "Gleason grading: Past, present and future," *Histopathology* **60**(1), 75–86 (2012).

32. J. Epstein, W. Allsbrook, and M. Amin, "2005 Iinternational Society of Urological Pathology (ISUP): Consensus conference on Gleason grading of prostatic carcinoma," *Am. J. Surg. Pathol.* **29**(9), 1228–1242 (2005).

33. R. Farjam, H. Soltanian-Zadeh, K. Jafari-Khouzani, and R. A. Zoroofi, "An image analysis apporach for automatic malignancy determination of prostate pathological images," *Cytometry Part B (Clinical Cytometry)* **72B**(4), 227–240 (2007).

34. J. Monaco, J. E. Tomaszewski, M. Feldman, M. Moradi, P. Mousavi, A. Boag, C. Davidson, P. Abolmaesumi, and A. Madabhushi, "Probabilistic pairwise Markov models: Application to prostate cancer detection," *Proc. SPIE* **7259**, 725903 (2009) [doi: 10.1117/12.812462].

35. A. W. Wetzel, R. Crowley, S. J. Kim, R. Dawson, L. Zheng, Y. Joo, Y. Yagi, J. Gilberston, C. Gadd, D. Deerfield, and M. J. Becich, "Evaluation of prostate tumor grades by content-based image retrieval," *Proc. SPIE* **3584**, 244–252 (1999) [doi: 10.1117/12.339826].

36. C. Wittke, J. Mayer, and F. Schweiggert, "On the classification of prostate carcinoma with methods from spatial statistics," *IEEE Trans. Inform. Technol. Biomed.* **11**(4), 406–414 (2007).

37. S. Naik, S. Doyle, S. Agner, A. Madabhushi, M. Feldman, and J. Tomaszewski, "Automated gland and nuclei segmentation for grading of prostate and breast cancer histopathology," *Fifth IEEE Int. Symp. Biomed. Imaging: From Nano to Macro (ISBI)* (2008).

38. K. Nguyen, A. Jain, and R. Allen, "Automated gland segmentation and classification for Gleason grading of prostate tissue images," *Proc. of the International 20th Conference on Pattern Recognition (ICPR)* (2010).

39. P. Khurd, L. Grady, A. Kamen, S. Giggs-Strauss, and E. M. Genega, "Network cycle features: Application to computer-aided Gleason grading of prostate cancer histopathological images," *Proc. IEEE International Symposium on Biomedical Imaging: From Nano to Macro* (2011).

40. X. Sun, S. H. Chuang, J. Li, and F. McKenzie, "Automatic diagnosis for prostate cancer using run-length matrix method," *Proc. SPIE* **7260**, 72603H (2009) [doi: 10.1117/12.811414].

41. E. Yu, J. Monaco, J. Tomaszewski, N. Shih, M. Feldman, and A. Madabhushi, "Detection of prostate cancer on histopathology using color fractals and Probabilistic Pairwise Markov models," *Proc. Annual International Conference of the IEEE Engineering in Medicine and Biology Society (EMBC)* (2011).
42. S. Doyle, M. Feldman, J. Tomaszewski, and A. Madabhushi, "A boosted Bayesian multiresolution classifier for prostate cancer detection from digitized needle biopsies," *IEEE Trans. Biomed. Eng.* **59**(5), 1205–1218 (2012).
43. K. Jafari-Khouzani and H. Soltanian-Zadeh, "Multiwavelet grading of pathological images of prostate," *IEEE Trans. Biomed. Eng.* **50**(6), 697–704 (2003).
44. E. Alexandratou, D. Yova, D. Gorpas, P. Maragos, G. Agrogiannis, and N. Kavantzas, "Texture analysis of tissue in Gleason grading of prostate cancer," *Proc. SPIE* **6859**, 685904 (2008) [doi: 10.1117/12.763377].
45. P. Huang and C. Lee, "Automatic classification for pathological prostate images based on fractal analysis," *IEEE Trans. Med. Imaging* **28**(7), 1037–1050 (2009).
46. S. Tai, C. Y. Li, Y. C. Wu, Y. J. Jan, and S. C. Lin, "Classification of prostatic biopsy," *Proc. IEEE Conference on Digital Content, Multimedia Technology and its Applications (IDC)* (2010).
47. P. Khurd, C. Bahlmann, P. Maday, A. Kamen, S. Gibbs-Strauss, E. Genega, and J. Frangioni, "Computer-aided gleason grading of prostate cancer histopathological images using texton forests," *Proc. 2010 IEEE International Symposium on Biomedical Imaging: From Nano to Macro* (2010).
48. H. J. Yoon, L. Ching-Chung, C. Christudass, R. Veltri, J. Epstein, and Z. Zhen, "Cardinal Multiridgelet-based prostate cancer histological image classification for Gleason grading," *Proc. 2011 IEEE International Conference on Bioinformatics and Biomedicine (BIBM)* (2011).
49. A. Almuntashri, S. Agaian, I. Thompson, D. Rabah, O. Zin Al-Abdin, and M. Nicolas, "Gleason grade-based automatic classification of prostate cancer pathological images," *Proc. 2011 IEEE International Conference on Systems, Man, and Cybernetics (SMC)* (2011).
50. C. Mosquera-Lopez and S. Agaian, "A new set of wavelet- and fractals-based features for Gleason grading of prostate cancer histopathology images," *Proc. SPIE* **8655**, 865516 (2013) [doi: 10.1117/12.1000193].
51. D. Diamond, S. J. Berry, C. Umbricht, H. Jewett, and D. S. Coffey, "Computerized image analysis of nuclear shape as a prognostic factor for prostatic cancer," *Prostate* **3**(4), 321–332 (1982).

52. J. I. Epstein, S. Berry, and J. C. Eggleston, "Nuclear roundness factor. A predictor of progression in untreated stage A2 prostate cancer," *Cancer* **54**, 666–671 (1984).

53. R. A. Stephenson, D. J. W. K. L. Zahniser, and M. L. Hutchinson, "An image analysis method for assessment of prognostic risk in prostate cancer: a pilot study," *Anal. Cell Pathol.* **3**(4), 243–248 (1991).

54. S. Vesalainen, P. Lipponen, M. Talja, J. Kasurinen, and K. Syrjanen, "Nuclear morphometry is of independent prognostic value only in T1 prostatic adenocarcinomas," *Prostate* **27**, 110–117 (1995).

55. J. M. Martinez-Jabaloyas, J. L. Ruiz-Cerda, M. Hernandez, A. Jimenez, and F. Jimenez-Cruz, "Prognostic value of DNA ploidy and nuclear morphometry in prostate cancer treated with androgen deprivation," *Urology* **59**(5), 715–720 (2002).

56. A. Buhmeida, T. Kuopio, and Y. Collan, "Nuclear size and shape in fine needle aspiration biopsy samples of the prostate," *Anal. Quant. Cytol. Histol.* **22**(4), 291–298 (2000).

57. K. Nguyen, A. K. Jain, and B. Sabata, "Prostate cancer detection: Fusion of cytological and textural features," *J. Pathol. Inform.* **2**(3), 10p (2011).

58. B. Djavan and M. Margreiter, "Biopsy standards for detection of prostate cancer," *World J. Urology* **25**(1), 11–17 (2007).

59. A. Angelucci, G. Pace, P. Sanita, C. Vicentini, and M. Bologna, "Tissue print of prostate biopsy: A novel tool in the diagnostic procedure of prostate cancer," *Diagnostic Pathol.* **6**(1), 34 (2008).

60. V. Ravery, S. Dominique, X. Panhard, M. Toublanc, L. Boccon-Gibod, and L. Boccon-Gibod, "The 20-core prostate biopsy protocol - a new gold standard?" *J. Urology* **179**(2), 504–507 (2008).

61. M. Moradi, P. Mousavi, and P. Abolmaesumi, "Computer-aided diagnosis of prostate cancer with emphasis on ultrasound-based approaches: A review," *Ultrasound Med. Biol.* **33**(7), 1010–1028 (2007).

62. I. M. Thompson et al., "Prevalence of Prostate Cancer among Men with a Prostate-Specific Antigen Level $\leq$4.0 ng per Milliliter," *N. Engl. J. Med.* **350**, 2239–2246 (2004).

63. Johns Hopkins Medicine, "Health library—prostate biosy," www.hopkinsmedicine.org/healthlibrary/test_procedures/urology/prostate_biopsy_92,P07710.

64. The University of Utah, "Histotechniques," library.med.utah.edu/WebPath/HISTHTML/HISTOTCH/HISTOTCH.html.

65. A. H. Fisher, K. A. Jacobson, J. Rose, and R. Zeller, "Hematoxylin and Eosin staining of tissue and cell sections," *CSH Protocols* **2008**(5) (2008).

66. S. Al-Janabi, A. Huisman, and P. J. Van Diest, "Digital pathology: Current status and future perspectives," *Histopathology* **61**, 1–9 (2012).
67. M. García-Rojo, G. Bueno-García, C. Peces-Mateos, J. González-García, and M. Carbajo-Vicente, "Critical comparison of 31 commercially available digital slide systems in pathology," *Int. J. Surgical Pathol.* **14**(4), 285–305 (2006).
68. M. Roula, J. Diamond, A. Bouridane, P. Miller, and A. Amira, "A multispectral computer vision system for automatic grading of prostatic neoplasia," *Proc. IEEE Int. Symp. Biomed. Imaging* (2002).
69. A. K. Jain and R. P. W. Duin, "Introduction to pattern recognition," *The Oxford Companion to the Mind*, Oxford University Press, Oxford, UK, pp. 698–703 (2004).
70. A. K. Jain, R. Duin, and M. Jianchang, "Statistical pattern recognition: A review," *IEEE Trans. Pattern Anal. Machine Intelligence* **22**(1), 4–37 (2000).
71. C. Smith, S. Agaian, and D. Akopian, "A wavelet-nenoising approach using polynomial threshold operators," *IEEE Signal Process. Lett.* **15**, 906–909 (2008).
72. B. Smith and S. Agaian, "Optimization of a parametric threshold in the wavelet domain," *SIAM Image Processing Conference*, Salt Lake City, Utah (2004).
73. C. B. Smith and S. Agaian, "Nonlinear noise suppression using a parametric class of wavelet shrinkage functions," *Proc. SPIE* **5439**, 41–50 (2004) [doi: 10.1117/12.542103].
74. D. Akopian, V. O., S. Agaian and A. J., "Processors for generalized stack filters," *IEEE Trans. Signal Process.* **43**(6), 1541–1546 (1995).
75. D. Z. Gevorkian, K. O. Egiazarian, S. S. Agaian, J. T. Astola, and O. Vainio, " Parallel algorithms and VLSI architectures for stack filtering using fibonacci p-codes," *IEEE Trans. Signal Processing*, **43**(1), 286–295 (1995).
76. S. Agaian, J. Astola, and K. Egiazarian, *Binary Polynomial Transforms and Nonlinear Digital Filters*, Marcel Dekker, Inc., New York (1995).
77. S. Agaian, "Visual Morphology," *Proc. SPIE* **3646**, 139–150 (1999) [doi: 10.1117/12.341081].
78. A. Almuntashri and S. Agaian, "Human visual system-based edge detection using image contrast enhancement and logarithmic ratio," *Proc. SPIE* **7708**, 770813 (2010) [doi: 10.1117/12.849609].
79. K. Panetta, S. Agaian, Y. Zhou, and E. Wharton, "Parameterized Logarithmic Framework for Image Enhancement Systems," *IEEE Trans. Systems, Man, and Cybernetics, Part B: Cybernetics* **41**(2), 460–473 (2011).

80. Z. Rahman, G. A. Woodell, and D. J. Jobson, "A Comparison of the Multiscale Retinex With Other Image Enhancement Techniques," *Proc. IS&T 50th Anniversary Conference* (1997).

81. K. Panetta, E. Wharton, and S. Agaian, "Human visual System Based Image Enhancement and Measure of Image Enhancement," *IEEE Trans. Systems, Man, and Cybernetics, Part A: Systems and Humans* **38**(1), 174–188 (2008).

82. E. Wharton, S. Agaian, and K. Panetta, "Adaptive Multi-Histogram Equalization using Human Vision Thresholding," *Proc. SPIE* **6497**, 64970G (2007).

83. S. Agaian, K. Panetta, and A. Grigoryan, "Transform Based Image Enhancement Algorithms with Performance Measure," *IEEE Trans. Image Process.* **10**(3), 367–380 (2001).

84. K. Panetta, J. Xia, and S. Agaian, "Color image enhancement based on the discrete cosine transform coefficient histogram," *J. Electron. Imaging* **21**(2), 021112-1–021112-17 (2012).

85. H. Greenspan, C. H. Anderson, and S. Akbar, "Image enhancement by nonlinear extrapolation in frequency space," *IEEE Trans. Image Process.* **9**(6), 1035–1048 (2000).

86. A. Grigoyan and S. Agaian, "Image Enhancement," *Advances in Imaging and Electron Physics*, Academic Press, San Diego, CA, pp. 165–243 (2004).

87. S. Agaian and S. A. McClendon, "Novel medical image enhancement algorithms," *Proc. SPIE* **7532**, 75320W (2010) [doi: 10.1117/12.839003].

88. T. Zong, H. Lin, and T. Kao, "Adaptive local contrast enhancement method for medical images displayed on a video monitor," *Med. Eng. Phys.* **22**, 79–87 (2000).

89. Y. Zhou, K. Panetta, and S. Agaian, "CT baggage image enhancement using a combination of alpha-weighted mean separation and histogram equalization," *Proc. SPIE* **7708**, 77080G (2010) [doi: 10.1117/12.849587].

90. H. D. Cheng and H. Xu, "A novel fuzzy logic approach to contrast enhancement," *Pattern Recogn.* **33**, 809–819 (2000).

91. E. Wharton, K. Panetta, and S. Agaian, "Edge Preserving Image Enhancement Using Anisotropic Diffusion," *Proc. SPIE* **6812**, 681218 (2008) [doi: 10.1117/12.766893].

92. G. R. Polesel and V. J. Mathews, "Image enhancement via adaptive unsharp masking," *IEEE Trans. Image Process.* **9**, 505–510 (2000).

93. A. Abdelwahab, M. Ahmed, and S. Hashem, "Image Enhancement Using a Contrast measure in the Discrete Wavelet Transform," *Radio Sci. Conf.* (2007).

94. J. Tang, E. Peli, and S. Acton, "Image Enhancement Using a Contrast measure in the Compressed Domain," *IEEE Signal Process. Lett.* **10**(10), 289–292 (2003).

95. S. C. Nercessian, K. A. Panetta, and S. S. Agaian, "Multiscale image fusion using an adaptive similarity-based sensor weighting scheme and human visual system-inspired contrast measure," *J. Electron. Imaging* **21**(2), 1–13 (2012).

96. E. Wharton, K. Panetta, and S. Agaian, "Human visual system based similarity metrics," *Proc. IEEE Systems, Man, and Cybernetics (SMC)*, Singapore (2008).

97. S. Agaian, B. Silver, and K. Panetta, "Transform coefficient histogram-based image enhancement algorithms using contrast entropy," *IEEE Trans. Signal Process.* **16**(3), 741–758 (2007).

98. K. Huang, Q. Wang, and Z. Wu, "Color image enhancement and evaluation algorithm based on human visual system," *Proc. IEEE ICASSP* (2004).

99. E. A. Silva, K. A. Panetta, and S. S. Agaian, "Quantify similarity with measurement of enhancement by entropy," *Proc. SPIE* **6579**, 65790U (2007) [doi: 10.1117/12.720087].

100. E. Wharton, S. Agaian, and K. Panetta, "A logarithmic measure of image enhancement," *Proc. SPIE* **6250**, 62500P (2006) [doi: 10.1117/12.665693].

101. E. Wharton, S. Agaian, and K. Panetta, "Comparative study of logarithmic enhancement algorithms with performance measure," *Proc. SPIE* **6064**, 606412 (2006) [doi: 10.1117/12.647489].

102. K. Panetta, Y. Zhou, S. Agaian, and H. Jia, "Nonlinear unsharp masking for mammogram enhancement," *IEEE Trans. Information Tech. Biomed.* **15**(6), 918–928 (2011).

103. S. DelMarco, and S. Agaian, "The design of wavelets for image enhacement and target detection," *Proc. SPIE* **7351**, 735103 (2009) [doi: 10.1117/12.816135].

104. S. Agaian, K. Tourshan, and J. P. Noonan Tang, "Parametric Slant-Hadamard transforms with applications," *IEEE Signal Process. Lett.* **9**(11), 375–377 (2002).

105. S. C. Nercessian, K. Panetta, and S. Agaian, "Multiscale image fusion using an adaptive similarity-based sensor weighting scheme and human visual system-inspired contrast measure," *J. Electron. Imaging* **21**(2), 1–13 (2012).

106. S. C. Nercessian, K. Panetta, and S. Agaian, "Nonlinear direct multiscale image enhancement based on the luminance and contrast masking characteristics of the human visual system," *IEEE Trans. Image Process.* **22**(9), 3549–3561(2013).

107. K. A. Panetta, S. S. Agaian, S. C. Nercessian, and A. A. Almunstashri, "Shape-dependent canny edge detector," *Opt. Eng.* **50**(8), 087008 (2011) [doi: 10.1117/1.3613941].

108. C. Smith and S. Agaian, "Mismatched wavelets: Information hiding and edge detection," *Cyber Journals: Multidisciplinary Journals in Science and Technology, Journal of Security* **2011**, 56–66 (2011).

109. S. S. Agaian, K. A. Panetta, S. C. Nercessian, and E. E. Danahy, "Boolean derivatives with application to edge detection for imaging systems," *IEEE Trans. Systems, Man, and Cybernetics, Part B: Cybernetics* **4**(2), 371–382 (2010).

110. K. A. Panetta, E. J. Wharton, and S. S. Agaian, "Logarithmic edge detection with applications," *J. Comp - JCP* **3**(9), 11–19 (2008).

111. Y. Yagi, "Color standardization and optimization in whole slide imaging," *Diagn. Pathol.* **6**(1), S15 (2011).

112. C. Mosquera-Lopez and S. Agaian, "Iterative local color normalization using fuzzy image clustering," *Proc. SPIE* **8755,** 875518 (2013) [doi: 10.1117/12.2016051].

113. D. Magee, D. Treanor, D. Crellin, M. Shires, K. Smith, K. Mohee, and P. Quirke, "Colour normalisation in digital histopathology images," *Proc. Optical Tissue Image Analysis in Microscopy, Histopathology, and Endoscopy (MICCAI Workshop)* (2009).

114. A. Hafiane, F. Bunyak, and K. Palaniappan, "Fuzzy clustering and active contours for histopathology image segmentation and nuclei detection," *Lecture Notes Comp. Sci.* **5259**, 903–914 (2008).

115. C. M. Mosquera-Lopez, S. Agaian, I. Sanchez, A. Almuntashri, O. Zinalabdin, A. Rikabi, and I. Thompson, "Exploration of efficacy of gland morphology and architectural features in prostate cancer gleason grading," *2012 IEEE Int. Conf. Systems, Man, and Cybernetics (SMC)*, Seoul (2012).

116. S. Naik, S. Doyle, M. Feldman, J. Tomaszewski, and A. Madabhushi, "Gland segmentation and computerized Gleason grading of prostate histology by integrating low-, high-level and domain specific information," *Proc. 2nd Int. Workshop Microscopic Image Anal. Appl. Biol. (MIAAB)*, Piscataway, NJ (2007).

117. C. Li, C. Xu, C. Gui, and M. Fox, "Level set evolution without re-initialization: a new variational formulation," *Proc. IEEE Comp. Soc. Conf. Comp. Vision Pattern Recogn. (CVPR)* (2005).

118. K. Nguyen, B. Sabatab, and A. K. Jain, "Prostate cancer grading: Gland segmentation and structural features," *Pattern Recogn. Lett.* **33**(7), 951–961 (2012).

119. J. Xu, R. Sparks, A. Janowcyzk, J. E. Tomaszewski, M. D. Feldman, and A. Madabushi, "High-throughput prostate cancer gland detection,

segmentation, and classification from digitized needle core biopsies," *Prostate Cancer Imaging: Computer-Aided Diagnosis, Prognosis, and Intervention*, Springer-Verlag, Berlin, pp. 77–88 (2010).

120. M. Ivanovici and N. Richard, "Fractal dimension of color fractal images," *IEEE Trans. Image Process.* **20**(1), 227–235 (2011).

121. R. M. Haralick, K. Shanmugam, and I. Dinstein, "Textural features for image classification," *IEEE Trans. Systems, Man, and Cybernetics* **3**(6), 610–621 (1973).

122. J. Diamond, N. Anderson, P. Bartels, R. Montironi, and P. Hamilton, "The use of morphological characteristics and texture analysis in the identification of tissue composition in prostatic neoplasia," *Human Pathol.* **35**, 1121–1131 (2004).

123. A. Tabesh, M. Teverovskiy, H. Pang, V. Kumar, D. Verbel, A. Kotsianti, and O. Saidi, "Multifeature prostate cancer diagnosis and Gleason grading of histological images," *IEEE Trans. Med. Imaging* **26**(10), 1366–1378 (2007).

124. M. M. Galloway, "Texture analysis using gray level run lengths," *Comp. Graphics and Image Process.* **4**, 172–179 (1975).

125. A. Chu, C. M. Sehgal, and J. Greenleaf, "Use of gray value distribution of run-length for texture analysis," *Pattern Recogn. Lett.* **11**, 415–420 (1990).

126. A. Tosun and C. Gunduz-Demir, "Graph run-length matrices for histopathological image segmentation," *IEEE Trans. Med. Imaging* **30**(3), 721–732 (2011).

127. B. Tso and M. P. M., "Pattern recognition principles," in *Classification Methods for Remotely Sensed Data,* 2nd edition, CRC Press, Boca Raton, FL, pp. 41–75 (2009).

128. A. Martinez and A. Kak, "PCA versus LDA," *IEEE Trans. Pattern Anal. Mach. Intel.* **23**(2), 228–233 (2001).

129. A. Hyvarinen and E. Oja, "Independent component analysis: algorithms and applications," *Neural Networks* **13**(4–5), 411–430 (2000).

130. R. Sparks and A. Madabhushi, "Gleason grading of prostate histology utilizing manifold regularization via statistical shape model of manifolds," *Proc. SPIE* **8315**, 83151J (2012) [doi: 10.1117/12.912887].

131. E. Alexandratou, V. Atlamazoglou, T. Thireou, G. Agrogiannis, D. Togas, N. Kavantzas, E. Patsouris, and D. Yova, "Evaluation of machine learning techniques for prostate cancer diagnosis and Gleason grading," *Int. J. Comp. Intel. Bioinformatics Systems Biol.* **1**(3), 297–315 (2010).

132. S. Doyle, M. Feldman, J. Tomaszewski, N. Shih, and A. Madabhushi, "Cascaded multi-class pairwise classifier (CascaMPa) for normal, cancerous, and cancer confounder classes in prostate histology," *Proc. 2011 IEEE Int. Symp. Biomed. Imaging: From Nano to Macro* (2011).

133. D. J. Hand, "Assessing the performance of classification methods," *Int. Stat. Rev.* **80**(3), 400–414 (2012).

134. P. Refaeilzadeh, L. Tang, and H. Liu, "Cross-validation," in *Encyclopedia of Database Systems*, Springer, Berlin (2009).

135. A. Fielding and J. F. Bell, "A review of methods for the assessment of prediction errors in conservation presence/absence models," *Environmental Conservation,* vol. 24, no. 1, pp. 38-49, 1997.

136. A. G. Lalkhen and A. McCluskey, "Clinical tests: Sensitivity and specificity," *Continuing Education in Anaesthesia, Critical Care, & Pain* **8**(6), 221–223 (2008).

137. D. J. Hand and R. J. Till, "A simple generalisation of the area under the ROC curve for multiple class classification problems," *Mach. Learning* **45**(2), 171–186 (2001).

138. S. Bouatmane, M. A. Roula, A. Bouridane, and S. Al-Maadeed, "Round-Robin sequential forward selection algorithm for prostate cancer classification and diagnosis using multispectral imagery," *Mach. Vision Appl.* **55**(5), 865–878 (2011).

139. M. Teverovskiy, V. Kumar, J. Ma, A. Kotsianti, D. Verbel, A. Tabesh, H. Pang, Y. Vengrenyuk, S. Fogarasi, and O. Saidi, "Improved prediction of prostate cancer recurrence based on an automated tissue image analysis system," *Proc. 2004 IEEE Int. Symp. Biomed. Imaging: Nano to Macro* (2004).

140. I. M. Thompson, D. P. Ankerst, C. Chi, P. J. Goodman, C. M. Tangen, M. Lucia, Z. Feng, H. L. Parnes, and C. A. Coltman, "Assessing Prostate Cancer Risk: Results from the Prostate Cancer Prevention Trial," *JNCI J. Nat. Cancer Inst.* **98**(8), 529–534 (2006).

141. D. S. Coffey, "Similarities of prostate and breast cancer: Evolution, diet, and estrogens," *J. Urology* **57**(4), 31–38 (2001).

Clara Mosquera-Lopez received her B.S. degree in Electronics Engineering and her M.S. degree in Management of Technology from the Universidad Pontificia Bolivariana, Medellin, Colombia in 2007 and 2010 respectively. Clara is currently a doctoral student in the Department of Electrical and Computer Engineering at the University of Texas at San Antonio. Her research interests include a broad spectrum of topics in signal processing and machine learning. Her present research is focused on the development of image processing algorithms for quantitative analysis of biomedical images and pattern recognition methods applied to computer-aided cancer diagnosis.

Sos S. Agaian is the Peter T. Flawn Professor of Electrical and Computer Engineering at the University of Texas, San Antonio, and a professor at the University of Texas Health Science Center, San Antonio. Dr. Agaian received his M.S. degree (summa cum laude) in Mathematics and Mechanics from Yerevan State University, Armenia; his Ph.D. in Mathematics and Physics from the Steklov Institute of Mathematics, Russian Academy of Sciences (RAS), Moscow; and his Eng.Sc.D. from the Institute of Control Sciences, RAS. He has authored more than 500 scientific papers, seven books, and holds 14 patents. He is a Fellow of SPIE, the American Association for the Advancement of Science (AAAS), and Imaging Sciences and Technology (IS&T). He also serves as a foreign member of the Armenian National Academy. He is the recipient of MAEStro Educator of the Year, sponsored by the Society of Mexican American Engineers. The technologies that Dr. Agaian invented have been adopted by multiple institutions, including the U. S. government, and commercialized by industry. He is an Editorial Board Member for the *Journal of Pattern Recognition and Image Analysis* and an Associate Editor for several journals, including the *Journal of Electronic Imaging* (SPIE, IS&T) and *System* (Elsevier). His research interests include multimedia processing, imaging systems, information security, artificial intelligence, computer vision, 3D imaging sensors, fusion, and biomedical and health informatics.

Alejandro Velez-Hoyos is an Associate Professor at the Universidad de Antioquia and the Universidad Pontificia Bolivariana, Medellin, Colombia. He is also part of the Department of Pathology at the Pablo Tobon Uribe Hospital and Dinamica IPS. Dr. Velez-Hoyos received his M.D. from the Universidad Pontificia Bolivariana, and after a residency in pathology, he received his specialization degree from the Universidad de Antioquia. He has authored several dozen scientific papers, six book chapters, and he has edited a textbook on pathology. Dr. Velez-Hoyos has attended several training sessions in the U.S. on cytology, uropathology, dermatopathology, immuno-histochemestry, and endocrine and soft-tissue pathology. He is the vice president of the Colombian Pathology Society and the vice president of the Colombian Cytology Society. His research interests include methods for tissue sampling, and diagnosis of various types of cancerous lesions through microscopic analysis of biopsy specimens. He actively participates in clinical studies on skin, prostate, and breast cancer, among other diseases.

Ian Thompson is the director of the Cancer Therapy and Research Center and a Professor in the Department of Urology in the School of Medicine at the University of Texas Health Science Center in San Antonio. Dr. Thompson received his undergraduate degree from West Point and his M.D. from Tulane University, New Orleans, LA. After a residency in urology in San Antonio, he completed a fellowship in Urologic Oncology at the Memorial Sloan Kettering Cancer Center in New York. He has published over 400 scientific papers, several dozen book chapters, and has edited five textbooks in medicine and surgery. He previously served as chair of the Residency Review Committee for Urology and currently serves as vice chair of the Early Detection Research Network of the National Cancer Institute. He previously served as the president of the Society of Urologic Oncology. He currently serves as chair of the Genitourinary Committee of the Southwest Oncology Group, the largest clinical trials organization supported by the National Cancer Institute. Dr. Thompson has served as a Visiting Professor at most major academic institutions in the U.S. as well as at many leading cancer centers in Europe, Asia, Central and South America, as well as Australia.

Chapter 10
Analysis of Breast Masses in Mammograms Using the Fractal Dimension and Shape Factors*

Grazia Raguso and Loredana Chieppa
Department of Mathematics, University of Bari, Italy

Antonietta Ancona
Radiology Unit, San Paolo Hospital of Bari, Italy

Samuela L'Abbate
Department of Statistics, University of Bari, Italy

Maria Luisa Pepe
Department of Radiology, University Hospital - Policlinico of Bari, Italy

Fabio Mangieri
Radiology Unit, San Paolo Hospital of Bari, Italy

Shantanu Banik and Rangaraj M. Rangayyan
Department of Electrical and Computer Engineering, Schulich School of Engineering, University of Calgary, Alberta, Canada

* This work was supported by the Fondazione Cassa di Risparmio di Puglia, Italy and the Natural Sciences and Engineering Research Council of Canada. We thank Thanh M. Nguyen and Naga R. Mudigonda for their contributions to this work.

10.1 Introduction

The practical utility of computational biology is far-reaching, and it involves various aspects concerned with understanding biological phenomena in general, and gaining insights in the areas of health, biotechnology, and the environment, among others.[1–3] In recent years, several research projects have been directed toward the health sector, with many of them studying mathematical and computational methods for computer-aided detection or diagnosis (CAD) of various diseases. At present, an effective screening technique for breast cancer is mammography, which helps to identify significant variations in the glandular tissue and appearance of tumoral lesions in the breast.[4] Several CAD systems have been developed to support radiologists in the area of mammography;[5–7] such systems can identify anomalous regions and masses with unusual morphological structure and abnormal density patterns.

When assessing an opacity in a mammogram, the following characteristics are taken into consideration: shape, definition or sharpness of the edges, roughness of the contour, variations in density or texture, and size. The regularity of the contour of a mass is the first parameter assessed: benign masses are often smooth, rounded, well-circumscribed, and surrounded by a halo of fairly low-density fat, whereas opacities with irregular contours and ill-defined edges are more likely to be malignant tumors.[4,8] A malignant tumor is often characterized by the presence of spicules (a stellate lesion typical of infiltrating ductal carcinoma, for example) and by a poorly defined irregular contour, one that could be considered to be a fractal pattern. The term carcinoma, from the Greek word "καρκιυος" (crab), was coined by Hippocrates; it indicates the infiltrative characteristics of a tumor as well as its ability to attack neighboring structures. Regardless, there are cases of malignant tumors with regular contours and benign masses with fractal-like contours; such cases are challenging for a physician to diagnose and make it difficult to build a classification model, leading to false negatives and false positives.[9] These observations have led to the idea of applying the concept of fractal dimension (FD) to analyze the contours of breast lesions.[8,10–12] As will be demonstrated in this chapter, fractal analysis can characterize the degree of complexity of a contour or shape, and can provide parameters to discriminate between benign masses and malignant tumors.[13]

Based on the differences observed in the forms of benign masses and malignant tumors, shape factors such as compactness (cf), fractional concavity (f_{cc}), spiculation index (SI), and a Fourier-descriptor-based factor (ff) have been proposed for their classification.[8,9,14] Guliato et al.[15,16] defined several shape factors based on the turning-angle function of a contour and demonstrated their usefulness in the classification of breast masses. Subtle textural differences have also been observed between benign masses and malignant tumors, with the former being mostly homogeneous, and the latter showing heterogeneous texture. Several studies have proposed measures of texture and edge sharpness to discriminate between benign masses and malignant tumors.[8,11,14,17–19] Sahiner et al.[20] and Alto et al.[19] explored several combinations of morphological and texture measures to classify breast masses. The notion of fractal analysis[21,22] is useful in studying the complexity of 1D functions, 2D contours, as well as grayscale or color images. A few studies have examined the application of fractals to classify breast masses based on the irregularity exhibited in their contours.[10–12] Matsubara et al.[23] reported 100% accuracy in the classification of 13 benign masses and malignant tumors using FD; their method involved computing a series of FD values for several contours of a given mass obtained by thresholding the mass at many levels; the variation in the FD of the given mass was used to categorize it as benign or malignant.

A study by Pohlman et al.[24] obtained greater than 80% classification accuracy with fractal analysis of signatures of contours of breast masses. However, the signature of a contour was derived as a function of the radial distance from the centroid to the contour versus the angle of the radial line over the range [0°, 360°], which could lead to a multivalued function in the case of an irregular or spiculated contour; the signature computed in this manner would also have ranges of undefined values in the case of a contour for which the centroid falls outside the region enclosed by the contour.

Dey and Mohanty[25] used fractal geometry to study breast lesions on cytology smears and found that the FD may be useful in discriminating between benign and malignant cells.

Fractal analysis can also be used to characterize the complexity of grayscale variations associated with texture. Zheng and Chan[26] used fractal analysis in a preprocessing step to select abnormal regions in mammograms. Guo et al.[27] computed the FD to characterize the complexity of regions of interest (ROIs) in mammograms, and used a support vector machine for the detection of abnormal regions related to breast masses. Guo et al.[28] also studied the Hausdorff FD to characterize architectural distortion.

Tourassi et al.[29] investigated the use of FD to distinguish between normal tissue patterns and architectural distortion in mammographic ROIs. The FD was estimated using the circular average power spectrum technique.[30,31] The method was applied to a dataset of 1,500 ROIs, including 112 ROIs with architectural distortion and 1,388 ROIs exhibiting normal tissue patterns. The best performance achieved was 0.89, in terms of the area (AUC) under the receiver operating characteristic (ROC) curve. Tourassi et al. observed that the presence of architectural distortion disrupts the self-similarity properties of breast parenchyma and that the average FD of the ROIs with architectural distortion was significantly lower than that of normal ROIs.

Eltonsy et al.[32] developed a method to detect masses and architectural distortion by locating points surrounded by concentric layers of image activity. A set of 80 images was used to evaluate the technique, including 13 masses, 38 masses with architectural distortion, and 29 images with only architectural distortion. An overall sensitivity of 91.3% with 9.1 false positives per image was obtained. A sensitivity of 93.1% was obtained in the detection of architectural distortion at the same rate of false positives.

Rangayyan et al.[33] used Gabor filters, phase portraits, FD, and texture features for the detection of architectural distortion in prior mammograms of screen-detected cancer and achieved a sensitivity of 79% at 8.4 false positives per image with a set of 14 prior mammograms. The methods were extended for the detection of architectural distortion in mammograms of interval-cancer cases taken prior to the detection of breast cancer;[34,35] a sensitivity of 80% was obtained at 5.8 false positives per image using the FD in combination with a number of measures of texture and angular spread of power, derived from a set of 106 prior mammograms of interval-cancer cases and 52 normal mammograms.[35]

Caldwell et al.[36] and Byng et al.[37] computed the FD of breast tumors by applying a modified box-counting method that represents grayscale values of the surfaces of the tumors as boxes of variable height. Such a fractal measure can be used to represent the complexity of density variations and texture in breast tissue. Byng et al.[37] showed that a grayscale-based fractal measure may be used to complement histogram skewness to relate breast density to the risk of development of breast cancer.

Other works have reported on the use of FD as a feature for the classification of tumors in magnetic resonance images of the brain,[38] ultrasonographic images of the liver,[39] and images related to colonic cancer.[40] Lee et al.[41] compared several shape factors, including the FD, in a study on the irregularity of the borders of melanocytic lesions. Kikuchi et al.[42] investigated the change in FD at different stages of ovarian tumor growth. Nam and Choi[43] computed the FD of regions in mammograms by using the box-counting method, and found that regions with higher FD indicated the presence of calcification.

Klonowski et al.[44] applied Higuchi's method for fractal analysis of the texture in histological images and for the analysis of the shape of breast masses. Tambasco et al.[45] applied fractal analysis for quantitative analysis of the pattern complexity seen in microscopic images of histological specimens of prostate and breast cancer; the FD demonstrated statistically highly significant differences between specimens of normal and cancerous tissue.

Li et al.[46] studied four approaches to estimate FD, including conventional and modified box counting, box counting using linear discriminant analysis (LDA), global Minkowski dimension, and a modified Minkowski technique using LDA. FD-based texture features were computed to characterize breast tissue patterns seen in mammograms. It was shown that the features could yield radiographic markers to assess the risk of development of breast cancer. Li et al.[47] derived a parameter via power spectral analysis that demonstrated a statistically significant difference between the high-risk and low-risk groups of women in the study.

Rangayyan and Nguyen[10] presented a study of four methods to compute the FD of the contours of breast masses, including the ruler method and the box-counting method applied to 1D and 2D representations of the contours; see also Cabral and Rangayyan.[11] The methods were applied to a dataset comprising 111 contours of breast masses in mammograms, including 65 contours of benign masses and 46 contours of malignant tumors. The FD was observed to complement other shape factors, in particular f_{cc}, in the representation of the complexity of the contours. The combination of FD with f_{cc} yielded the highest AUC value of 0.93; the two measures, on their own, resulted in AUC values of 0.89 and 0.88, respectively.

The aim of the present study is to employ fractal analysis and several shape factors for the classification of breast masses by using only their contours.[13] Even though fractal analysis has been widely used in the analysis of biomedical images, only a few studies have specifically applied the method to study and classify mammographic masses (as reviewed above). The FD may be used as a quantitative measure of the complexity of the contour or boundary of an object. Benign masses and malignant tumors differ significantly in shape complexity, and therefore, it should be possible to differentiate between them by using the FD in combination with shape factors.

The present work is an extended investigation based on our related initial work,[13] and it represents a thorough follow-up study and further evaluation of the methods proposed by Rangayyan and Nguyen.[10] This work presents the results of extensive comparative analysis and evaluation of several methods for shape analysis, and the results of classification experiments including cross-validation with three different datasets of breast-mass contours from multiple sources and patient populations, with the contours drawn by four radiologists. The results demonstrate strong performance of fractal analysis and shape factors in the classification of breast masses based on their contours.

10.2 Methods

10.2.1 Fractal analysis

The term *fractal* comes from the Latin word *fractus*, meaning fractionated, broken, irregular, or chaotic. In his pioneering work on *The Fractal Geometry of Nature*,[21] the mathematician Benoît Mandelbrot states that "the world around us is full of fractals." The genius of Mandelbrot lies in his understanding that many natural objects can consist of portions similar to the object itself; and therefore, "the whole is similar to one or more of the parts," a phenomenon known as self-similarity. Fractals are capable of representing a wide variety of objects and phenomena in nature: not only a stretch of coastline, the branches or roots of a tree, and clouds, but also the ramifications of a lightning bolt, the bronchial and vascular systems, and, we believe, the contours of breast masses and tumors.

Fractal geometry is, as Mandelbrot says, the "geometry of nature," because it is suitable for describing the *complexity* and *variety* of shapes in the world around us. Fractal geometry investigates the morphology of the amorphous. The correspondence between *fractals* and *chaos* is by no means accidental—rather, it is the sign of a profound relationship: *fractal geometry is the geometry of chaos*. The most-important fractals include the Cantor set, the Von Koch curve, the Sierpinski triangle, the Mandelbrot set, and the Lorenz attractor.[11,21,48]

Fractals are irregular figures that can be generated by the iteration of linear or nonlinear functions (Julia and Mandelbrot sets).[21,48] They are sometimes self-similar and have a fine structure that reveals new details at every level of magnification.[21] In order to measure the degree of complexity or irregularity of a fractal, the concept of FD was introduced; this concept is derived from the more-general notion of the Hausdorff dimension.[49] For a subset F of the plane, the *s-dimensional Hausdorff measure* is defined as follows:

$$H^s(F) = \lim_{\delta \to 0} H^s_\delta \tag{10.1}$$

with

$$H^s_\delta = inf\left\{\sum_{j=1}^{\infty} \delta_j^s\right\}, \tag{10.2}$$

where s, δ, and δ_j are positive real numbers, and $\delta_j \leq \delta$ are the diameters of a family of circles that makes up a δ-cover (countable) of F. The Hausdorff dimension of F is the number D such that

$$H^s(F) = \begin{cases} \infty, & \text{if } s < D; \\ 0, & \text{if } s > D. \end{cases}$$

The measure of F depends on the dimension of the space in which F is imagined to be immersed, whereas its dimension is a number, even a fraction, intrinsic to F. Such a definition can be applied to a *fractal figure*, and hence the Hausdorff dimension is also called the fractal dimension.

Now consider the self-similarity dimension and its relationship with the Hausdorff dimension. Without entering into the mathematical complexities, it is commonly asserted that the (topological) dimension of a point is 0, that of a line is 1, that of a rectangle is 2, and that of a parallelepiped is 3. If we consider, therefore, a rectangle made up of m copies of itself, reduced by a scale factor of $\frac{1}{s}$, the power law that links the dimension $D(=2)$ of the figure with the number of its parts $m(=9)$ and the scale factor $\frac{1}{s}(=3)$ is

$$m = \left(\frac{1}{s}\right)^{D}. \tag{10.3}$$

Therefore, in agreement with the above, it is reasonable to define the *self-similarity dimension* D_{sim}[49] of a self-similar figure consisting of m copies of itself reduced by a scale factor $\frac{1}{s}$ as

$$D_{sim} = \frac{\log m}{\log \frac{1}{s}}. \tag{10.4}$$

When s varies (and therefore, so does m), the set of points on the plane, on a log–log scale, is approximated by the regression straight line with a slope that provides an estimate of D_{sim} via Eq. (10.4). For self-similar figures, the self-similarity dimension coincides with the Hausdorff dimension.

The Hausdorff dimension generalizes the concept of the self-similarity dimension in the sense that it is applicable to any set of the plane and, therefore, to a fractal set that is not strictly self-similar. The difficulties involved in defining the Hausdorff dimension have led many authors to find alternative methods to estimate the FD. The common alternatives are the box-counting method and the ruler method, which have been extensively described in the literature.[10,11,49] The box-counting method consists of partitioning the plane and, therefore, the object into square boxes of side s; counting the minimum number $N(s)$ of the boxes that have at least one point (image pixel) in common with the object; and representing the values on a log–log scale of points $[\log N(s), \log \frac{1}{s}]$ for the variations (decreasing) in s. An estimate of the FD of the object is given by the slope of the regression straight line (best fit), i.e.,

$$D = \frac{\log N(s)}{\log \frac{1}{s}}. \tag{10.5}$$

The ruler method, also known as the compass or divider method, consists of providing the approximate length of the contour using a segment or ruler of length s. The smaller the value of s, the better the approximation. Once the length of the contour has been described as a function $u(s)$, the power law, in this case, is given by

$$u(s) = c\frac{1}{s^{d}}, \tag{10.6}$$

where c is the proportionality constant, and the exponent d is linked to the FD by the relation

$$D = 1 + d. \tag{10.7}$$

Applying the logarithm to Eq. (10.6), we obtain

$$\log u(s) = \log c + d \log \left(\frac{1}{s}\right), \tag{10.8}$$

in which d coincides with the slope of the straight line in the plane [$\log u(s)$, $\log \frac{1}{s}$]; thus, $D = 1 + d$ is an estimate of the FD. The self-similarity dimension, the box-counting dimension, and the ruler dimension are particular forms of Mandelbrot's FD.[49]

10.2.2 Shape factors

Each point p of a contour or curve C can be represented, in the plane, either with its Cartesian coordinate pair (x, y) or by indexing the points as pairs $[p, f(p)]$, in which f is a function defined on the indices of the contour point p. An example of $f(p)$ is the Euclidean distance between the point, p, under consideration and a reference, such as the centroid of the object enclosed by the contour. The first representation of C mentioned above is 2D, whereas the second representation is 1D and is known as a *signature* of C. These two representations are shown in Figs. 10.1 and 10.2.

In the present work, estimates of the FD were calculated using four methods: the ruler and box-counting methods applied to each 2D contour and its 1D signature. In the rest of this chapter, these features are referred to as FD-ruler 2D, FD-ruler 1D, FD-box 2D, and FD-box 1D.

Various measures can be associated with a contour or curve: these are the so-called shape factors, which have proven to be effective in describing shapes in many research fields, in particular in the medical field.[8,10,11] In general, it is desirable that a shape factor is invariant to translation, rotation, starting point, and contour size; in the present application, it is also desirable for a shape factor to increase in value as the shape of a contour gets to be more complex and rough. The shape factors used in the present work are cf, f_{cc}, SI, ff, and four estimates of the FD (mentioned earlier); these measures have been proven to be effective in the classification of breast masses.[8–11,14] In addition, a measure of shape known as convex deficiency (CD) is used, defined as the difference between the areas of a given contour and its convex hull, expressed as a fraction of the area of the contour. See Rangayyan[8] for details on the shape factors.

Compactness is defined as[8]

$$cf = 1 - \frac{4\pi A}{P^2}, \tag{10.9}$$

where P and A are the perimeter and the area of the contour, respectively. A high compactness value indicates a long perimeter enclosing a small area.

Fractional concavity is defined as[8,9]

$$f_{cc} = \frac{CC}{L}, \tag{10.10}$$

where CC represents the sum of the concave segment lengths of the contour, and L is the total length of the contour. Malignant tumors with significant

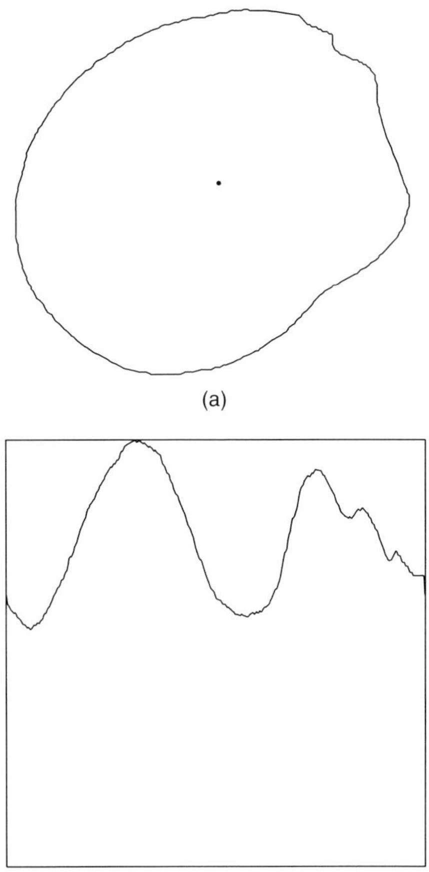

(a)

(b)

Figure 10.1 (a) The contour of a benign breast mass (the dot represents the centroid of the contour) and (b) the corresponding signature. The contour and signature have been normalized. FD-ruler 1D = 1.1643, FD-ruler 2D = 1.0107, FD-box 1D = 1.0349, FD-box 2D = 1.0480, cf = 0.1493, CD = 0.0155, f_{cc} = 0.1478, SI = 0.0277, and ff = 0.1490.

concavities in their contours are expected to have higher values of f_{cc} (as well as CD) than benign masses with smooth and oval shapes.

In order to define the spiculation index (SI), it is necessary to consider the notion of the length and narrowness angle of a spicule. Consider a spicule (a strongly concave or convex arc) of a contour, and suppose that, upon applying a polygonal modeling method, the polygon obtained contains M segments s_i that define $M - 1$ angles θ_i.[8,9,16] The narrowness angle θ of the spicule θ, is defined as the arithmetic mean of those angles θ_j that are less than or equal to the mean of all the angles θ_i for $i = 1, 2, \ldots, M - 1$. The length of the spicule is defined as

$$S = \sum_{i=1}^{M} s_i. \tag{10.11}$$

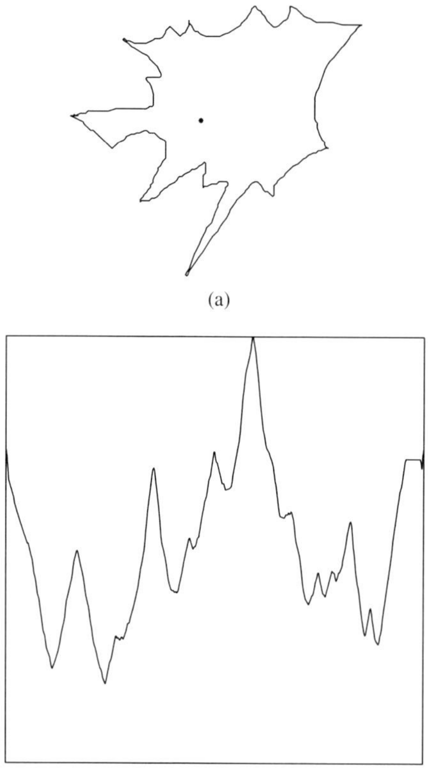

Figure 10.2 (a) The contour of a malignant tumor (the dot represents the centroid of the contour) and (b) the corresponding signature. The contour and signature have been normalized. FD-ruler 1D = 1.2698, FD-ruler 2D = 1.1658, FD-box 1D = 1.1580, FD-box 2D = 1.1612, *cf* = 0.7914, CD = 0.5872, f_{cc} = 0.5952, SI= 0.4883, and *ff* = 0.4602.

For reasons of simplicity, s_i indicates both the segment and its length. Let N be the number of significant spicules in the contour, and let θ_n and S_n $(n = 1, 2, \ldots, N)$ be the narrowness angle and the spicule length, respectively. The value of SI of the contour is defined as[8,9,16]

$$SI = \frac{\sum_{n=1}^{N} (1 + \cos\theta_n) S_n}{\sum_{n=1}^{N} S_n}. \tag{10.12}$$

The shape factor *ff* based on Fourier descriptors is a measure related to the presence of roughness or high-frequency components in contours.[8,50,51] The measure is derived by taking the sum of the normalized Fourier descriptors of the coordinates of the contour pixels divided by the corresponding indices, dividing it by the sum of the normalized Fourier descriptors, and subtracting the result from unity, as follows:[8,50]

$$ff = 1 - \frac{\sum_{k=-N/2+1}^{N/2} |Z_o(k)|/|k|}{\sum_{k=-N/2+1}^{N/2} |Z_o(k)|}. \tag{10.13}$$

Here, $Z_o(k)$ are the normalized Fourier descriptors, defined as

$$Z_o(k) = \begin{cases} 0, & k = 0; \\ \dfrac{Z(k)}{Z(1)}, & \text{otherwise.} \end{cases}$$

The Fourier descriptors themselves are defined as

$$Z(k) = \frac{1}{N} \sum_{n=0}^{N-1} z(n) \ \exp\left[-j \frac{2\pi}{N} nk\right], \tag{10.14}$$

$k = -\frac{N}{2}, \ldots, -1, 0, 1, 2, \ldots, \frac{N}{2} - 1$, where $z(n) = x(n) + jy(n), n = 0, 1, \ldots, N - 1$, represents the sequence of contour pixel coordinates. Contours of malignant tumors are expected to be more rough with increased high-frequency content, in general, than the contours of benign masses; hence, the value of ff is expected to be higher for the former than the latter.[8,9,14,20]

Rangayyan and Oloumi[12] applied the approach of power spectral analysis to signatures of contours of breast masses to estimate the FD. The approach is based on fractional Brownian motion and processes with their power spectra varying in proportion to $1/f^{\beta}$, where f is the frequency, and β, known as the spectral exponent, is related to the FD. The results were comparable to those provided by other estimates of the FD obtained using the box-counting and ruler methods. This method is not included in this chapter.

10.2.3 Feature analysis, selection, and classification

In the present study, nine features (shape factors) were extracted for each contour, as described in Section 10.2.2; each feature or shape factor is a single-valued quantity. The dimension of the feature set or vector for each breast mass, at the initial stage of analysis, is $N_f = 9$. The capability of each feature to discriminate between benign masses and malignant tumors was assessed by analyzing the associated ROC curve.[52] The ROC curve shows the variation in the sensitivity versus the false-positive rate (complement of the specificity) for various values of the decision threshold. The area under the curve (AUC) serves as a measure of the performance of the decision-making system or classifier.

Feature analysis and selection[53–56] was performed by evaluating the performance of each feature or combinations of several features based on their p-values,[57] AUC values,[52] and other measures of performance. In the present work, the stepwise logistic regression procedure was used, which includes regression models in which the choice of the predictive variables is performed by an automatic procedure.[58] Each step consists of one step of forward selection and one step of backward elimination. These two tasks were repeated until no features were added or removed. Feature selection was also performed using stepwise regression, which iteratively varies the number of features used in the classification step by entering features into or removing features from the group of selected features based on a selection criterion using the F-statistic.[54]

Classification was performed using the selected sets of features with LDA, quadratic discriminant analysis (QDA),[53] and an artificial neural network with radial basis functions (RBFs).[59] The leave-one-out (LOO) procedure was used in the feature-selection step as well as the training and testing steps of each classifier. The inclusion of the feature selection step within the LOO procedure reduces the bias introduced by the training sample; however, the number of selected features varies for each trial (for each mass being tested). ROC curves were obtained based on the discriminant values derived from each classifier for each contour.

For further cross-validation of the features and the classification procedures, feature selection and training of each of the classifiers mentioned above were performed with one of the three datasets of contours used in the present work, and the classifier so obtained was tested on the remaining two datasets. In this case, the number of features selected depends upon the training dataset and remains constant for all masses in the test datasets.

10.3 Datasets of Contours of Breast Masses

Dataset A: Mammograms of 20 cases with breast masses were obtained from Screen Test: the Alberta Program for the Early Detection of Breast Cancer.[19,60] Ethics approval of the project was obtained from the Conjoint Health Research Ethics Board, Office of Medical Bioethics, University of Calgary, and the Calgary Regional Health Authority. The mammograms were digitized using the Lumiscan 85 scanner at a resolution of 50 μm with 12 bits/pixel. The dataset includes 57 images, 37 of which contain benign masses and 20 of which contain malignant tumors.[19] The diagnostic classification of the masses was based upon biopsy. The areas of the benign masses vary from 39 to 423 mm^2, with an average and standard deviation of 163 ± 87 mm^2. The areas of the malignant tumors vary from 34 to 1122 mm^2, with an average and standard deviation of 265 ± 283 mm^2. The contour of each mass was manually drawn on the corresponding mammographic image by an expert radiologist specialized in mammography and verified independently by another radiologist. Contours were drawn meticulously so as to include fine details related to microlobulations and spicules. Most of the benign masses in this dataset are well circumscribed (circumscribed benign or CB), whereas most of the malignant tumors are spiculated (spiculated malignant or SM), as typically encountered in mammographic images. BI-RADS ranking[61] was not used in the Screen Test program.

Dataset B: In an earlier study, a set of 54 mammographic images containing masses was obtained from the original Mammographic Image Analysis Society (MIAS) database[62] and the teaching library of the Foothills Hospital in Calgary.[9] The original MIAS images were digitized at a resolution of 50 μm, whereas the Foothills Hospital images were digitized at a resolution of 62 μm. The diagnostic classification of the masses was based upon biopsy. The areas of the benign masses vary from 32 to 1207 mm^2, with an average and standard deviation of 281 ± 288 mm^2. The areas of the malignant tumors vary from 46 to 1244 mm^2, with an average and standard deviation of 286 ± 292 mm^2. The contours of the masses in the mammographic dataset were drawn manually on the digitized images by an expert radiologist specialized in mammography.[9,10,19] Contours were drawn meticulously so as to include fine details related to microlobulations and spicules. This dataset includes circumscribed and spiculated cases in both the benign and malignant categories. Spiculated benign (SB) masses and circumscribed malignant (CM) tumors are unusual, and they tend to cause difficulties in pattern-classification studies.[9] This dataset has a total of 54 contours, including 16 CB, 12 SB, 19 SM, and seven CM types. BI-RADS ranking[61] is not provided by the MIAS database and was not used at the time of preparation of the Foothills Hospital cases.

Dataset C: A set of 192 mammograms was obtained from 192 patients at the Senology Unit, San Paolo Hospital, Bari, Italy, ASL Ba/4 (medical group). The study was approved by the San Paolo Hospital, and informed consent was obtained from patients for anonymous use of mammographic and related data for scientific purposes. The patients were diagnosed to have breast disease via screen-film mammography and confirmed histologically; 163 of the cases were malignant, and 29 were benign. The most-useful mammographic projections were selected to analyze the contours of the lesions. The areas of the benign masses vary from 13 to 927 mm^2, with an average and standard deviation of 249 ± 233 mm^2. The areas of the malignant tumors vary from 13 to 1263 mm^2, with an average and standard deviation of 333 ± 244 mm^2.

During an initial phase, contours of the lesions present on the film images were manually traced by a team of three radiologists specialized in mammography, using a colored grease pencil. The contoured image was then placed on a Wacom Intuos3 A4 graphic tablet, and using an optical pen, the part of interest was retraced so as to obtain a digital representation of the contour. Each contour was saved in a text file as a list of points, corresponding to the (x, y) coordinates of the points on the contour. The contour was then resampled so as to achieve a uniform distribution of points.

In the second phase, a custom CAD environment was implemented in MATLAB®, with a graphical user interface consisting of a full-screen window and a drop-down menu for draw and save commands. The commands available in the menu were also activated by the optical pen to make the process of manual drawing of the contours and the related data acquisition simpler and faster.

10.4 Results and Discussion

Figure 10.3 shows the ROC curves for the four estimates of the FD listed in Table 10.1 with Dataset A; Figure 10.4 shows the ROC curves for the shape factor *cf* with the three datasets. Table 10.1 shows the AUC values obtained for each of the shape factors studied with each of the datasets used in the present work. Several of the shape factors have provided high classification performance across the highly variable datasets of contours of breast masses, with an AUC in the range of $[0.7308, 0.9973]$. All of the shape factors have performed very well with Dataset A, with an AUC in the range of $[0.9135, 0.9973]$, due to the fact that most of the benign masses in the dataset have smooth contours whereas most of the malignant tumors have spiculated contours. The performance of the shape factors is not as high with Dataset B, albeit good, being in the range of $[0.7308, 0.8448]$, due to the presence of a

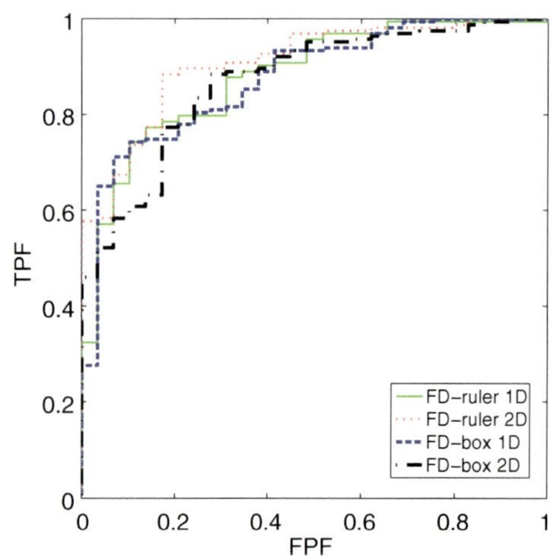

Figure 10.3 ROC curves indicating the classification performance of FDs obtained with the ruler method and the box-counting method with Dataset C. The values of the area AUC under the ROC curves are listed in Table 10.1. TPF: true-positive fraction; FPF: false-positive fraction.

Table 10.1 List of the nine shape factors and their individual AUC values for each dataset used in the present study.

Shape Factor	Dataset A	Dataset B	Dataset C
FD-ruler 1D	0.9419	0.8228	0.8794
FD-ruler 2D	0.9743	0.8448	0.9084
FD-box 1D	0.9230	0.8173	0.8752
FD-box 2D	0.9135	0.7761	0.8695
cf	0.9851	0.7967	0.9175
CD	0.9824	0.7308	0.9135
f_{cc}	0.9973	0.7527	0.8367
SI	0.9662	0.8118	0.8887
ff	0.9878	0.8173	0.9040

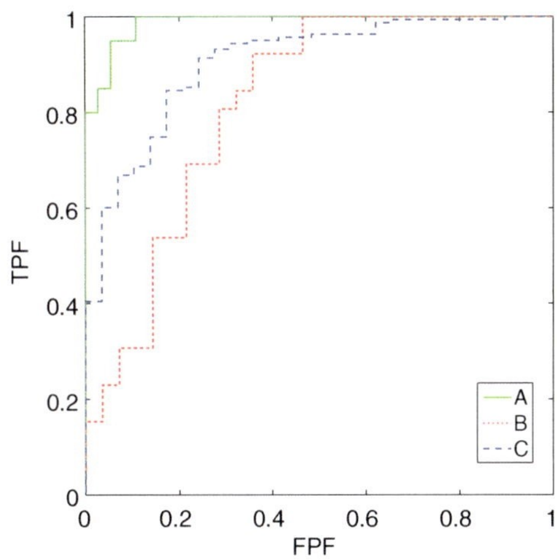

Figure 10.4 ROC curves representing the classification performance of *cf* with the three datasets used in the study. The values of the area under the ROC curves are listed in Table 10.1.

large number of SB and CM cases. The results with Dataset C are very good, with an AUC in the range of [0.8367, 0.9175].

By selecting the point closest to the upper-left corner with the coordinates [0.0, 1.0] in the ROC curve shown in Fig. 10.4 for the shape factor *cf*, a high performance is achieved by using Dataset A with a sensitivity of 95.2% at a specificity of 94.4%. The performance of the same feature with Dataset C is slightly poorer, with a sensitivity of 84.4% at a specificity of 83%. Other values of sensitivity and specificity may be obtained by choosing other operating points on the various ROC curves obtained in the study.

The results of feature selection with the stepwise logistic regression procedure and the stepwise regression procedure with the F-statistic are shown in Table 10.2 for the three datasets used in the study and their combinations. The set of selected features varies in each step of the LOO procedure (for each

Table 10.2 List of the shape factors selected and the AUC values with various classifiers for the datasets used in the present study and combinations thereof. The rows indicated with an asterisk represent the features selected most often in the LOO procedure for each dataset listed. The set of selected features and the dimension of the feature vector (N_f) varies in each step of the LOO procedure (for each mass being tested). The initial set of features has a dimension of 9.

Feature Selection	Classifier	Dataset A	Dataset B	Dataset {A, B}	Dataset C	Dataset {A, B, C}
All	LDA	0.9797	0.7390	0.9117	0.8877	0.9267
features	QDA	0.9797	0.7885	0.9154	0.8500	0.9084
in Table 10.1	RBF	0.9919	0.7981	0.9348	0.9162	0.9309
Logistic regression	*	f_{cc}	FD-ruler 2D, FD-box 2D	f_{cc}, SI, CD	FD-ruler 1D, CD	FD-ruler 1D, f_{cc}
	LDA	0.9973	0.8448	0.9247	0.9243	0.9327
	QDA	0.9973	0.8393	0.9177	0.8982	0.9283
	RBF	0.9973	0.8599	0.9324	0.9264	0.9393
Stepwise regression	*	f_{cc}, SI	FD-ruler 2D, FD-box 2D	f_{cc}, SI, CD	FD-ruler 1D, cf, f_{cc}	FD-ruler 1D, cf, f_{cc}, SI, CD
	LDA	0.9919	0.8448	0.9247	0.9076	0.9297
	QDA	0.9920	0.8393	0.9177	0.8944	0.9044
	RBF	0.9973	0.8599	0.9324	0.9156	0.9362

mass being tested). The features selected most often with the two procedures and various datasets are listed in Table 10.2; these features may be considered to be stronger than the other features that were not selected as often. The AUC values obtained with the three classifiers used in the study indicate that some combinations of the shape factors can lead to slightly higher classification performance than the individual features (compare the AUC values in Table 10.2 with those in Table 10.1).

Table 10.3 summarizes the results of cross-validation of the selected features and classifiers with the various datasets used in the study. The results indicate the importance of training a classifier with a large dataset including a collection of diverse cases. The best results have been obtained by training the

Table 10.3 Results of cross-validation of the selected features and classifiers with various datasets (in terms of AUC). The list of selected features in each case was obtained using the stepwise regression procedure with the F-statistic. The dimension of the feature vector (N_f) is equal to the number of features selected. The initial set of features has a dimension of 9.

Training Set	Selected Features	Test Set	LDA	QDA	RBF
A	$N_f = 2$: f_{cc}, SI	B	0.7816	0.8063	0.7971
A	$N_f = 2$: f_{cc}, SI	C	0.8864	0.8551	0.8866
B	$N_f = 2$: FD-ruler 2D, FD-box 2D	A	0.9487	0.9487	0.9608
B	$N_f = 2$: FD-ruler 2D, FD-box 2D	C	0.8718	0.8788	0.8972
{A, B}	$N_f = 3$: f_{cc}, SI, CD	C	0.8877	0.8132	0.8540
C	$N_f = 3$: FD-ruler 1D, cf, f_{cc}	A	0.9932	0.9932	0.9978
C	$N_f = 3$: FD-ruler 1D, cf, f_{cc}	B	0.8049	0.8338	0.8387
C	$N_f = 3$: FD-ruler 1D, cf, f_{cc}	{A, B}	0.9127	0.9234	0.9170

classifiers with Dataset C and testing on Datasets A and B. Whereas the results with the LOO procedure listed in Table 10.2 may be biased to some extent, the results in Table 10.3 are more reliable due to the complete independence between the training and testing datasets. The results with high AUC values in Table 10.3 indicate the robustness of the shape factors used in the present study to variations in the datasets, feature selection methods, pattern classification techniques, and cross-validation methods.

The present study has performed comparative analysis and extensive evaluation of several measures of shape for the purpose of classification of breast masses based on the roughness of their contours. Although most typical benign masses have smooth contours and most malignant tumors have rough or spiculated contours, such a distinction may not be present always. It is desirable to use a combination of several measures of shape that characterize various notions of shape roughness. The following points are noteworthy regarding the several measures of shape used in the present study:

- Fractal dimension characterizes the complexity of a contour's shape in terms of the presence of nesting patterns with self-similarity and its space-filling nature; this helps distinguish between macrolobulated benign masses and microlobulated malignant tumors.
- Compactness captures the relationship between the perimeter of a contour and the area contained, and helps identify contours with narrow or oblong parts; this helps discriminate between benign masses with oval contours and malignant tumors with finger-like protrusions or excursions.
- Fractional concavity and convex deficiency help recognize contours with several concave parts, indentations, or incursions; they help separate well-circumscribed benign masses from malignant tumors with ill-defined and undulating contours.
- The Fourier factor quantifies the relative amount of high-frequency power associated with rapid fluctuations in a contour, and helps classify benign masses with smooth contours as being distinct from malignant tumors with rough contours.
- Spiculation index yields a large value in the presence of long and narrow spicules with small internal angles that are expected in malignant tumors; its value will be small for benign masses with round contours and no spicules.

The results of the present study demonstrate the effectiveness of the proposed shape factors in the classification of breast masses and tumors. The various shape factors proposed capture different distinguishing characteristics of benign masses and malignant tumors. The multiple notions of shape roughness described above and captured by the different shape factors may find other applications in computer vision and image analysis.

10.5 Conclusion

The results obtained in the present study support the hypothesis that contours of breast masses possess fractal properties and that the FD can help discriminate benign masses from malignant tumors. The results also indicate that several other shape factors, such as cf, CD, f_{cc}, ff, and SI, can provide comparable performance and also augment the performance of the FD. The variations in the performance of the features across the datasets used in the study may be related to the specifics of the datasets, in terms of the smooth-versus-rough nature of the benign masses and malignant tumors. Regardless, the results of cross-validation across three different sets of contours from different sources and with annotations by multiple radiologists indicate that shape analysis can make substantial contributions to the analysis of breast masses in mammograms. Further studies are in progress on detailed evaluation of the shape features with contours of breast masses in mammograms obtained with digital mammographic imaging systems. It is desirable to evaluate the shape features with automatically detected contours of breast masses[20] and a large database. Given the difficulty in obtaining accurate contours of masses in mammograms, it is also desirable to evaluate the classification performance of combinations of shape factors with other features related to texture and edge sharpness.[20] The methods described in the present work are expected to make important contributions to CAD of breast cancer.

References

1. T. F. Nonnenmacher, G. A. Losa, and E. R. Weibel, *Fractals in Biology and Medicine, Vol.I*, Birkhäuser, Basel, Switzerland (1994).
2. T. F. Nonnenmacher, G. A. Losa, and E. R. Weibel, *Fractals in Biology and Medicine, Vol.II*, Birkhäuser, Basel, Switzerland (1998).
3. M. M. Novak, *Fractals in the Natural and Applied Sciences*, North Holland, Amsterdam (1994).
4. M. J. Homer, *Mammographic Interpretation: A Practical Approach*, 2nd Ed., McGraw-Hill, New York (1997).
5. K. Doi, "Diagnostic imaging over the last 50 years: research and development in medical imaging science and technology," *Physics in Medicine and Biology* **51**, R5–R27 (2006).
6. K. Doi, "Computer-aided diagnosis in medical imaging: historical review, current status and future potential," *Computerized Medical Imaging and Graphics* **31**, 198–211 (2007).
7. J. Tang, R. M. Rangayyan, J. Xu, I. E. Naqa, and Y. Yang, "Computer-aided detection and diagnosis of breast cancer with mammography: Recent advances," *IEEE Transactions on Information Technology in Biomedicine* **13**(2), 236–251 (2009).

8. R. M. Rangayyan, *Biomedical Image Analysis*, CRC Press, Boca Raton, FL (2005).
9. R. M. Rangayyan, N. R. Mudigonda, and J. E. L. Desautels, "Boundary modelling and shape analysis methods for classification of mammographic masses," *Medical and Biological Engineering and Computing* **38**, 487–496 (2000).
10. R. M. Rangayyan and T. M. Nguyen, "Fractal analysis of contours of breast masses in mammograms," *Journal of Digital Imaging* **20**(3), 223–237 (2007).
11. T. M. Cabral and R. M. Rangayyan, *Fractal Analysis of Breast Masses in Mammograms*, Morgan & Claypool, San Francisco, CA (2012).
12. R. M. Rangayyan and F. Oloumi, "Fractal analysis and classification of breast masses using the power spectra of signatures of contours," *Journal of Electronic Imaging* **21**(2), 023018:1–9 (2012).
13. G. Raguso, A. Ancona, L. Chieppa, S. L'Abbate, M. L. Pepe, F. Mangieri, M. De Palo, and R. M. Rangayyan, "Application of fractal analysis to mammography," *Proceedings of the 32nd Annual International Conference of the IEEE Engineering in Medicine and Biology Society 2010*, pages 3182–3185, Buenos Aires, Argentina (2010).
14. R. M. Rangayyan, N. M. El-Faramawy, J. E. L. Desautels, and O. A. Alim, "Measures of acutance and shape for classification of breast tumors," *IEEE Transactions on Medical Imaging* **16**(6), 799–810 (1997).
15. D. Guliato, J. D. de Carvalho, R. M. Rangayyan, and S. A. Santiago, "Feature extraction from a signature based on the turning angle function for the classification of breast tumors," *Journal of Digital Imaging* **21**(2), 129–144 (2008).
16. D. Guliato, R. M. Rangayyan J. D. Carvalho, and S. A. Santiago, "Polygonal modeling of contours of breast tumors with the preservation of spicules," *IEEE Transactions on Biomedical Engineering* **55**(1), 14–20 (2008).
17. D. Guliato, R. M. Rangayyan, W. A. Carnielli, J. A. Zuffo, and J. E. L. Desautels, "Segmentation of breast tumors in mammograms using fuzzy sets," *Journal of Electronic Imaging* **12**(3), 369–378 (2003).
18. N. R. Mudigonda, R. M. Rangayyan, and J. E. L. Desautels, "Gradient and texture analysis for the classification of mammographic masses," *IEEE Transactions on Medical Imaging* **19**(10):1032–1043, 2000.
19. H. Alto, R. M. Rangayyan, and J. E. L. Desautels, "Content-based retrieval and analysis of mammographic masses," *Journal of Electronic Imaging* **14**(2), 023016:1–17 (2005).
20. B. S. Sahiner, H. P. Chan, N. Petrick, M. A. Helvie, and L. M. Hadjiiski, "Improvement of mammographic mass characterization using spiculation

measures and morphological features," *Medical Physics* **28**(7), 1455–1465 (2001).

21. B. B. Mandelbrot, *The Fractal Geometry of Nature*, W. H. Freeman, San Francisco, CA (1983).
22. A. L. Goldberger, D. R. Rigney, and B. J. West, "Chaos and fractals in human physiology," *Scientific American* **262**, 42–49 (1990).
23. T. Matsubara, H. Fujita, S. Kasai, M. Goto, Y. Tani, T. Hara, and T. Endo, "Development of new schemes for detection and analysis of mammographic masses," *Proceedings of the 1997 IASTED International Conference on Intelligent Information Systems (IIS 97)*, pages 63–66, Grand Bahama Island, Bahamas (1997).
24. S. Pohlman, K. A. Powell, N. A. Obuchowski, W. A. Chilcote, and S. Grundfest-Broniatowski, "Quantitative classification of breast tumors in digitized mammograms," *Medical Physics* **3**(8), 1337–1345 (1996).
25. P. Dey and S. K. Mohanty, "Fractal dimensions of breast lesions on cytology smears," *Diagnostic Cytopathology* **29**(2), 85–86 (2003).
26. L. Zheng and A. K. Chan, "An artificial intelligent algorithm for tumor detection in screening mammogram," *IEEE Transactions on Medical Imaging* **20**(7), 559–567 (2001).
27. Q. Guo, V. Ruiz, J. Shao, and F. Guo, "A novel approach to mass abnormality detection in mammographic images," *Proceedings of the IASTED International Conference on Biomedical Engineering*, pages 180–185, Innsbruck, Austria (2005).
28. Q. Guo, J. Shao, and V. Ruiz, "Characterization and classification of tumor lesions using computerized fractal-based texture analysis and support vector machines in digital mammograms," *International Journal of Computer Assisted Radiology and Surgery* **4**(1), 11–25 (2009).
29. G. D. Tourassi, D. M. Delong, and C. E. Floyd, Jr., "A study on the computerized fractal analysis of architectural distortion in screening mammograms," *Physics in Medicine and Biology* **51**(5), 1299–1312 (2006).
30. E. Anguiano, M. A. Pancorbo, and M. Aguilar, "Fractal characterization by frequency analysis: I. Surfaces," *Journal of Microscopy* **172**, 223–232 (1993).
31. M. Aguilar, E. Anguiano, and M. A. Pancorbo, "Fractal characterization by frequency analysis: II. A new method," *Journal of Microscopy* **172**, 233–238 (1993).
32. N. Eltonsy, G. Tourassi, and A. Elmaghraby, "Investigating performance of a morphology-based CAD scheme in detecting architectural distortion in screening mammograms," *Proceedings of the 20th International Congress and Exhibition on Computer Assisted Radiology and Surgery (CARS 2006)*, pages 336–338, Osaka, Japan (2006).

33. R. M. Rangayyan, S. Prajna, F. J. Ayres, and J. E. L. Desautels, "Detection of architectural distortion in mammograms acquired prior to the detection of breast cancer using Gabor filters, phase portraits, fractal dimension, and texture analysis," *International Journal of Computer Assisted Radiology and Surgery* **2**(6), 347–361 (2008).
34. R. M. Rangayyan, S. Banik, and J. E. L. Desautels, "Computer-aided detection of architectural distortion in prior mammograms of interval cancer," *Journal of Digital Imaging* **23**(5), 611–631 (2010).
35. S. Banik, R. M. Rangayyan, and J. E. L. Desautels, "Detection of architectural distortion in prior mammograms," *IEEE Transactions on Medical Imaging* **30**(2), 279–294 (2011).
36. C. B. Caldwell, S. J. Stapleton, D. W. Holdsworth, R. A. Jong, W. J. Weiser, G. Cooke, and M. J. Yaffe, "Characterization of mammographic parenchymal pattern by fractal dimension," *Physics in Medicine and Biology* **35**(2), 235–247 (1990).
37. J. W. Byng, N. F. Boyd, E. Fishell, R. A. Jong, and M. J. Yaffe, "Automated analysis of mammographic densities," *Physics in Medicine and Biology* **41**, 909–923 (1996).
38. K. M. Iftekharuddin, W. Jia, and R. Marsh, "Fractal analysis of tumor in brain MR images," *Machine Vision and Applications* **13**, 352–362 (2003).
39. C. M. Wu, Y. C. Chen, and K. S. Hsieh, "Texture features for classification of ultrasonic liver images," *IEEE Transactions on Medical Imaging* **11**(2), 141–152 (1992).
40. A. N. Esgiar, R. N. G. Naguib, B. S. Sharif, M. K. Bennett, and A. Murray, "Fractal analysis in the detection of colonic cancer images," *IEEE Transactions on Information Technology Biomedicine* **6**(1), 54–58 (2002).
41. T. K. Lee, D. I. McLean, and M. S. Atkins, "Irregularity index: a new border irregularity measure for cutaneous melanocytic lesions," *Medical Image Analysis* **7**, 47–64 (2003).
42. K. Kikuchi, S. Kozuma, K. Sakamaki, M. Saito, G. Marumo, T. Yasugi, and Y. Taketani, "Fractal tumor growth of ovarian cancer: sonographic evaluation," *Gynecologic Oncology* **87**, 295–302 (2002).
43. S. H. Nam and J. Y. Choi, "A method of image enhancement and fractal dimension for detection of microcalcifications in mammogram," *Proceedings of the 20th Annual International Conference of the IEEE Engineering in Medicine and Biology Society*, pages 1009–1011, Hong Kong (1998).
44. W. Klonowski, R. Stepien, and P. Stepien, "Simple fractal method of assessment of histological images for application in medical diagnostics," *Nonlinear Biomedical Physics* **4**(7), www.nonlinearbiomedphys.com/content/4/1/7 (2010).

45. M. Tambasco, B. M. Costello, A. Kouznetsov, A. Yau, and A. M. Magliocco, "Quantifying the architectural complexity of microscopic images of histology specimens," *Micron* **40**(4), 486–494 (2009).

46. H. Li, M. L. Giger, O. I. Olopade, and L. Lan, "Fractal analysis of mammographic parenchymal patterns in breast cancer risk assessment," *Academic Radiology* **14**, 513–521 (2007).

47. H. Li, M. L. Giger, O. I. Olopade, and M. R. Chinander, "Power spectral analysis of mammographic parenchymal patterns for breast cancer risk assessment," *Journal of Digital Imaging* **21**(2), 145–152 (2008).

48. H. O. Peitgen and P. H. Richter, *The Beauty of Fractals—Images of Complex Dynamical Systems*, Springer, Berlin (1986).

49. H. O. Peitgen, H. Jurgens, and D. Saupe, *Chaos and Fractals: New Frontiers of Science*, Springer, New York (2004).

50. L. Shen, R. M. Rangayyan, and J. E. L. Desautels, "Detection and classification of mammographic calcifications," *International Journal of Pattern Recognition and Artificial Intelligence* **7**(6), 1403–1416 (1993).

51. L. Shen, R. M. Rangayyan, and J. E. L. Desautels, "Application of shape analysis to mammographic calcifications," *IEEE Transactions on Medical Imaging* **13**(2), 263–274 (1994).

52. C. E. Metz, "ROC methodology in radiologic imaging," *Investigative Radiology* **21**, 720–733 (1986).

53. R. O. Duda, P. E. Hart, and D. G. Stork, *Pattern Classification*, 2nd ed., Wily-Interscience, New York (2001).

54. B. Sahiner, H.-P. Chan, N. Petrick, R. F. Wagner, and L. Hadjiiski, "Feature selection and classifier performance in computer-aided diagnosis: the effect of finite sample size," *Medical Physics* **27**(7), 1509–1522 (2000).

55. R. J. Nandi, A. K. Nandi, R. M. Rangayyan, and D. Scutt, "Classification of breast masses in mammograms using genetic programming and feature selection," *Medical and Biological Engineering and Computing* **44**, 683–694 (2006).

56. S. Banik, R. M. Rangayyan, and J. E. L. Desautels, *Computer-aided Detection of Architectural Distortion in Prior Mammograms of Interval Cancer*, Morgan & Claypool, San Francisco, CA (2013).

57. J. H. Ware, F. Mosteller, F. Delgado, C. Donnelly, and J. A. Ingelfinger, "P values," in *Medical Uses of Statistics*, 2nd ed., J. C. Bailar III and F. Mosteller, Eds., NEJM Books, Boston, MA, pages 181–200 (1992).

58. F. L. Ramsey and D. W. Schafer, *The Statistical Sleuth: A Course in Methods of Data Analysis*, Duxbury Press, Belmont, CA (1997).

59. S. Haykin, *Neural Networks: A Comprehensive Foundation*, 2nd ed., Prentice Hall, Englewood Cliffs, NJ (1999).

60. Alberta Health Services, *Screen Test and the Alberta Breast Cancer Screening Program*, http://www.albertahealthservices.ca/services.asp?pid=service&rid=1002353, accessed September 2012.

61. American College of Radiology (ACR), *Illustrated Breast Imaging Reporting and Data System (BI-RADS)*, 4th ed., Reston, VA (2003).

62. J. Suckling, J. Parker, D. R. Dance, S. Astley, I. Hutt, C. R. M. Boggis, I. Ricketts, E. Stamakis, N. Cerneaz, S.-L. Kok, P. Taylor, D. Betal, and J. Savage, "The Mammographic Image Analysis Society Digital Mammogram Database," in *Digital Mammography: Proceedings of the 2nd International Workshop on Digital Mammography*, A. G. Gale, S. M. Astley, D. D. Dance, and A. Y. Cairns, Eds., Elsevier, York, UK, pages 375–378 (1994).

Grazia Raguso graduated in Mathematics and is a Professor of Geometry at the Faculty of Sciences of the University of Bari, Italy since 1988. She has been a member of the Board of Directors of the university since 2005, as well as a member and coordinator of numerous university committees and a reviewer for the journal *Mathematical Reviews*. She has published many papers in journals and conference proceedings. Her research interests include combinatorial structures and their applications, and integral geometry and geometric probability. In recent years, she has been involved in studies of dynamical systems and attractors, with particular attention devoted to the study of fractal geometry and its applications in biomedicine, specifically in mammography.

Antonietta Ancona graduated in Medicine and Surgery and obtained a diploma in professional qualification at the University of Siena, Italy in 1977. In 1981, she specialized in Radiology at the University of Bari, Italy. She received her primary diploma in Diagnostic Radiology in 1989. She worked at the Civil Hospital in Molfetta and served in the USL BA/9 - USL BA/4 Policlinico Hospital Consortium. In 1999, she moved to the ASL BA/4 - ASL BA Radiology Hospital "S. Paul," where she was appointed Head of Unit Simple - Section of Breast - Radiology Service. Since 1983, she has been a member of FONCAM (National Breast Cancer Task Force). From 1986 to date, she has taught at the Italian School of Breast directed by Prof. Umberto Veronesi. Since 1999, she has worked in business/organization for the Section of Breast - Radiology Hospital "S. Paul" certified by Health Management. From 2008 to present, she is the Medical Director of UOC Radiology - P.O. Sao Paulo - ASL BA. Her research interests are in biomedicine, specifically in mammography.

Loredana Chieppa graduated in Mathematics and received her Ph.D. in Computational Mathematics from the University of Bari, Italy in 2010. Her research interests are in applied mathematics, biomedical signal and image analysis, computer graphics, and computer-aided design; she has published related papers in journals and conference proceedings.

Samuela L'Abbate graduated in Mathematics and received her Ph.D. in Statistics from the Statistics Department at the University of Bari, Italy in May 2013. Her research interests are in fractal geometry and its applications in biomedicine, specifically in mammography. In the statistics field, she was involved in data mining, in particular, cluster analysis, a technique of unsupervised data mining. She has published papers in journals and conference proceedings on these topics.

Maria Luisa Pepe graduated as a medical doctor in 2006 and as a radiologist in 2011, with her B.S. and specialization gained cum laude at the University of Bari, Italy. Her research interests are in breast imaging and in other fields of imaging.

Fabio Felice Mangieri graduated in Medicine and Surgery in 1997. He specialized in Diagnostic Radiology in 2005. Since 2001, he has been working in the breast section of the Hospital San Paolo in Bari, Italy.

Shantanu Banik received his B.Sc. in 2005 in Electrical and Electronic Engineering from the Bangladesh University of Engineering and Technology (BUET), Dhaka, Bangladesh, and his M.Sc. and Ph.D. in 2008 and 2011, respectively, from the Department of Electrical and Computer Engineering, University of Calgary, Canada. He has coauthored several journal papers, a number of conference papers, three book chapters, and two books. He has received many awards and scholarships as a graduate student and postdoctoral fellow at the University of Calgary, including the Institute of Cancer Research (ICR), Canada Publication Prize for significant contributions to cancer research; a postdoctoral fellowship with the Natural Sciences and Engineering Research Council (NSERC) of Canada and Collaborative Research and Training Experience (CREATE); J. B. Hyne Research Innovation Award for outstanding research activity at the University of Calgary; Robert B. Paugh Memorial Award; Graduate Student Productivity Award; the Queen Elizabeth II Graduate (Doctoral) Scholarship; the Graduate Faculty Council Scholarship (Doctoral); the University Technologies International, Inc. (UTI)

Fellowship; the University of Calgary Alumni Association Graduate Scholarship; and the Schulich School of Engineering Teaching Assistant Excellence Award. He is currently working as a Research and Development Engineer at Circle Cardiovascular Imaging, Calgary, Alberta, Canada. His research interests include medical signal and image processing and analysis, development of computer-aided diagnosis techniques for the detection of cancer, segmentation and analysis of cardiac abnormalities, landmarking and segmentation of medical images, pattern recognition and classification, medical imaging, and automatic segmentation and analysis of tumors in CT and MRI images.

Rangaraj M. Rangayyan is a Professor of Electrical and Computer Engineering, and an Adjunct Professor of Surgery and Radiology, at the University of Calgary, Canada. He received his B.Eng. degree in Electronics and Communication from the University of Mysore at the People's Education Society College of Engineering, Mandya, Karnataka, India in 1976, and his Ph.D. in Electrical Engineering from the Indian Institute of Science, Bangalore, Karnataka, India in 1980. He has published more than 150 papers in journals and 250 papers in conference proceedings. He has been recognized with the 1997 and 2001 Research Excellence Awards of the Department of Electrical and Computer Engineering, the 1997 Research Award of the Faculty of Engineering, and by appointment as "University Professor" (2003–2013) at the University of Calgary. He is the author of two textbooks: *Biomedical Signal Analysis* (IEEE/Wiley, 2002) and *Biomedical Image Analysis* (CRC, 2005). He has coauthored and coedited several other books, including *Color Image Processing with Biomedical Applications* (SPIE, 2011). He has been recognized with the IEEE Third Millennium Medal in 2000 and the 2013 IEEE Canada Outstanding Engineer Medal, and elected as a Fellow of the following organizations: IEEE (2001), Engineering Institute of Canada (2002), American Institute for Medical and Biological Engineering (2003), SPIE (2003), Society for Imaging Informatics in Medicine (2007), Canadian Medical and Biological Engineering Society (2007), and Canadian Academy of Engineering (2009). His research interests include digital signal and image processing, biomedical signal and image analysis, and computer-aided diagnosis.

Chapter 11
Another Step towards Successful Tomographic Imaging in Cancer: Solving the Problem of Image Reconstruction

Artyom M. Grigoryan
Department of Electrical and Computer Engineering, University of Texas at San Antonio, USA

11.1 Introduction

The problem of image reconstruction from a finite number of projections is important in computed tomography (CT), which is used in diagnostic medicine[1] and electron microscopy.[2,3] The technology of CT scanners is advancing, and new scanner generations provide high-quality pictures, e.g., third-generation scanners wherein the detector array covers the entire FOV, and fan-beam geometry is used instead of parallel-beam geometry.[4,5] Computed tomography, which is also called computerized axial tomography (CAT), is a diagnostic procedure that uses special x-ray equipment to obtain cross-sectional pictures of the human body.

CT images are called tomograms; they show structures inside the body, including internal organs (such as the kidneys, liver, or spleen), blood vessels, bones, tumors, and other tissues. Tomograms are used to detect or confirm the presence of a tumor (including brain tumors); to provide information about the size and location of the tumor, and whether it has spread; to guide a biopsy (the removal of cells or tissues for examination under a microscope); and to help plan radiation therapy or surgery.

11.1.1 CT images and lung cancer

A wide variety of medical conditions that are not visible in conventional radiographs can be seen in CT images. According to the American College of Radiology, CT can be used for lung-cancer screening, and recent results from the National Lung Screening Trial have shown that CT lung-cancer screening can significantly reduce the number of people who die each year from lung cancer when CT is performed in the context of careful patient selection and follow-up. Also, new research published in the January issue of *The Journal of Nuclear Medicine* shows that CT and positron emission tomography (PET) imaging offer significant prognostic stratification information during initial staging for patients with locally advanced breast cancer. When compared to conventional imaging, 18F-FDG PET/CT more-accurately showed lesions in the chest, abdomen, and bones. The study by Iagaru et al.[6] suggests that 18F-FDG PET/CT should be considered as complimentary imaging tools in the pre- and postoperative work-up of patients diagnosed with breast cancer.

11.1.2 CT images and breast cancer

As reported by Pettigrew and Berry,[7] breast cancer is a pathology that causes 40,000 deaths each year in the United States, and it is the second leading cause of death in women.[8,9] Early detection is necessary to treat patients and improve their chances for survival. Over the last 15 years, the number of breast-cancer-related deaths has been drastically reduced (by 30–50%) because of technological advancements in mammography. Recent studies show the potential benefits of the use of computed tomography technology for breast imaging. The breast CT becomes another alternative to mammography and magnetic resonance imaging for use in breast cancer screening. There are a few complications in mammography, and the major one is the lack of sensitivity with mammography, which makes it impossible to detect some abnormalities, for instance, breast lesions and breast cancer.[10,11] Studies show that 30% of cancers are not detected by mammography, and 70–90% of biopsies conducted as a result of suspicious mammograms are negative, driving up medical costs and causing patients unnecessary stress. Other complications include not being able to visualize the breast in a 3D view and how the superimposition of breast tissues with mammography makes it harder to detect abnormalities.[12,13] It is currently recognized that CT may potentially play a major role in breast imaging by providing advanced treatment options for patients. Experimental and research studies have addressed using breast CT as another diagnostic tool for breast cancer.[14] Studies indicate that breast CT can be as valuable as mammography, if not more so.

Using cone-beam technology for breast CT has been shown to benefit image quality. A simulation study conducted by Chen and Ning[13] showed the feasibility of cone-beam dedicated breast CT systems in identifying lesions only a few millimeters in size and calcifications 100 micrometers in size. The breast CT can produce better low-contrast images that aid in detecting breast masses. A precise measurement of the location and volume visualization of lesions is possible, as well as optimal resolution for calcification with high contrast. This is an important aspect of breast CT because physicians can now localize the lesions, which will lead to advanced treatment options with CT, such as using breast CT for biopsies.

11.1.3 Breast cancer with CT, mammography, and MRI

The breast CT may become another alternative to magnetic resonance imaging (MRI) for screening women who are high at risk for breast cancer and for estimating the extent of their disease.[15–17] The cost of breast CT will be less expensive than MRI, and the potential interpretation time for CT may be shorter than that for MRI. The potential advantages of breast CT systems far outweigh the disadvantages. These advantages include the ability to image the entire volume of the breast, the high spatial resolution, the superior contrast resolution, the elimination of superimposed breast tissues and structures (resulting in improved tumor detection), and the fine anatomical detail produced with breast CT. Although dedicated breast CT utilizes ionizing radiation, the radiation dose is equal to or less than the dose used for mammography. Also, unlike mammography, breast CT does not need to use breast compression, which is a significant benefit for patients.

Although breast CT systems are still in development, the study of breast CT is moving in a positive direction, and it has the potential to gradually change the field of breast imaging. If breast CT becomes commercially available in the future, it will likely play an important role in screening and early diagnosis. Will breast CT replace mammography or MRI? Breast CT may have a vital role in evaluating women with dense breasts, implants, and small cancers in dense breasts. Translational research should help provide evidence-based guidelines for clinical implementation of breast CT. Lowering patient dose while increasing the resolution needed to detect angiogenesis at the ductal level will help increase survival rates among women.

11.1.4 Algorithms in CT

With new and still-developing generations of CT scanners and their effective implementation in tomographic imaging, the complexity of the mathematical methods and approaches for image reconstruction from projections increases. The problem solved by Radon in 1917[18,19] for reconstructing a function from its projections has not yet been solved for practical applications because an infinite set of line integrals (measurements) is not available. Many of the existing methods of image reconstruction from a finite number of projections process the projection data on the polar grid of coordinates and then transform the data into the Cartesian lattice without fully covering it, which results in low-quality reconstructions. This is the traditional way to calculate the approximation of the reconstructed image, which is shown by the Fourier slice theorem.[1,2]

This chapter presents a new approach for solving the problem of image reconstruction, which differs from the known methods of back-projection, iterative reconstruction, Fourier filtering, Radon filtering, and convolution filtering.[20–28] The reconstruction is performed by calculating the tensor or paired transform[29–33] of the image from the projection data. In the tensor representation, a 2D discrete image is considered as the sum of direction images, each of which is determined by the corresponding signal. These signals are called the splitting-signals; they also carry the spectral information of the image at frequency points of different subsets that cover the whole domain of frequencies. The discrete image can be reconstructed by its splitting-signals by calculating the 2D DFT or by directly performing the inverse 2D tensor transform. In addition, the tensor representation possesses the following important property. This transform is in the discrete space, and all components of the transform are defined as the ray sums of the discrete image in the Cartesian lattice. However, it is very important to note that these sum rays can be exactly calculated from the ray integrals. In other words, the splitting-signals of the discrete image $f_{n,m}$ can be calculated from the projection of the image $f(x,y)$. Each projection is processed by a system of linear equations or linear convolutions to calculate the corresponding part of the 2D tensor or paired representation of the image,[33,45,49] and then the inverse transform is calculated to obtain the discrete image. The model described for image reconstruction is simple, and the reconstruction is exact. The proposed method was implemented in MATLAB and C++, and the experimental results of image reconstruction are illustrated. Preliminary results also show high-quality images when applying the proposed method with an incomplete set of projections for the angular range scanned down to 10 deg.

11.2 Model of the Image

Consider the simple discrete model of image reconstruction when the parallel scanning scheme is used (although the proposed algorithms can be generalized to the fan scanning scheme, too). Figure 11.1 illustrates the problem of image

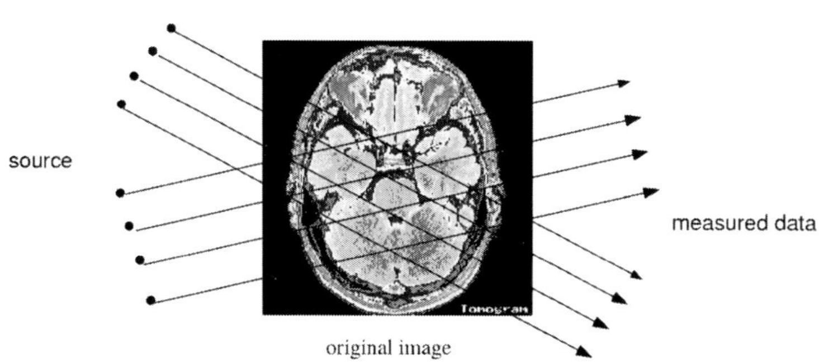

Figure 11.1 The image and two sets of parallel rays.

reconstruction when assuming that the radiation source and detector represent themselves—the points and the rays spreading between them are straight.

To simplify the calculations, consider the 2D image $f(x,y)$ to be reconstructed in the unit square $[0,1] \times [0,1]$ with the Cartesian lattice $N \times N$ on it. The integer $N > 1$ is the size of the lattice. The lattice is denoted by $X = X_{N,N} = \{(n,m); n,m = 0 : (N-1)\}$. The square is divided by small pieces or image elements (IE) of size $\Delta x = \Delta y = 1/N$ each, and the knots of the lattice are in the centers of IE. Image elements are numbered by (n,m). In each image element, the value of the image is considered to be constant:

$$f_d(x,y) = \frac{1}{(\Delta x)^2} \int_{IE} f(x,y)dxdy, \quad \text{if} \quad (x,y) \in IE.$$

Thus, the image $f(x,y)$ is represented by $f_d(x,y)$, and the discrete image to be reconstructed on the lattice from $f_d(x,y)$ is $f_{n,m}$. The values of the discrete image are defined by

$$f_{n,m} = \int_{(n,m)\text{th} \in IE} f(x,y)dxdy = (\Delta x)^2 f_d(x,y), \qquad \forall (x_0, y_0) \in (n,m)\text{th IE},$$

and they are placed in the centers of the corresponding (n,m)th IE. When N is large and the image $f(x,y)$ is presented by its small elements, the earlier equations describe a good model of the image on the Cartesian lattice. Such a model is considered in the well-known series-expansion methods for solving the system of line integrals in the discrete case.[20–25]

11.2.1 Line integrals and ray sums

The sum of the discrete image along a ray l passing through the knots of the Cartesian lattice is denoted by v_l, and they are called ray sums:

$$v_l = \sum_{(n,m)\in l} f_{n,m}. \tag{11.1}$$

In Eq. (11.1), the line integral along a ray l in the (n,m)th IE is defined as follows:

$$w_l^{n,m} = \int_{(n,m)\text{th} \in IE} f(x,y)dl = (\Delta l) f_d(x_0, y_0) = \Delta l \frac{1}{(\Delta x)^2} f_{n,m} = (\Delta l) N^2 f_{n,m}, \tag{11.2}$$

where (x_0, y_0) is a point in the (n,m)th IE, and $\Delta l = \Delta l_{n,m}$ denotes the length of the ray l in the (n,m)th IE. The ray is referred to as the line, i.e., the ray with zero width. The line integral along the ray l can be written as

$$w_l = \int f(x,y)dl = \sum_{(n,m)\in l} w_l^{(n,m)} = N^2 \sum_{(n,m)\in l} (\Delta l_{n,m}) f_{n,m}, \tag{11.3}$$

where the summation is performed by the image elements through which the ray l passes. Thus, there is a large system of linear equations describing the relationship between the line integrals w_l and the values $f_{n,m}$ of the discrete image. The solution of this system is unknown; it is difficult to find this solution because the values of $(\Delta l_{n,m})$ are generally different even for the parallel rays of the same projection. However, it is important to say that there exist sets of parallel rays with the same value of length in each IE through which the rays pass (these rays are described in a moment). The rays l of such a set may differ from the rays defined by the ray sums in Eq. (11.1) depending on the projection, but in all cases the ray sums can be uniquely calculated from line integrals.

11.2.2 Two types of parallel rays

This section describes the rays that pass the unit square and the rays that pass through the knots of the Cartesian lattice. Therefore, two coordinate systems on the plane are considered. The first system of coordinates (x, y) is for the image $f(x, y)$ on the square $[0, 1] \times [0, 1]$, as shown in Fig. 11.2(a), for the image with 13 ellipses, each with constant intensity. The second coordinate system (n, m), where n and m are integers, is for the lattice $X_{N,N}$ in the square. This system is used for the discrete image $f_{n,m}$, as shown in Fig. 11.2(b). Parameters x and n run from left to right, and parameters y and m run from top to bottom.

Consider the parallel lines on the unit square and lattice, which are parameterized by coordinates of the frequency points. Given the frequency point $(p, s) \in X_{N,N}$, such that g.c.d. $(p, s) = 1$, consider the lines

$$l(t) = l_{p,s}(t) = \{(n, m); pn + sm = t\}, \quad t = 0 : (p + s)(N - 1)$$

on the square lattice $X_{N,N}$. These lines are referred to as the *arithmetical rays*.

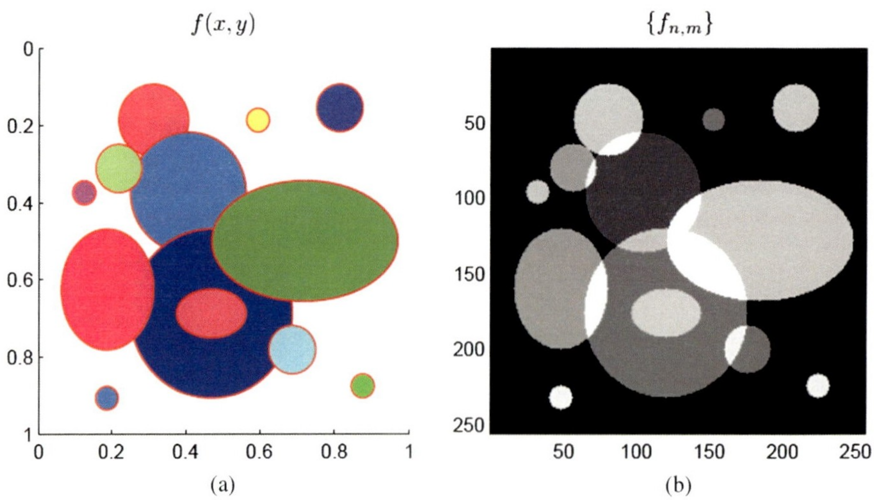

Figure 11.2 Two coordinate systems for images $f(x, y)$ and $f_{n,m}$.

The equations of these lines on the square $[0,1] \times [0,1]$ are

$$l(t) = l_{p,s}(t) = \left\{(x,y); px + sy = \frac{t}{N} + \frac{p+s}{2N}\right\}, \quad t = 0 : (p+s)(N-1).$$

These lines are referred to as the *geometrical rays* to distinguish the discrete and continuous cases. These two types of rays are denoted by $l(t)$, and the same set of t is considered for the rays, $t = 0 : (p+s)(N-1)$. The generator (p,s) defines the slope $-\tan^{-1}(p/s)$ of these rays. The set of line integrals

$$W = W_{p,s} = \{w_{l(t)}; l(t) = l_{p,s}(t), \quad t = 0 : (p+s)(N-1)\}$$

is called the (p,s)-projection of the image. One should consider other sets of $(p+s)(N-1)+1$ shifted lines that are parallel to $l(t)$ and call them the geometric rays, too.

11.3 The Image and the Set of Splitting-Signals

This section describes the images and their 2D discrete Fourier transforms (2D DFTs) by the unique sets of 1D signals and 1D DFTs, respectively. The images are considered in the tensor and paired representations, which are the 2D frequency and 1D time representations of the image.[29–35] These representations are unique and invertible, and fast algorithms exist for them. Both tensor and paired transforms of images can be used effectively to calculate the 2D DFT and other transforms, including 2D Hartley, Hadamard, and cosine transforms.[34–36,48] The tensor transform can be used in image enhancement,[40–42] image denoising,[43,49,50] and discrete image reconstruction.[44,45]

In the tensor representation, the discrete image $f_{n,m}$ is the set of 1D splitting-signals:

$$\chi\colon \{f_{n,m}\} \to \{f_{T_{p,s}} = \{f_{p,s,t}; \quad t = 0 : (N-1)\}\}_{(p,s)\in J_{N,N}}. \tag{11.4}$$

Here, $J_{N,N}$ is a set of frequency points (p,s) or generators of the splitting-signals that are selected in a way to cover the lattice $X_{N,N}$ of frequency points (p,s) with a minimum number of the following subsets:

$$T_{p,s} = \{(k mod\ N, ks\ modN); \quad k = 0 : (N-1)\}.$$

These sets are cyclic groups with generators (p,s). When N is a power of two, $N = 2^r$, $r > 1$, the set of generators (p,s) contains $3N/2$ elements and can be defined as

$$J_{N,N} = \{(p,1); \quad p = 0 : (N-1)\} \cup \{(1,2s); \quad s = 0 : (N/2-1)\}. \tag{11.5}$$

The components $f_{p,s,t}$ of the splitting-signals $f_{Tp,s}$ are the sums of the image $f_{n,m}$ along the parallel lines on the lattice

$$f_{p,s,t} = \chi_{p,s,t}{^\circ}f = \sum\nolimits_{(n,m)\in X}\{f_{n,m}; np + ms = t\ mod\ N\}, \quad t = 0 : (N-1). \tag{11.6}$$

The binary function $\chi_{p,s,t}(n,m)$ takes value 1, when $np + ms = t \ mod \ N$ and 0, otherwise. Given (p, s), the components $f_{p,s,t}$ are periodic by t, i.e., $f_{p,s,t+N} = f_{p,s,t}$ for $t = 0 : (N - 1)$.

Figure 11.3(a) shows the image, and 11.3(b) shows the tensor transform. The tensor transform is redundant, and the set of all splitting-signals can be divided by two parts, or matrices. The first part is with the generators $\{(p, 1); p = 0 : (N - 1)\}$, and the second part is for the splitting-signals with the remaining generators $\{(1, 2s); s = 0 : (N/2 - 1)\}$ of the set $J_{256,256}$. The splitting-signals are written along the rows in these two matrices.

Figure 11.4 shows the splitting-signal of the image, which is generated by the frequency point (4,1) in (a). This splitting-signal is written in row number 5 in the first part of the tensor transform in Fig. 11.3.

The splitting-signal $f_{T_{p,s}}$ carries the spectral information of the 2D DFT at N frequency points of the set $T_{p,s}$, i.e., the following holds:

$$F_{kp \ modN, \ ks \ modN} = \sum_{t=0}^{N-1} f_{p,s,t} W^{kt}, \quad k = 0 : (N - 1), \tag{11.7}$$

where $W = W_N = \exp(-2\pi j/N)$.

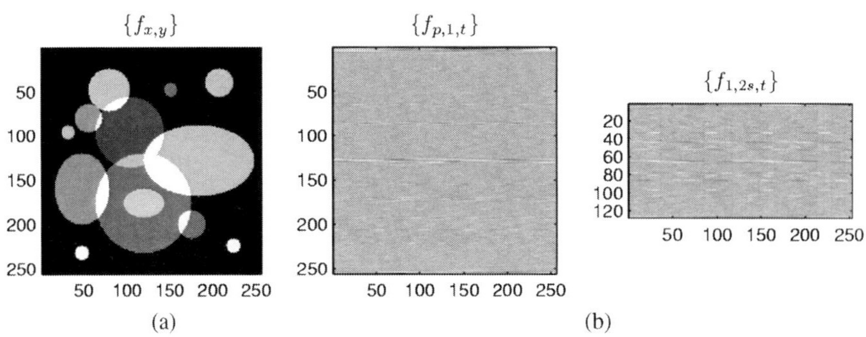

Figure 11.3 (a) Image of size 256 × 256 and (b) two parts of the tensor transform of the image.

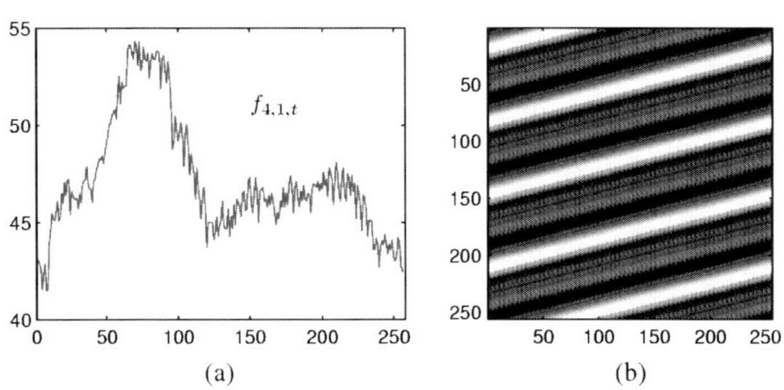

Figure 11.4 (a) Splitting-signal $\{f_{4,1,t}; t = 0 : 255\}$ of the image and (b) its direction image.

As an example, Figure 11.5 illustrates the 256×256 case. The splitting-signal or image-signal $f_{T2,1}$ of length 256 is shown in (a), along with the magnitude of the 1D DFT of the signal in (c) and the spectrum of the image in (d). Two bright parallel lines on the spectrum show the samples at points of the set $T_{2,1}$ at which the 2DFT of the image is the 1DFT of the image signal. The 2D DFT at frequency points of this set has been amplified in order to see the location of the group and the directions of the projection along which the components of the tensor are calculated as ray sums. The image after amplifying the 2DFT at frequency points of the set $T_{2,1}$ is shown in Fig. 11.5(b). The effect of amplifying those spectral components shows the contribution of the direction image. The projections are calculated at angle $\psi = 63.4349$ deg, and the 1D DFT is filled by the 2D DFT along three lines at angle $\theta = 90 - \psi = 26.5651$ deg.

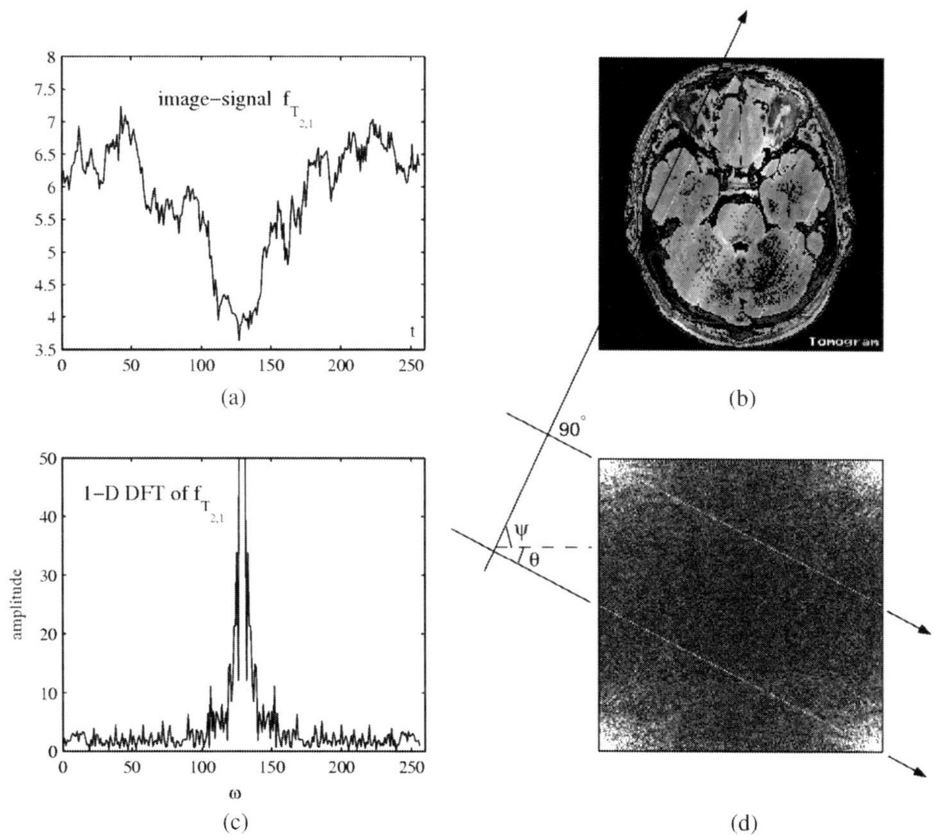

Figure 11.5 (a) The image signal corresponding to the set $T_{2,1}$. (b) The image after amplifying the 2DFT at frequency points of $T_{2,1}$. (c) Magnitude of the 1D DFT of the image signal (the zero component is shifted to the center and truncated). (d) The 2D DFT of the image with amplified samples of the set $T_{2,1}$.

To remove the redundancy of the tensor transform, the paired representation of the image is used when the set of splitting-signals is defined as[30,31,33]

$$\chi': \{f_{n,m}\} \rightarrow \{f_{T'_{p,s}} = \{f'_{p,s,t}; t = 0 : (N/2^k - 1), \text{g.c.d.}(p,s) = 2^k\}\}_{(p,s)\in J'_{N,N}} \tag{11.8}$$

Here, $J'_{N,N}$ is a set of $3N - 2$ frequency points (p,s) or generators, and it is defined as $J'_{N,N} \cup 2J'_{N/2,N/2} \cup 4J'_{N/4,N/4} \cup \ldots \cup \{(N/2,N/2)\} \cup \{(0,0)\}$. The components of the paired splitting-signals are defined from the tensor transforms as $f'_{p,s,t} = f_{p,s,t} - f_{p,s,t+N/2}$, when $t = 0 : (N/2^k - 1)$. The splitting-signals carry information of the 2D DFT of the image at frequency points of the subset $T'_{p,s} = \{(2m+1)(p,s); m = 0 : (N/2^k - 1)\}$ of $T_{p,s}$. The 2D paired transform for the general case—when N is a power of a prime—is defined similarly to the case when such a prime is 2.[32,49,50]

The family of subsets $T_{p,s}$ occurs when $(p,s) \in J_{N,N}$ covers the entire lattice $X_{N,N}$ and $3N/2$ 1D DFTs of splitting-signals $f_{Tp,s}$ define completely the 2D DFT of the image. The tensor representation is unique, and the image can be defined through the 2D DFT calculated by Eq. (11.7) or directly from the tensor transform. Indeed, according to the principle of superposition,[33,46–48] the image $f_{n,m}$ can be composed from splitting-signals as follows:

$$f_{n,m} = \frac{1}{2N}\sum_{k=0}^{r-1}\frac{1}{2^k}\sum_{(p,s)\in J_{N/2^k,N/2^k}}\left[f_{2^k p,2^k s,t} - f_{2^k p,2^k s,t+\frac{N}{2}}\right] + \frac{1}{N^2}f_{0,0,0}, \tag{11.9}$$

where $t = t(2^k p, 2^k s; n,m) = (n2^k p + m2^k s)\ mod\ N$. All components $f_{2^k p,2^k s,t}$ in this equation are defined by Eq. (11.6), but they can also be calculated by using the following recursive formula:

$$f_{2^k p,2^k s,2^k t_1} = f_{2^{k-1}p,2^{k-1}s,2^{k-1}t_1} + f_{2^{k-1}p,2^{k-1}s,2^{k-1}t_1+N/2}, \quad k = 1,2,\ldots,r-1, \tag{11.10}$$

where p or s equals 1, and $t_1 = 0 : (N/2^k + 1 - 1)$. In other words, these components can be calculated from $3N/2$ splitting-signals $f_{Tp,s}$ generated by the frequencies $(p,s) \in J_{N,N}$. Each image $d_{n,m} = f_{p,s,(np+ms) mod\ N}$, $n,m = 0 : (N-1)$ is the direction image with N values that are located along the parallel lines $(np + ms)$ mod N that pass through the knots of the lattice. For the case when $(p,s) = (3,1)$, Fig. 11.4(b) shows the direction image that corresponds to the splitting-signal in (a).

Thus, the image $f_{n,m}$ is the linear combination of direction images. This composition of the image can be used to reconstruct the image from $3N/2$ projections or an incomplete set of projections. It should be mentioned that the inverse formula in Eq. (11.9) can also be used for other sets of $3N/2$ generators (p,s) when the covering of the lattice $N \times N$ by the groups $T_{p,s}$ is irreducible. For instance, $\{(2p,1); p = 0 : (N/2 - 1)\} \cup \{(1,s); s = 0 : (N-1)\}$.

In the case where N is prime, the inverse tensor transform of the $N \times N$ image $f_{n,m}$ can be composed from $(N+1)$ tensor splitting-signals, as follows:

$$f_{n,m} = \frac{1}{N}\sum_{s=0}^{N-1}\left[f_{1,s,(n+ms)\,mod\,N} - f_{0,1,m}\right] - NE[f], \quad n,m = 0:(N-1). \quad (11.11)$$

The set of generators (p,s) for cyclic groups $T_{p,s}$ is defined as

$$J_{N,N} = \{(1,0),(1,1),(1,2),(1,3),\cdots,(1,N-1)\} \cup \{(0,1)\}. \quad (11.12)$$

This set contains $(N+1)$ frequency generators, and this number shows the number of splitting-signals in the tensor representation of the image.

11.4 Geometry of the Projections on the Lattice

This section describes the method of transferring the geometry from the image plane to the Cartesian lattice, i.e., a way to calculate the ray sums of the discrete image from the line integrals of the image $f(x,y)$. The sets of rays for these two geometries, i.e., the sets of arithmetical and geometrical rays, may be the same or different, and that will depend on the angle of projections. Consider a few examples of projections in the case when N is a prime.

Example 1 (7 × 7)

Consider the (1,2)-projection and the $N = 7$ case. Figure 11.6 shows three parallel lines $l(2)$, $l(9)$, and $l(16)$ in the unit square with the Cartesian

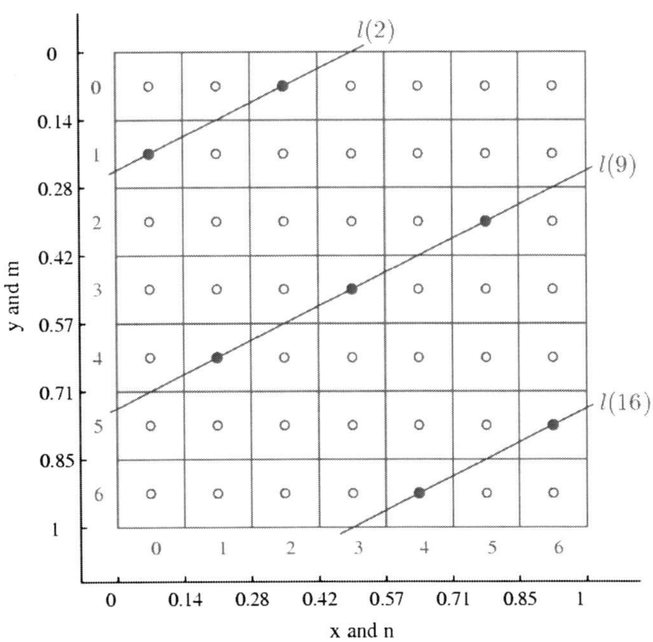

Figure 11.6 Parallel rays on the square $[0,1] \times [0,1]$ and the 7×7 lattice.

lattice 7×7. The generator (p, s) is (1,2); it determines the slope, $\arctan(p/s)$, of the lines. The equations of these three lines on the $[0,1] \times [0,1]$ square are

$$l(2) = l_{1,2}(2) = \left\{(x,y); x + 2y = \frac{2}{7} + \frac{3}{14} = \frac{7}{14}\right\},$$

$$l(9) = l_{1,2}(9) = \left\{(x,y); x + 2y = \frac{9}{7} + \frac{3}{14} = \frac{21}{14}\right\},$$

$$l(16) = l_{1,2}(16) = \left\{(x,y); x + 2y = \frac{16}{7} + \frac{3}{14} = \frac{35}{14}\right\}.$$

In the discrete lattice 7×7, these lines are described by $l(2) = l_{1,2}(2) = \{(n,m); n + 2m = 2\}$, $l(9) = l_{1,2}(9) = \{(n,m); n + 2m = 9\}$, and $l(16) = l_{1,2}(16) = \{(n,m); n + 2m = 16\}$. The sums of the image along arithmetical rays are denoted by

$$v_{p,s}(t) = v_l(t) = \sum_{(n,m)\in l_{p,s}(t)} f_{n,m}.$$

The set of 19 geometrical rays that coincide with these arithmetical rays is defined as

$$l(t) = l_{1,2}(t) = \left\{(x,y); x + 2y = \frac{t}{7} + \frac{3}{14}\right\}, \quad t = 0:18,$$

where $(x,y) \in [0,1] \times [0,1]$. The parallel rays $n + 2m = t$, where $t = 0:18$, are shown in Figure 11.7 on the left. The number of rays is calculated as $(p+s)(N-1)+1 = 3 \cdot 6 + 1 = 19$. To number these rays, the set of control points that are also shown on the right are used.

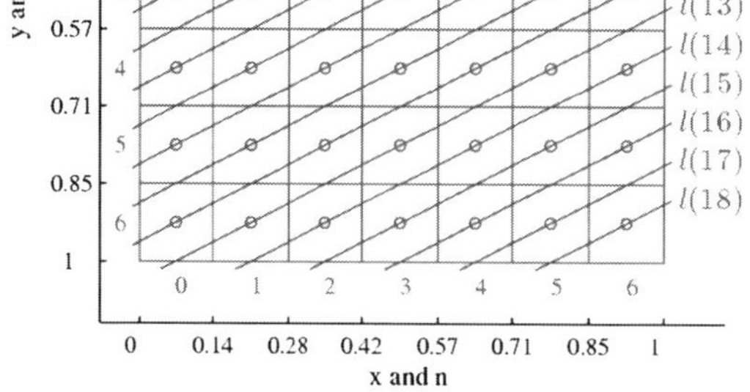

$$\begin{bmatrix} 0\bullet & 1\bullet & 2\bullet & 3\bullet & 4\bullet & 5\bullet & 6\bullet \\ \circ & \circ & \circ & \circ & \circ & 7\bullet & 8\bullet \\ \circ & \circ & \circ & \circ & \circ & 9\bullet & 10\bullet \\ \circ & \circ & \circ & \circ & \circ & 11\bullet & 12\bullet \\ \circ & \circ & \circ & \circ & \circ & 13\bullet & 14\bullet \\ \circ & \circ & \circ & \circ & \circ & 15\bullet & 16\bullet \\ \circ & \circ & \circ & \circ & \circ & 17\bullet & 18\bullet \end{bmatrix}$$

(for rays at angle -26.56^o).

Figure 11.7 19 arithmetical rays for the (1,2)-projection and the corresponding numbered control points.

Nineteen arithmetical rays of this projection are used to determine the components $\{f_{1,2,t}; t = 0:6\}$ of the tensor transforms. The masks of the basic functions $\chi_{1,2,t}(n,m)$ have coefficients equaling 1 on these parallel lines. Consider the masks of the first two basic functions:

$$[\chi 1,2,0] = \begin{bmatrix} 1 & 0 & 0 & 0 & 0 & 0 & 0 \\ 0 & 0 & 0 & 0 & 0 & 1 & 0 \\ 0 & 0 & 0 & 1 & 0 & 0 & 0 \\ 0 & 1 & 0 & 0 & 0 & 0 & 0 \\ 0 & 0 & 0 & 0 & 0 & 0 & 1 \\ 0 & 0 & 0 & 0 & 1 & 0 & 0 \\ 0 & 0 & 1 & 0 & 0 & 0 & 0 \end{bmatrix}, \quad [\chi 1,2,1] = \begin{bmatrix} 0 & 1 & 0 & 0 & 0 & 0 & 0 \\ 0 & 0 & 0 & 0 & 0 & 0 & 1 \\ 0 & 0 & 0 & 0 & 1 & 0 & 0 \\ 0 & 0 & 1 & 0 & 0 & 0 & 0 \\ 1 & 0 & 0 & 0 & 0 & 0 & 0 \\ 0 & 0 & 0 & 0 & 0 & 1 & 0 \\ 0 & 0 & 0 & 1 & 0 & 0 & 0 \end{bmatrix}$$

These masks and the parallel rays passing through coefficients 1 are shown in Fig. 11.8.

It is not difficult to see that seven components $f_{1,2,t}$, $t = 0:6$ of the tensor transform are calculated as

$$\begin{aligned} f_{1,2,0} &= v_0 + v_7 + v_{14}, \\ f_{1,2,1} &= v_1 + v_8 + v_{15}, \\ f_{1,2,2} &= v_2 + v_9 + v_{16}, \\ f_{1,2,3} &= v_3 + v_{10} + v_{17}, \\ f_{1,2,4} &= v_4 + v_{11} + v_{18}, \\ f_{1,2,5} &= v_5 + v_{12}, \\ f_{1,2,6} &= v_6 + v_{13}, \end{aligned} \tag{11.13}$$

where the variables $v_t = v_{1,2}(t)$ denote the sums of the discrete image along the arithmetical rays $l(t)$, $t = 0:18$. This system of equations can be

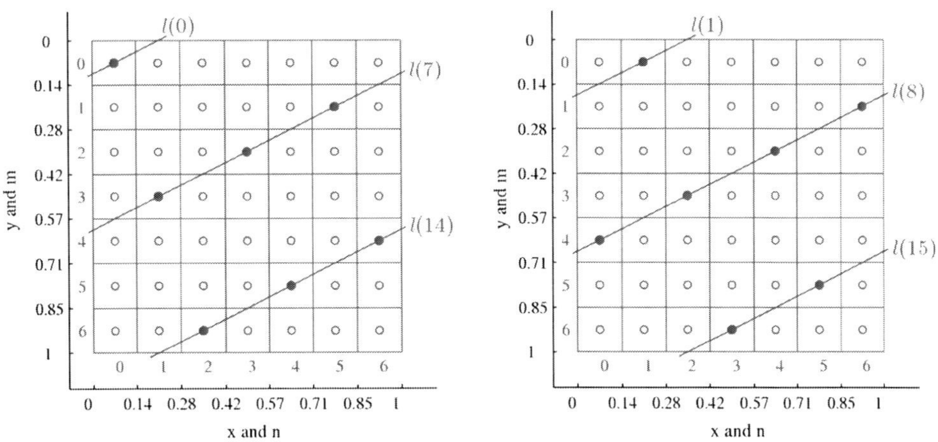

Figure 11.8 Masks of functions $\chi_{1,2,0}(n,m)$ and $\chi_{1,2,1}(n,m)$ with the parallel rays.

written as $f_{1,2,t} = v_t + v_{t+7} + v_{t+14}$, $t = 0:6$, where $v_t = 0$ if $t > 18$. To calculate the signal $f_{1,2,t}$ from the projection data, we will derive and then simplify the formula describing the linear relation between the integrals $w(t)$ and sums $v(t)$, $t = 0:18$. First, consider the mask of the tensor function $\chi_{1,2,t}(n,m)$ with coefficients 1 of the rays $l(0)$, $l(7)$, and $l(14)$, as shown in Fig. 11.9.

It is not difficult to see that the line integral along the ray $l = l(6)$ in image elements can be written as

$$w_l^{(0,3)} = \frac{\Delta l_{0,3}}{\Delta x} 7f_{0,3}, \quad w_l^{(1,3)} = \frac{\Delta l_{1,3}}{\Delta x} 7f_{1,3}, \quad w_l^{(1,2)} = \frac{\Delta l_{1,2}}{\Delta x} 7f_{1,2},$$
$$w_l^{(2,2)} = \frac{\Delta l_{2,2}}{\Delta x} 7f_{2,2}, \quad w_l^{(3,2)} = \frac{\Delta l_{3,2}}{\Delta x} 7f_{3,2}, \quad w_l^{(3,1)} = \frac{\Delta l_{3,1}}{\Delta x} 7f_{3,1},$$
$$w_l^{(4,1)} = \frac{\Delta l_{4,1}}{\Delta x} 7f_{4,1}, \quad w_l^{(5,1)} = \frac{\Delta l_{5,1}}{\Delta x} 7f_{5,1}, \quad w_l^{(5,0)} = \frac{\Delta l_{5,0}}{\Delta x} 7f_{5,0},$$
$$w_l^{(6,0)} = \frac{\Delta l_{6,0}}{\Delta x} 7f_{6,0}.$$

The lengths of intersection of the rays with image elements equal $\Delta l_{0,3} = \Delta l_{2,2} = \Delta l_{4,1} = \Delta l_{6,0} = (\sqrt{5}/7)/2$ and $\Delta l_{1,3} = \Delta l_{1,2} = \Delta l_{3,2} = \Delta l_{3,1} = \Delta l_{5,1} = \Delta l_{5,0} = (\sqrt{5}/7)/4$. Therefore, the line integral $w_1(6)$ can be written as

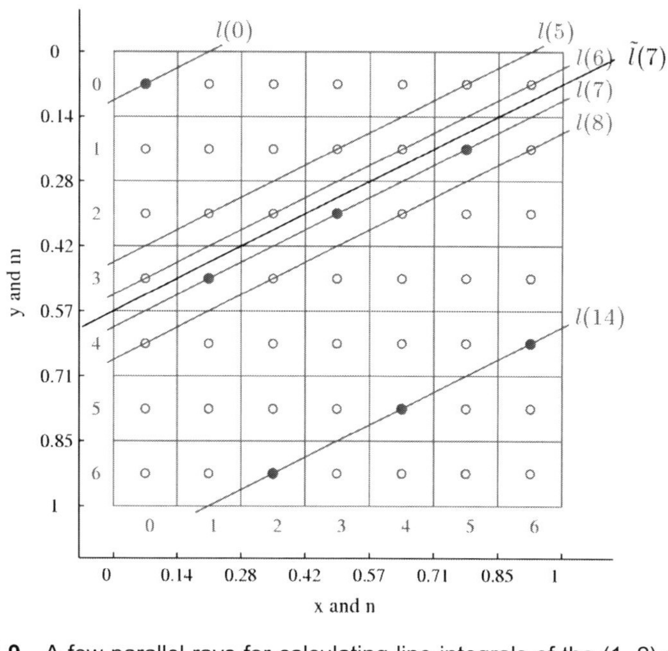

Figure 11.9 A few parallel rays for calculating line integrals of the (1, 2)-projection.

$$
\begin{aligned}
w_l(6) &= \sum_{\Delta l(n,m)\neq 0} w_l^{(n,m)} \\
&= \left[w_l^{(0,3)} + w_l^{(2,2)} + w_l^{(4,1)} + w_l^{(6,0)}\right] + \left[w_l^{(1,3)} + w_l^{(3,2)} + w_l^{(5,1)}\right] \\
&\quad + \left[w_l^{(1,2)} + w_l^{(3,1)} + w_l^{(5,0)}\right] \\
&= \frac{7\sqrt{5}}{2}[f_{0,3} + f_{2,2} + f_{4,1} + f_{6,0}] + \frac{7\sqrt{5}}{4}[f_{1,3} + f_{3,2} + f_{5,1}] \\
&\quad + \frac{7\sqrt{5}}{4}[f_{1,2} + f_{3,1} + f_{5,0}] \\
&= \frac{7\sqrt{5}}{2}\left[v_{1,2}(6) + \frac{1}{2}v_{1,2}(5) + \frac{1}{2}v_{1,2}(7)\right].
\end{aligned}
$$

A similar equation holds for other line integrals $w_{1,2}(t)$, as well. For instance,

$$
wl(7) = \frac{7\sqrt{5}}{2}\left[v_{1,2}(7) + \frac{1}{2}v_{1,2}(6) + \frac{1}{2}v_{1,2}(8)\right].
$$

Different coefficients at the line sums $v_{1,2}(t)$, $v_{1,2}(t-1)$, and $v_{1,2}(t+1)$ complicate the solution of the system of these equations with respect to $v_{1,2}(t)$. To simplify the calculation of the sums $v_{1,2}(t)$, a system of equations, wherein all coefficients at the sums are equal, is described. First, consider the geometrical ray $\tilde{l}$ that is located between the rays $l(6)$ and $l(7)$; this ray is denoted as $\tilde{l}(7)$ in the figure. One can notice that this ray equally intersects all image elements with numbers (0,3), (1,3), (2,2), (3,2), (4,1), (5,1), and (6,0). Therefore, the integral of the image along the geometrical ray $\tilde{l}$ can be calculated from the sums of the discrete image along two arithmetical rays by

$$
\begin{aligned}
w_{\tilde{l}} &= \sum_{\Delta\tilde{l}(n,m)\neq 0} w_{\tilde{l}}^{(n,m)} \\
&= w_l^{(0,3)} + w_l^{(1,3)} + w_l^{(2,2)} + w_l^{(3,2)} + w_l^{(4,1)} + w_l^{(5,1)} + w_l^{(6,0)} \\
&= \left[w_l^{(0,3)} + w_l^{(2,2)} + w_l^{(4,1)} + w_l^{(6,1)}\right] + \left[w_l^{(1,3)} + w_l^{(3,2)} + w_l^{(5,1)}\right] \\
&= \frac{7\sqrt{5}}{2}[f_{0,3} + f_{2,2} + f_{4,1} + f_{6,0}] + \frac{7\sqrt{5}}{2}[f_{1,3} + f_{3,2} + f_{5,1}] \\
&= \frac{7\sqrt{5}}{2}[v_{1,2}(6) + v_{1,2}(7)].
\end{aligned}
$$

A simple equation for the line integral along the geometrical ray $\tilde{l}$ is obtained. This ray is considered as the ray number 7 in the new set of parallel geometrical rays that are defined as

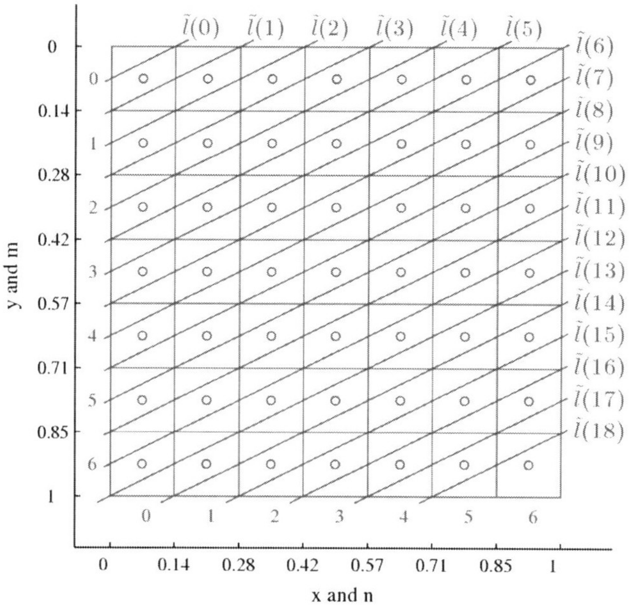

Figure 11.10 The set of shifted geometrical rays for calculating 19 line integrals of the (1,2)-projection.

$$\tilde{l}(t) = l_{1,2}(t-0.5) = \left\{(x,y); x+2y = \frac{t}{7}+\frac{1}{7}\right\}, \quad t = 0:18$$

and illustrated in Fig. 11.10.

The relationship between the integrals along the geometrical rays $\tilde{l}(t)$ and the sums along the arithmetical rays $l(t)$ is described by the following system of linear equations:

$$w_{1,2}(t) = w(\tilde{l}(t)) = \frac{7\sqrt{5}}{2}[v_{1,2}(t-1)+v_{1,2}(t)], \quad t=0:18,$$

where $v_{1,2}(-1) = 0$. These equations can be written in matrix form as

$$\mathbf{w} = \begin{bmatrix} w_{1,2}(0) \\ w_{1,2}(1) \\ w_{1,2}(2) \\ w_{1,2}(3) \\ w_{1,2}(4) \\ \vdots \\ w_{1,2}(17) \\ w_{1,2}(18) \end{bmatrix} = \frac{7\sqrt{5}}{2}\mathbf{A}\mathbf{v} = \frac{7\sqrt{5}}{2} \begin{bmatrix} 1 & 0 & 0 & 0 & 0 & \cdots & 0 & 0 \\ 1 & 1 & 0 & 0 & 0 & \cdots & 0 & 0 \\ 0 & 1 & 1 & 0 & 0 & \cdots & 0 & 0 \\ 0 & 0 & 1 & 1 & 0 & \cdots & 0 & 0 \\ 0 & 0 & 0 & 1 & 1 & \cdots & 0 & 0 \\ \vdots & \vdots & \vdots & \vdots & \vdots & \ddots & \vdots & \vdots \\ 0 & 0 & 0 & 0 & 0 & \cdots & 1 & 0 \\ 0 & 0 & 0 & 0 & 0 & \cdots & 1 & 1 \end{bmatrix} \begin{bmatrix} v_{1,2}(0) \\ v_{1,2}(1) \\ v_{1,2}(2) \\ v_{1,2}(3) \\ v_{1,2}(4) \\ \vdots \\ v_{1,2}(17) \\ v_{1,2}(18) \end{bmatrix}.$$

The above Toeplitz matrix $\mathbf{A}$ of size 19×19 has the triangle inverse matrix, and all sums along the arithmetical rays are calculated by

$$\begin{bmatrix} v_{1,2}(0) \\ v_{1,2}(1) \\ v_{1,2}(2) \\ v_{1,2}(3) \\ v_{1,2}(4) \\ \vdots \\ v_{1,2}(17) \\ v_{1,2}(18) \end{bmatrix} = \frac{2}{7\sqrt{5}} \begin{bmatrix} 1 & 0 & 0 & 0 & 0 & \cdots & 0 & 0 \\ -1 & 1 & 0 & 0 & 0 & \cdots & 0 & 0 \\ 1 & -1 & 1 & 0 & 0 & \cdots & 0 & 0 \\ -1 & 1 & -1 & 1 & 0 & \cdots & 0 & 0 \\ 1 & -1 & 1 & -1 & 1 & \cdots & 0 & 0 \\ \vdots & \vdots & \vdots & \vdots & \vdots & \ddots & \vdots & \vdots \\ -1 & 1 & -1 & 1 & -1 & \cdots & 1 & 0 \\ 1 & -1 & 1 & -1 & 1 & \cdots & -1 & 1 \end{bmatrix}.$$

The required sums $v(t)$ of the discrete image $f_{n,m}$ can be calculated directly by $\mathbf{v} = 2/(7\sqrt{5})\mathbf{A}^{-1}\mathbf{w}$ or by the method of direct substitution in the following recursive form:

$$\begin{aligned} b_0 &= w(0), \\ b_1 &= w(1) - b_0, \\ b_t &= w(t) - b_{t-1}, \quad t = 2, 3, \ldots, 18, \end{aligned} \tag{11.14}$$

and, then, $v(t) = 2/(7\sqrt{5})\ b_t$, where $t = 0 : 18$. Substituting these values in (11.13), we obtain all components of the splitting-signal $\{f_{1,2,t};\ t = 0 : 7\}$.

Example 2 (7 $\times$ 7)

Consider the $N = 7$ case and $(p, s) = (1, 5)$. The masks of matrices of the tensor functions $\chi_{1,5,t}(n, m)$ with the first two triplet numbers of the subset $\{(1, 5, t); t = 0 : 6\}$ are

$$[\chi_{1,5,0}] = \begin{bmatrix} 1 & 0 & 0 & 0 & 0 & 0 & 0 \\ 0 & 0 & 1 & 0 & 0 & 0 & 0 \\ 0 & 0 & 0 & 0 & 1 & 0 & 0 \\ 0 & 0 & 0 & 0 & 0 & 0 & 1 \\ 0 & 1 & 0 & 0 & 0 & 1 & 0 \\ 0 & 0 & 0 & 1 & 0 & 0 & 0 \\ 0 & 0 & 0 & 0 & 0 & 1 & 0 \end{bmatrix}, [\chi_{1,5,1}] = \begin{bmatrix} 0 & 1 & 0 & 0 & 0 & 0 & 0 \\ 0 & 0 & 0 & 1 & 0 & 0 & 0 \\ 0 & 0 & 0 & 0 & 0 & 1 & 0 \\ 1 & 0 & 0 & 0 & 0 & 0 & 0 \\ 0 & 0 & 0 & 0 & 1 & 0 & 0 \\ 0 & 0 & 0 & 0 & 0 & 0 & 1 \\ 0 & 0 & 0 & 0 & 0 & 1 & 0 \end{bmatrix}.$$

The direction of parallel rays in these masks are defined by the angle arctan(5), as shown in Fig. 11.11(a) for the first mask. There are six such parallel rays that pass through the ones in the mask. Instead of parallel rays of the (1,5)-projection, consider that the $(1, -2)$-projection also defines the same set of masks as the tensor functions. Indeed, the equation $1n + 5m = t$ mod 7 can be written as $1n + (7 - 2)m = t$ mod 7 or $1n - 2m = t$ mod 7.

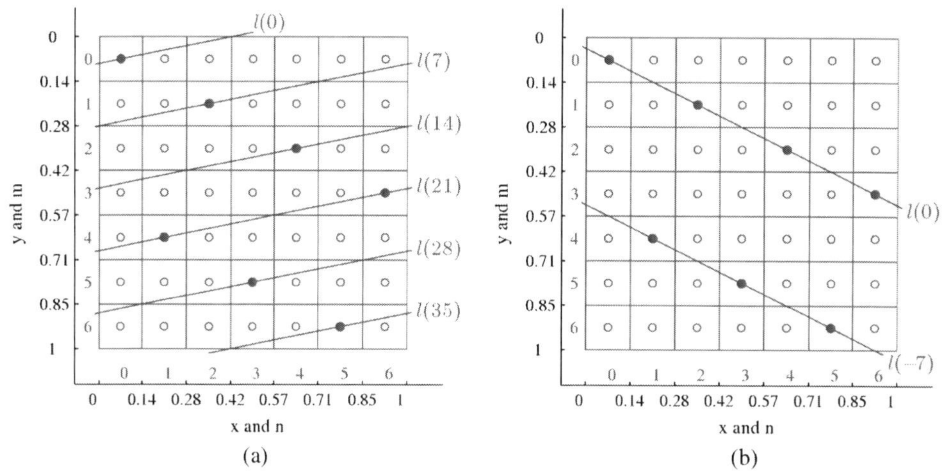

Figure 11.11 The mask $[\chi_{1,5,0}]$ with two different sets of parallel rays.

The same mask with two parallel rays of the (1,−2)-projection is illustrated in Fig. 11.11(b).

The set of all rays for the (1, −2)-projection is shown in Fig. 11.12 together with the control points numbering these rays.

Such a change of projection allows one to reduce the number of rays from $(1+5)6+1=37$ to $(1+|-2|)6+1=19$. The set of 19 arithmetical rays of this projection is described as

$$l(t)=l_{1,-2}(t)=\left\{(x,y);x-2y=\frac{t}{7}-\frac{1}{14}\right\},\quad t=6:-1:12, \tag{11.15}$$

where $(x, y)\in[0,1]\times[0,1]$.

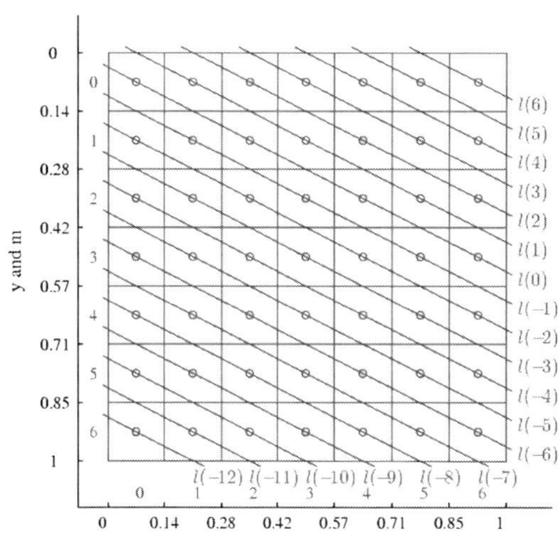

$$\begin{bmatrix} 0\bullet & 1\bullet & 2\bullet & 3\bullet & 4\bullet & 5\bullet & 6\bullet \\ -2\bullet & -1\bullet & \circ & \circ & \circ & \circ & \circ \\ -4\bullet & -3\bullet & \circ & \circ & \circ & \circ & \circ \\ -6\bullet & -5\bullet & \circ & \circ & \circ & \circ & \circ \\ -8\bullet & -7\bullet & \circ & \circ & \circ & \circ & \circ \\ -10\bullet & -9\bullet & \circ & \circ & \circ & \circ & \circ \\ -12\bullet & -11\bullet & \circ & \circ & \circ & \circ & \circ \end{bmatrix}$$

(for rays at angle 26.56^o).

Figure 11.12 The set of 19 arithmetical rays for the (1, −2)-projection and control points.

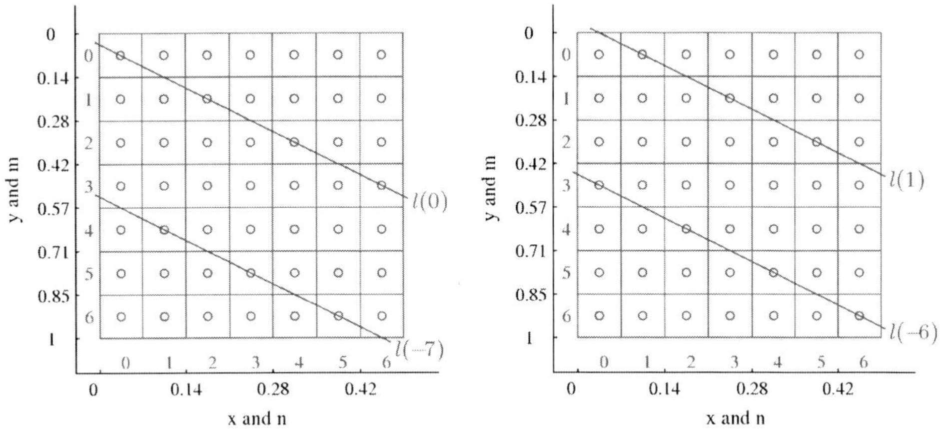

Figure 11.13 Two masks with parallel rays for the (1, −2)-projection.

The masks for the tensor functions $\chi_{1,5,0}(n, m)$ and $\chi_{1,5,1}(n, m)$ are shown in Fig. 11.13.

It is not difficult to see that the components of the tensor transform $f_{1,5,t}$ are calculated by

$$f_{1,5,0} = v_0 + v_{-7}, \quad f_{1,5,1} = v_1 + v_{-6}, \quad f_{1,5,2} = v_2 + v_{-5} + v_{-12},$$
$$f_{1,5,3} = v_3 + v_{-4} + v_{-11}, \quad f_{1,5,4} = v_4 + v_{-3} + v_{-10},$$
$$f_{1,5,5} = v_5 + v_{-2} + v_{-9}, \quad f_{1,5,6} = v_6 + v_{-1} + v_{-8}, \tag{11.16}$$

where the sums of the discrete image along the rays are denoted by $v(t) = v_{1,-2}(t)$, $t = 6 : -1 : -12$.

The system of linear equations in Eq. (11.16) can also be written as follows:

$$f_{1,5,t} = \sum_{m=0}^{2} v_{t-7m} = \sum_{m=0}^{2} v_{1,-2,t-7m}, \quad t = 0 : 6,$$

where $v_t = 0$ if $t < -7$. Consider the shifted set of the rays, $l(t) \to l(t - 0.5)$, as the set of 19 geometrical rays:

$$\tilde{l}(t) = l_{1,-2}(t - 0.5) = \left\{(x, y); x - 2y = \frac{t}{7} - \frac{1}{7}\right\}, \quad t = 6 : -1 : -12.$$

The set of these geometrical rays is shown in Fig. 11.14.

For this set of geometrical rays, the system of equations for the line integrals is defined as

$$w(t) = w(\tilde{l}(t)) = \frac{7\sqrt{5}}{2}[v(t) + v(t - 1)], \quad t = 6 : -1 : -12, \tag{11.17}$$

where $v(-13) = 0$. This system can be written in matrix form with the same Toeplitz matrix described for the (1,2)-projection. Therefore, the

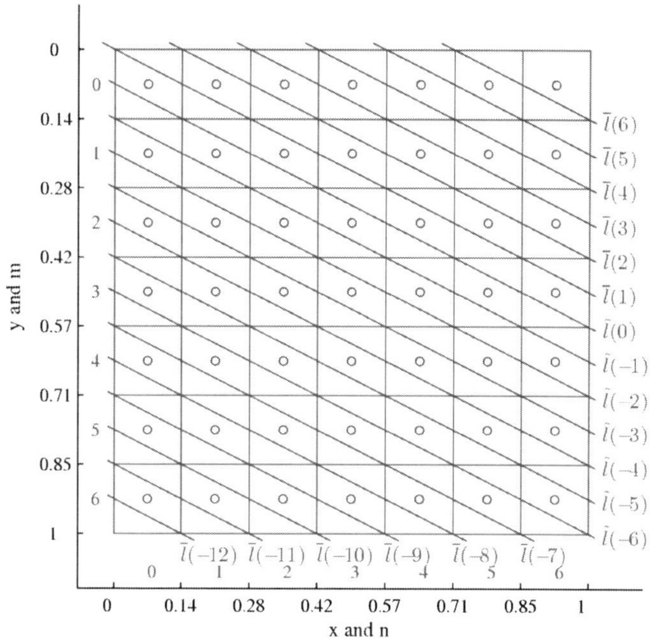

Figure 11.14 The set of geometrical rays for the (1, −2)-projection.

sums $v(t)$ of the discrete image $f_{n,m}$ can be determined by $v(t) = b_t/(7\sqrt{5}/2)$, where the components b_t are calculated recursively by

$$b_{-12} = w(-12), \quad b_{-11} = w(-11) - b_{-12}, \quad b_{-10} = w(-10) - b_{-11},$$
$$b_t = w(t) - b_{t-1}, \quad t = -9, -8, -7, \ldots 6. \tag{11.18}$$

It should be noted that the numbering of the control points can be changed by the transform $t \to 6 - t$. The set of 19 geometrical rays of this projection in Eq. (11.15) will then be written as

$$l(t) = l_{1,-2}(6-t) = \left\{(x,y); x - 2y = -\frac{1}{7} + \frac{11}{14}\right\},$$

and the system of equations in Eq. (11.17) can be written as $w(t) = \mathrm{w}(\tilde{l}(t)) = 7\sqrt{5}/2[v(t) + v(t+1)]$, $t = 0 : 18$, where $v(19) = 0$. Equation (11.18) for the coefficients $b_t = 7\sqrt{5}/2v(t)$ is considered as

$$b_{18} = w(18),$$
$$b_{17} = w(17) - b_{18},$$
$$b_{16} = w(16) - b_{17},$$
$$b_t = w(t) - b_{t+1}, \quad t = 15, 14, \ldots, 0.$$

Therefore, the components of the splitting signal $f_{1,5,t}$ in Eq. (11.16) can be calculated as follows:

$$f_{1,5,t} = \sum_{m=0}^{2} v_{6-t+7m}, \quad t = 0 : 6, (v_t = 0, t > 18).$$

11.4.1 Main equations for geometrical rays when *N* is prime

Generally, when N is prime, the image reconstruction from $N+1$ projections is described like the $N=7$ case. Table 11.1 provides the general equations for rays and for calculating the sums $v(t)$ from the line integrals $w(t)$ for all projections. Two subsets of generators (p,s) are considered separately. The first subset contains generators $(1,s)$, where $s \leq (N-1)/2$. The generators $(1,s)$, where $s > (N-1)/2$, compose the second subset. For each generators $(1,s)$ of the second subset, the $(1,s)$-projection is substituted by the $(1,s-N)$-projection. In other words, the splitting-signal $\{f_{1,s,t}; t=0:(N-1)\}$ is calculated from the line integrals of the $(1,s-N)$-projection, similar to the (1,5)-projection described earlier for the $N=7$ case. Each such "substitution" saves the number of line integrals used in the reconstruction.

Table 11. 1 General equations of rays $l(t)$, linear integrals $w(t)$, sums $v(t)$, and splitting-signals $f_{p,s,t}$.

	Equations	t
$l(t)$	$xp+ys=\frac{t}{N}+\frac{1}{N}$, where $s \leq \frac{N-1}{2}$	$0:(p+s)(N-1)-1$
	$xp+y(s-N)=\frac{t}{N}+\frac{s+1-N}{N}$, where $s > \frac{N-1}{2}$	$N-1:-1:(s-N)(N-1)$
$w(t)$	$w(t)-Nv(t)$, where $s=0$ or $p=0$,	$0:N-1$
	$w(t)=K\left[v(t)+v(t-1)+\ldots+v(t-(s-1))\right]$, where $s \leq \frac{N-1}{2}$ and $K=\frac{N}{s}\sqrt{p^2+s^2}$	$0:(p+s)(N-1)-1$
	$w(t)=K\left[v(t)+v(t-1)+\ldots+v(t-(s-1))\right]$, where $s > \frac{N-1}{2}$, and $K=\frac{N}{N-s}\sqrt{p^2+(N-s)^2}$	$-(N-s)(N-1):N-1$
$b(t)$	$b(0)=w(0), b_t=w(t)-(b_t+\cdots+b_{t-(N-s-1)})$, where $s \leq \frac{N-1}{2}$	$0:(p+s)(N-1)-1$
	$b(0)=w(0), b_t=w(t)-(b_t+\cdots+b_{t-(N-s-1)})$, where $s > \frac{N-1}{2}$	$-(N-s)(N-1):N-1$
$v(t)$	$v(t)=\frac{b_t}{K}$, where $s \leq \frac{N-1}{2}$ and $K=\frac{N}{8}\sqrt{p^2+s^2}$	$0:(p+s)(N-1)-1$
	$v(t)=\frac{b_t}{K}$, where $s\frac{N-1}{2}$, and $K=\frac{N}{N-s}\sqrt{p^2+(N-s)^2}$	$(s-N)(N-1):N-1$
$f_{p,s,t}$	$\sum_m v_{t+7m}$, where $s \leq \frac{N-1}{2}$	$0:N-1$
	$\sum_m v_{t-7m}$, where $s > \frac{N-1}{2}$	$0:N-1$

The number of parallel geometrical rays used in each projection is defined as

$$n(1,s) = (1+s)(N-1)+1 = N+s(N-1), \quad s = 0:(N-1)/2,$$
$$n(1,s) = n(1,N-s) = N^2 - s(N-1), \quad s = (N+1)/2 : N-1,$$
$$n(0,1) = n(1,0) = N. \tag{11.19}$$

11.4.2 Simulation results for modeled images

The proposed method of reconstruction is implemented in MATLAB and C+. For prime N, the block diagram of the reconstruction of the image on the Cartesian grid $N \times N$ from $N+1$ projections is shown in Fig. 11.15.

The following steps describe the proposed method of image reconstruction.

1. An image $N \times N$ is composed (for instance, from a few random rectangles or ellipses) on the square $[0,1] \times [0,1]$.
2. The set $J_{N,N}$ of generators (p,s) is calculated to define $N+1$ projections. The considered set is $J_{N,N} = \{(1,s); s = 0:(N-1)\} \cup \{(0,1)\}$.
3. Given the frequency point (p,s), the (p,s)-projection is calculated, i.e., all line integrals $w(t)$ along the set of geometrical rays $\widetilde{l}p, s(t)$.
4. The geometry of the geometrical rays of the (p,s)-projection is transformed to arithmetical rays, i.e., the set of sums $v(t)$ of the discrete image $f_{n,m}$ is calculated from the integrals $w(t)$ by solving the system of linear equations with the Toeplitz matrix $(M \times M)$, where $M = (p+s-1)(N-1)+1$, if $s \leq (N-1)/2$, and $M = (p+(N-s)-1)(N-1)+1$, if $s > (N-1)/2$. The Toeplitz matrix is triangular, and the transform of geometry of geometrical rays to arithmetical rays is fast.

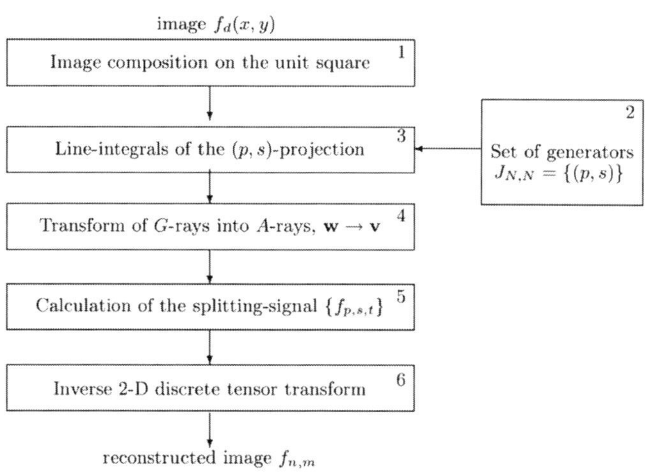

Figure 11.15 Block diagram of the reconstruction of the image $f(x,y) = f_d(x,y)$ composed of image elements with constant intensity each on the Cartesian grid $N \times N$ placed in the square $[0,1] \times [0,1]$.

5. The splitting-signal $\{f_{p,s,0}, f_{p,s,1}, \ldots, f_{p,s,N-1}\}$ is calculated from $v(t)$. This signal is written into the corresponding row of the matrix $N \times (N+1)$ of the 2D discrete tensor transform.
6. The inverse 2D tensor transform is calculated, which is the reconstructed discrete image $f_{n,m}$.

Figure 11.16 shows the image $f(x, y)$ on the square $[0, 1] \times [0, 1]$, which is generated by the random rectangles with coordinates on the lattice 131×131. The reconstruction $f_{n,m}$ of the image by 132 projections is also shown.

One can see from the reconstructed image that the reconstruction is exact. The experiments were simulated on the computer with an Intel® Core™ 2Duo E8400 3.00 GHz and 3.21 GB of RAM. The time for image reconstruction is 32.16 seconds. For the (1, 4)-projection, Fig. 11.17 shows

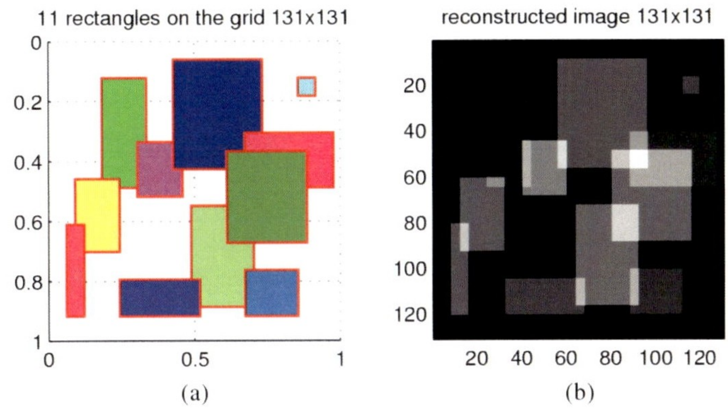

Figure 11.16 (a) Image with 13 rectangles on the lattice 131×131 and (b) its reconstruction.

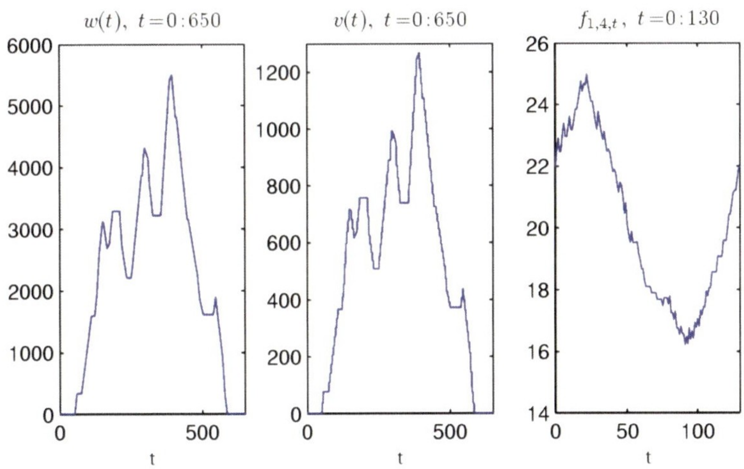

Figure 11.17 Line integrals $w(t)$, ray sums $v(t)$, and splitting-signal for the (1,4)-projection.

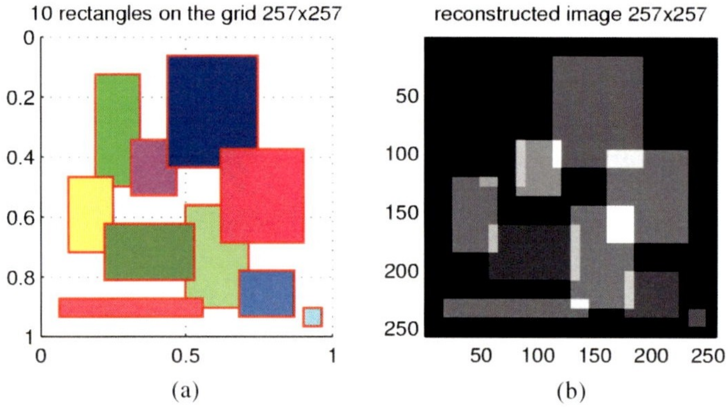

Figure 11.18 Image with ten rectangles on the lattice 257 × 257 and its reconstruction.

the line-integrals $w(t)$ on the left, the sums $v(t)$ in the middle, and the splitting-signal $\{f_{1,4,0}, f_{1,4,1}, \ldots, f_{1,4,130}\}$ on the right.

Consider the 257 × 257 example, where 258 projections are required for the image reconstruction. Figure 11.18(a) presents the image, and 11.18(b) shows the exact reconstruction.

Table 11.2 shows the total time of image processing, when the program was implemented in Ci+. The tested image $f(x, y)$ was composed of 15 random rectangles on the square $[0, 1] \times [0, 1]$. The time includes the calculation of all line integrals $w(t)$, line sums $v(t)$, and the calculation of the direct and inverse tensor transforms.

11.5 Geometry for the Lattice $N \times N$ when N is a Power of Two

When reconstructing the image $f(x, y)$ from its projections on the Cartesian lattice of size $N \times N$, when $N = 2^r$, $r > 1$ is a power of two, components of the tensor transform of the image from the line integrals can be defined in a way similar to that when N is a prime. Effective reconstruction involves the 2D paired transform because it removes the redundancy of the tensor transform. Therefore, this section considers the image in paired representation and describes the method of calculating the paired splitting-signals from line integrals.

Table 11.2 Time for the Ci-based program (the first version).

$N \times N$	Projection Data Processing and Reconstruction
67 × 67	00 : 00 : 00.08
127 × 127	00 : 00 : 00.50
257 × 257	00 : 00 : 04.21
521 × 521	00 : 00 : 37.24
1031 × 1031	00 : 04 : 55.69
2053 × 2053	00 : 39 : 40.44

Given generator $(p,s)\in J'_{N,N}$, the integer k is defined by $2^k=$ g.c.d. (p,s), and $k=r-1$ when $(p,s)=(0,0)$. The components of the 2D paired transform

$$f'_{p,s,2^kt}=\chi'_{p,s,2^kt}\circ f=f_{p,s,2^kt}-f_{p,s,2^kt+N/2},\quad t=0:(N/2^{k+1}-1)$$

are calculated by the system of orthogonal paired functions that are defined as

$$\chi'_{p,s,2^kt}(n,m)=\begin{cases}1 & \text{if } np+ms=2^kt \bmod N\\ -1 & \text{if } np+ms=(2^kt+N/2) \bmod N\\ 0 & \text{otherwise,}\end{cases}$$

where $(n,m)\in X_{N,N}$. The set of N^2 triplet-numbers of the paired functions is taken as

$$U_{N,N}=\bigcup_{k=0}^{r-1}\left\{(p,s,2^kt);(p,s)\in 2^kJ_{N/2^k,N/2^k},t=0:(N/2^{k+1}-1)\right\}\cup\{(0,0,0)\}.$$

To describe the method of the transferring geometry from the image plane to the Cartesian lattice, consider the $N=8$ case in the following example.

Example 3 (8 $\times$ 8)

Consider the $N=8$ case and $(p,s)=(4,1)$. Four masks of the paired functions $\chi'_{1,4,t}(n,m)$ with the triplet-numbers $(1,4,t)$, $t=0$: 3 are

$$[\chi'_{1,4,0}]=\begin{bmatrix}1&0&0&0&-1&0&0&0\\-1&0&0&0&1&0&0&0\\1&0&0&0&-1&0&0&0\\-1&0&0&0&1&0&0&0\\1&0&0&0&-1&0&0&0\\-1&0&0&0&1&0&0&0\\1&0&0&0&-1&0&0&0\\-1&0&0&0&1&0&0&0\end{bmatrix},\ [\chi'_{1,4,1}]=\begin{bmatrix}0&1&0&0&0&-1&0&0\\0&-1&0&0&0&1&0&0\\0&1&0&0&0&-1&0&0\\0&-1&0&0&0&1&0&0\\0&1&0&0&0&-1&0&0\\0&-1&0&0&0&1&0&0\\0&1&0&0&0&-1&0&0\\0&-1&0&0&0&1&0&0\end{bmatrix},$$

$$[\chi'_{1,4,2}]=\begin{bmatrix}0&0&1&0&0&0&-1&0\\0&0&-1&0&0&0&1&0\\0&0&1&0&0&0&-1&0\\0&0&-1&0&0&0&1&0\\0&0&1&0&0&0&-1&0\\0&0&-1&0&0&0&1&0\\0&0&1&0&0&0&-1&0\\0&0&-1&0&0&0&1&0\end{bmatrix},\ [\chi'_{1,4,3}]=\begin{bmatrix}0&0&0&1&0&0&0&-1\\0&0&0&-1&0&0&0&1\\0&0&0&1&0&0&0&-1\\0&0&0&-1&0&0&0&1\\0&0&0&1&0&0&0&-1\\0&0&0&-1&0&0&0&1\\0&0&0&1&0&0&0&-1\\0&0&0&-1&0&0&0&1\end{bmatrix}.$$

The set of the 36 parallel rays of this projection is shown in Fig. 11.19:

$$l(t)=l_{1,4}(t)=\left\{(x,y);x+4y=\frac{t}{8}+\frac{5}{16}\right\},t=0:35,(x,y)\in[0,1]\times[0,1].$$

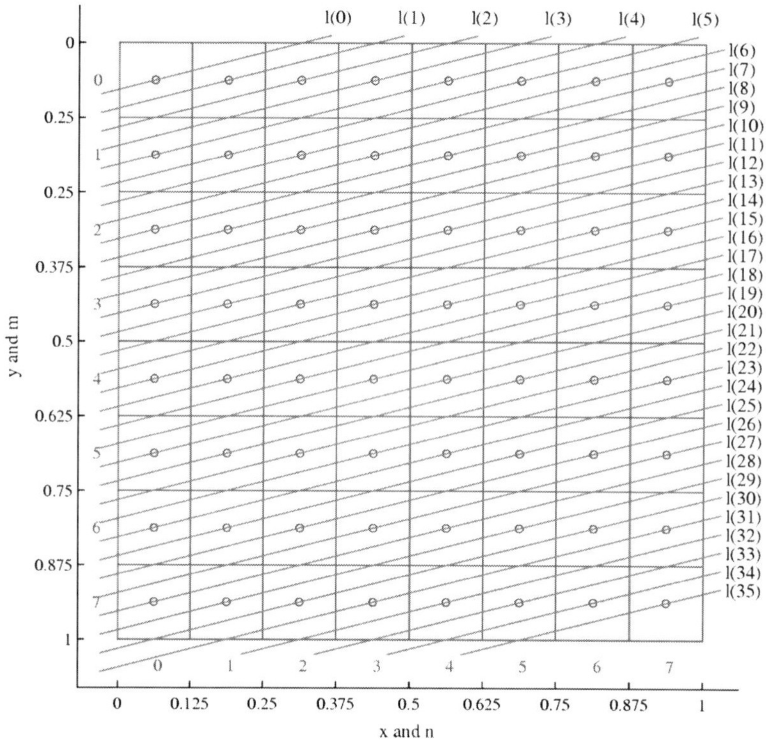

Figure 11.19 The sets of parallel rays for the (1,4)-projection.

$(1+4)7+1=36$ parallel rays are considered for the (1,4)-projection. The control points of the set of rays are defined as follows:

$$\begin{bmatrix} 0\bullet & 1\bullet & 2\bullet & 3\bullet & 4\bullet & 5\bullet & 6\bullet & 7\bullet \\ \cdot & \cdot & \cdot & \cdot & 8\bullet & 9\bullet & 10\bullet & 11\bullet \\ \cdot & \cdot & \cdot & \cdot & 12\bullet & 13\bullet & 14\bullet & 15\bullet \\ \cdot & \cdot & \cdot & \cdot & 16\bullet & 17\bullet & 18\bullet & 19\bullet \\ \cdot & \cdot & \cdot & \cdot & 20\bullet & 21\bullet & 22\bullet & 23\bullet \\ \cdot & \cdot & \cdot & \cdot & 24\bullet & 25\bullet & 26\bullet & 27\bullet \\ \cdot & \cdot & \cdot & \cdot & 28\bullet & 29\bullet & 30\bullet & 31\bullet \\ \cdot & \cdot & \cdot & \cdot & 32\bullet & 33\bullet & 34\bullet & 35\bullet \end{bmatrix} \text{(angle of rays is } -\tan^{-1}(1/4) = -14.036^\circ).$$

The rays are shown separately on the masks of the paired functions $\chi'_{1,4,t(n,m)}$, $t=0{:}3$, in Fig. 11.20.

We denote by $v_t = v_{1,4}(t)$, where $t = 0:35$, the sums of the discrete image $f_{n,m}$ along these rays. The components of the paired transform with the triplet-numbers $(1,4,t)$ can be calculated as follows:

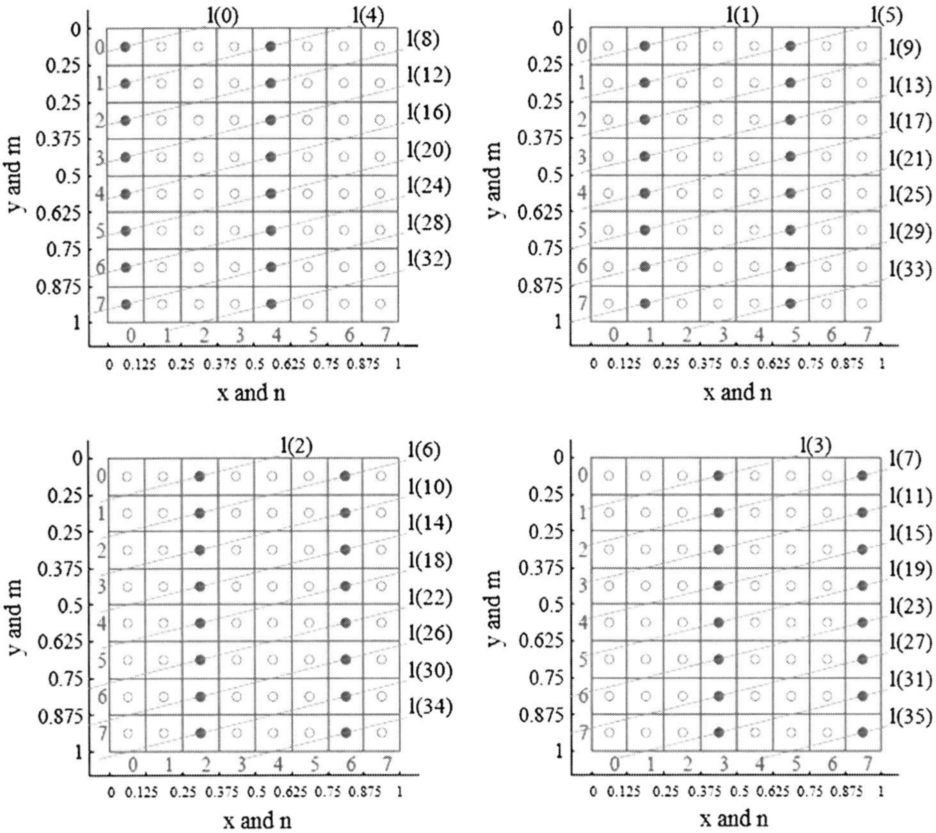

Figure 11.20 Four masks with the set of arithmetical rays for the paired functions with triplet numbers $(1, 4, t)$, $t = 0 : 3$.

$$
\begin{aligned}
f'_{1,4,0} &= v_0 - v_4 + v_8 - v_{12} + v_{16} - v_{20} + v_{24} - v_{28} + v_{32},\\
f'_{1,4,1} &= v_1 - v_5 + v_9 - v_{13} + v_{17} - v_{21} + v_{25} - v_{29} + v_{33},\\
f'_{1,4,2} &= v_2 - v_6 + v_{10} - v_{14} + v_{18} - v_{22} + v_{26} - v_{30} + v_{34},\\
f'_{1,4,3} &= v_3 - v_7 + v_{11} - v_{15} + v_{19} - v_{23} + v_{27} - v_{31} + v_{35}.
\end{aligned}
\tag{11.20}
$$

As follows from Eq. (11.10), from the ray sums of the (1,4)-projection, the components $f'_{2,0,0}, f'_{2,0,2}, f'_{4,0,0}$, and $f'_{0,0,0}$ can also be calculated. Two masks with rays for the paired functions $\chi'_{2,0,t}$, $t = 0, 2$ are shown in Fig. 11.21.

The corresponding components of the 2D paired transform are calculated by

$$
\begin{aligned}
f'_{2,0,0} = &\ (v_0 + v_4 + v_8 + v_{12} + v_{16} + v_{20} + v_{24} + v_{28} + v_{32})\\
&-(v_2 + v_6 + v_{10} + v_{14} + v_{18} + v_{22} + v_{26} + v_{30} + v_{34}),\\
f'_{2,0,2} = &\ (v_1 + v_5 + v_9 + v_{13} + v_{17} + v_{21} + v_{25} + v_{29} + v_{33})\\
&-(v_3 + v_7 + v_{11} + v_{15} + v_{19} + v_{23} + v_{27} + v_{31} + v_{35}).
\end{aligned}
\tag{11.21}
$$

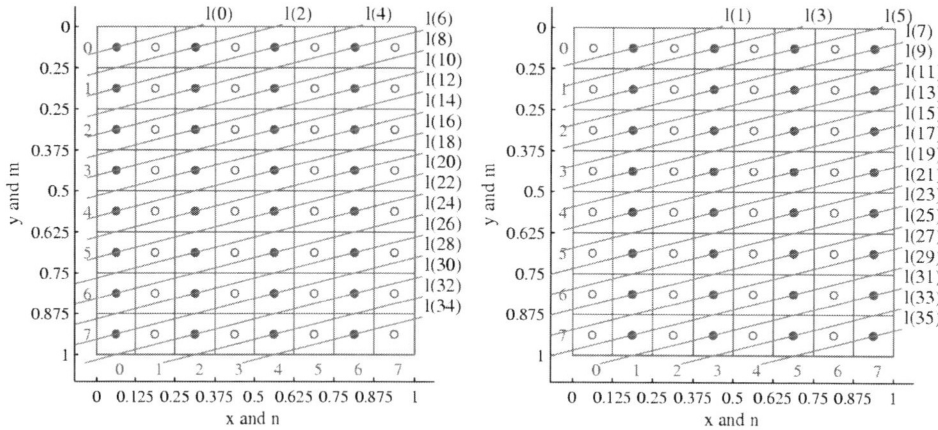

Figure 11.21 Two masks with the set of arithmetical rays for the paired functions with triplet numbers $(2, 0, t)$, $t = 0, 1$.

Similarly, for the remaining two components, there is the following: $f'_{4,0,0} = v_0 - v_1 + v_2 - v_3 + \ldots + v_{34} - v_{35}$ and $f'_{0,0,0} = v_0 + v_1 + v_2 + v_3 + \ldots + v_{34} + v_{35}$.

We consider the following set of triplet numbers:

$$U(1,4) = \{(1,4,0),\ (1,4,1),\ (1,4,2),\ (1,4,3),\ (2,0,0),\ (2,0,2),\ (4,0,0),\ (0,0,0)\},$$

and define by $\mathbf{P}(1,4) = \{f'_{p_1,s_1,t};\ (p_1, s_1, t) \in U(1,4)\}$ the components of the 2D paired transforms with these triplets. Introducing the new variables

$$a_n = \sum_{m=0}^{4} v_{n+8m} = \sum_{m=0}^{4} v_{1,4,n+8m}, \quad n = 0:7,$$

where $v_k = 0$ if $k > 35$, the complete system of linear equations for the eight components $f'_{p_1,s_1,t} \in \mathbf{P}(1,4)$ can be written as follows: $f'_{1,4,0} = a_0 - a_4$,

$$\begin{aligned}
f'_{1,4,1} &= a_1 - a_5,\\
f'_{1,4,2} &= a_2 - a_6,\\
f'_{1,4,3} &= a_3 - a_7,\\
f'_{2,0,0} &= (a_0 + a_4) - (a_2 + a_6),\\
f'_{2,0,2} &= (a_1 + a_5) - (a_3 + a_7),\\
f'_{4,0,0} &= (a_0 + a_4) - (a_1 + a_5) + (a_2 + a_6) - (a_3 + a_7),\\
f'_{0,0,0} &= (a_0 + a_4) + (a_1 + a_5) + (a_2 + a_6) + (a_3 + a_7). \qquad (11.22)
\end{aligned}$$

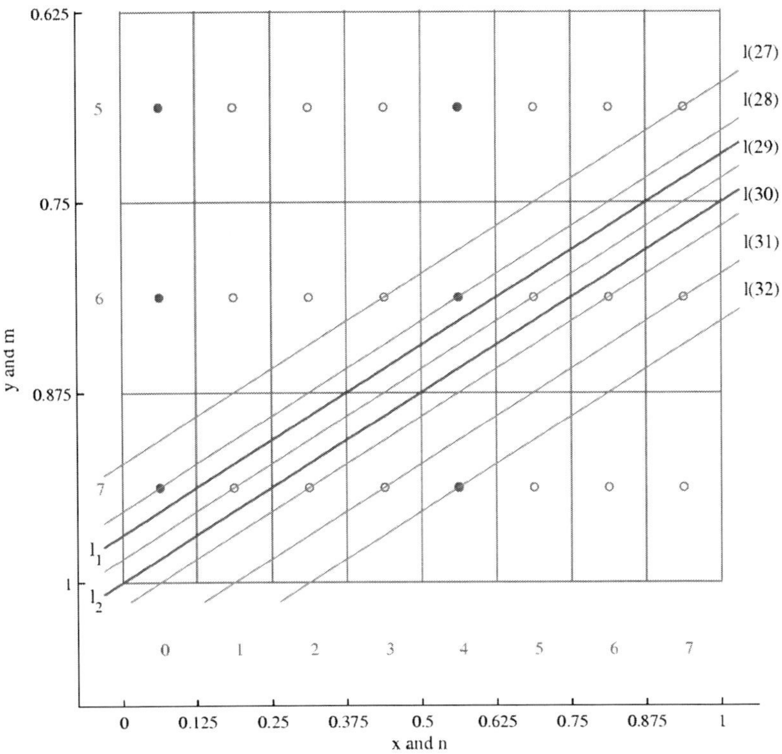

Figure 11.22 Set of geometrical rays for calculating line-integrals when $(p, s) = (1, 4)$.

These calculations correspond to the fast 8-point discrete paired transform (DPT)[33,34,50] of the vector $\mathbf{a} = (a_0, a_1, a_2, \ldots, a_7)'$, i.e., $\mathbf{P}(1,4) = [\chi_8']\mathbf{a}$.

To express the sums $v(t)$ of the discrete image $f_{n,m}$ through the line integrals $w(t) = w_{1,4}(t)$, consider the part of the mask of the paired function with number (1,4,0), which is shown in Fig. 11.22. The geometrical ray $l(29)$ between two points of the discrete image, (0,7) and (5,6), passes through six image elements of numbers (0,7), (1,7), (2,7), (3,7), (3,6), and (4,6). The length of the intersection $\Delta l_{0,7}$ of the ray with the IE of number (0,7) is twice the length of the intersection with each IE of number (3,7) and (3,6), but it is equal to the length of intersection with the IE of numbers (1,7) and (2,7). Because $\Delta l_{0,7} = \sqrt{17}/4\Delta x$, the following can be written:

$$w(29) = w_{1,4}(29) = \frac{\Delta l_{0,7}}{(\Delta x)^2}\left[v(28) + v(29) + v(30) + \frac{1}{2}v(31) + \frac{1}{2}v(27)\right].$$

Similar equations can be used to calculate other integrals $w(t)$, $t = 0 : 35$, and then the sums $v(t)$ can be found. Now consider the geometrical rays l_1

and l_2 between rays 28, 29 and 29, 30, respectively. The length of the intersection of ray l_1 with the image elements of numbers (0,7), (1,7), (2,7), and (3,6) is the same: $\sqrt{17}/4\Delta x$. The length of the intersection of ray l_2 with each image element of number (0,7), (1,7), (2,7), and (3,7) is also $\sqrt{17}/4\Delta x$.

Therefore, the integrals w_1 and w_2 of the image along the shifted rays l_1 and l_2, respectively, can be written as follows:

$$w_1 = \frac{\sqrt{17}/4\Delta x}{(\Delta x)^2}[v(27) + v(28) + v(29) + v(30)] = 2\sqrt{17}[v(27) + v(28) + v(29) + v(30)],$$

$$w_2 = 2\sqrt{17}[v(28) + v(29) + v(30) + v(31)].$$

These equations can be generalized if the integral w_1 is denoted by $w(30)$, and w_2 by $w(31)$. [Integral w_1 can also be denoted by $w(27)$ and w_2 by $w(28)$.] In this case, a simple solution for sums $v(t)$ can be obtained. Indeed, consider a new set of parallel rays defined by the shifting $t \to t - 3/2$,

$$\tilde{l}(t) = l_{1,4}(t - 3/2) = \left\{(x, y); x + 4y = \frac{t}{8} + \frac{1}{8}\right\}, \quad t = 0 : 35,$$

which is shown in Fig. 11.23.

For the integrals along the set of geometrical rays, we can write the following equations:

$$w(t) = 2\sqrt{17}[v(t) + v(t-1) + v(t-2) + v(t-3)], t = 0 : 35.$$

where $v(-1) = v(-2) = v(-3) = 0$. This system of linear equations can be written in matrix form as

$$\mathbf{w} = 2\sqrt{17}\mathbf{A}\mathbf{v} = 2\sqrt{17}\begin{bmatrix} 1 & 0 & 0 & 0 & 0 & 0 & \dots & 0 & 0 & 0 & 0 & 0 \\ 1 & 1 & 0 & 0 & 0 & 0 & \dots & 0 & 0 & 0 & 0 & 0 \\ 1 & 1 & 1 & 0 & 0 & 0 & \dots & 0 & 0 & 0 & 0 & 0 \\ 1 & 1 & 1 & 1 & 0 & 0 & \dots & 0 & 0 & 0 & 0 & 0 \\ 0 & 1 & 1 & 1 & 1 & 0 & \dots & 0 & 0 & 0 & 0 & 0 \\ 0 & 0 & 1 & 1 & 1 & 1 & \dots & 0 & 0 & 0 & 0 & 0 \\ \cdot & \cdot & \cdot & \cdot & \cdot & \cdot & \dots & \cdot & \cdot & \cdot & \cdot & \cdot \\ \cdot & \cdot & \cdot & \cdot & \cdot & \cdot & \dots & \cdot & \cdot & \cdot & 0 & 0 \\ 0 & 0 & 0 & 0 & 0 & 0 & \dots & 1 & 1 & 1 & 1 & 0 \\ 0 & 0 & 0 & 0 & 0 & 0 & \dots & 0 & 1 & 1 & 1 & 1 \end{bmatrix}\mathbf{v}, \quad (11.23)$$

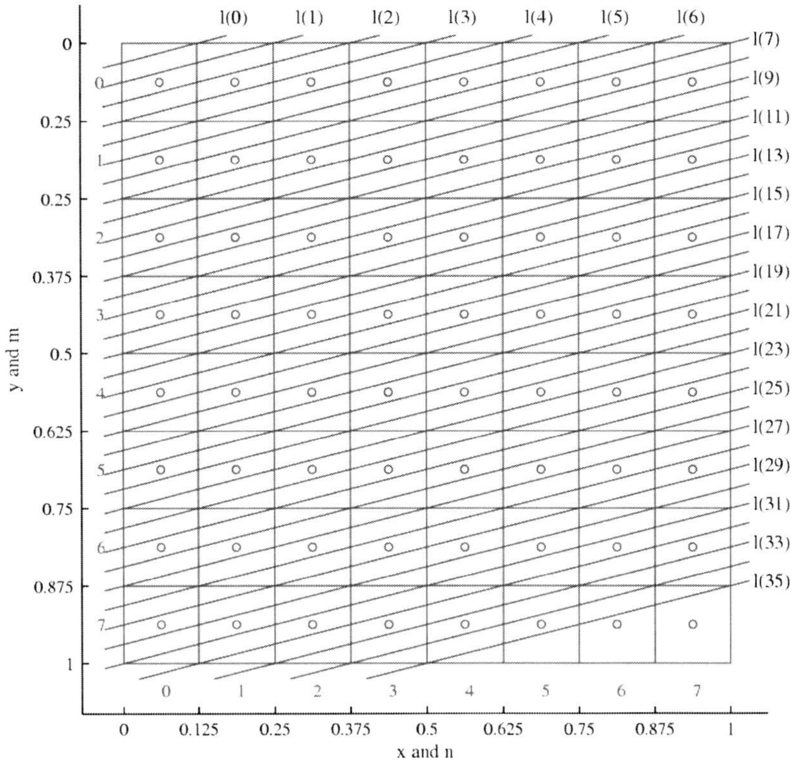

Figure 11.23 The set of geometrical rays for the (1, 4)-projection.

where $\mathbf{w} = (w_0, w_1, \ldots, w_{35})'$, and $\mathbf{v} = (v_0, v_1, \ldots, v_{35})'$. The above Toeplitz matrix 36 × 36 has the following inverse lower-triangle Toeplitz matrix:

$$\mathbf{A}^{-1} = \begin{bmatrix} 1 & 0 & 0 & 0 & 0 & 0 & 0 & \ldots & 0 & 0 & 0 & 0 \\ -1 & 1 & 0 & 0 & 0 & 0 & 0 & \ldots & 0 & 0 & 0 & 0 \\ 0 & -1 & 1 & 0 & 0 & 0 & 0 & \ldots & 0 & 0 & 0 & 0 \\ 0 & 0 & -1 & 1 & 0 & 0 & 0 & \ldots & 0 & 0 & 0 & 0 \\ 1 & 0 & 0 & -1 & 1 & 0 & 0 & \ldots & 0 & 0 & 0 & 0 \\ -1 & 1 & 0 & 0 & -1 & 1 & 0 & \ldots & 0 & 0 & 0 & 0 \\ 0 & -1 & 1 & 0 & 0 & -1 & 1 & \ldots & 0 & 0 & 0 & 0 \\ 0 & 0 & -1 & 1 & 0 & 0 & -1 & \ldots & 0 & 0 & 0 & 0 \\ . & . & . & . & . & . & . & \ldots & . & . & . & . \\ 1 & 0 & 0 & -1 & 1 & 0 & 0 & \ldots & 1 & 0 & 0 & 0 \\ -1 & 1 & 0 & 0 & -1 & 1 & 0 & \ldots & -1 & 1 & 0 & 0 \\ 0 & -1 & 1 & 0 & 0 & -1 & 1 & \ldots & 0 & -1 & 1 & 0 \\ 0 & 0 & -1 & 1 & 0 & 0 & -1 & \ldots & 0 & 0 & -1 & 1 \end{bmatrix} \quad (11.24)$$

From the solution of the equation $\mathbf{b} = \mathbf{A}^{-1}\mathbf{w}$, the sums $\mathbf{v} = \mathbf{b}/(2\sqrt{17})$ are obtained. The vector $\mathbf{b}$ can be calculated by using the following recurrent procedure:

$$\begin{aligned} b_0 &= w(0), \\ b_1 &= w(1) - b_0, \\ b_2 &= w(2) - (b_1 + b_0), \\ b_3 &= w(3) - (b_2 + b_1 + b_0), \\ b_4 &= w(4) - (b_3 + b_2 + b_1), \\ b_t &= w(t) - (b_{t-1} + b_{t-2} + b_{t-3}), \quad t = 5, 6 \ldots, 35. \end{aligned} \tag{11.25}$$

Similar calculations hold for other projections, and the ray sums can be calculated from the line integrals by solving the system of linear equations described by the triangle Toeplitz matrices (see Grigoryan and Grigoryan[45] for more details).

11.5.1 Algorithm of image reconstruction

For the $N = 8$ case, all $3N/2 = 12$ projections are required to reconstruct the image $N \times N$. The number of parallel geometrical rays used in the (p, s)-projections is defined by Eq. (11.19). In the case where N is a power of two, the line integrals are used to calculate the splitting-signals in the paired or tensor representation of the discrete image. Therefore, the block diagram given in Eq. (11.15) can be used for reconstructions by the tensor transform. In order to reduce the complexity of calculations due to redundancy of the tensor transform, the 2D paired transform can be calculated, as shown in the above $N = 8$ example, and then the reconstruction $f_{n,m}$ is calculated by the inverse paired transform.

Each triplet number (p,s,t) of masks of the paired functions in this table has the form $2^k(p_1,s_1,t_1)$, where $2^k = \text{g.c.d}(p, s)$, and $t_1 = 0 : N/2^{k+1} - 1$, and its multiplicity (i.e., the number of subsets U covering this triplet number) equals 2^k. Therefore, there are two ways to use this property in image reconstruction.

1. One can use the complete 1D paired transforms over all vectors $\mathbf{a}$, whose components are defined as

$$a_n = \sum_m v_{n \pm Nm} = \sum_m v_{p,s,n \pm Nm}, \quad n = 0 : N - 1.$$

   The components $f'_{p,s,t}$ of the 2D paired transform, which have been calculated by 1D paired transform $\mathbf{P}(p, s) = [\chi'_N]\mathbf{a}$, are then normalized by the factor of $2^k = \text{g.c.d.}(p, s)$, when $(p, s) \neq (0, 0)$, and N, when $(p, s) = (0, 0)$.
2. One can use the incomplete complete 1D paired transforms over vectors $\mathbf{a}$ to avoid repeating the calculation of components of the 2D paired

transform of the image. For instance, when $N = 8$, the complete 1D paired transform can be used for the first (1,0)-projection, but only six outputs can be used for the transform when calculating $\mathbf{P}(1,2) = [\chi'_8]\mathbf{a}$ for the (1,2)-projection, only half of the calculations when calculating $\mathbf{P}(1,4) = [\chi'_8]\mathbf{a}$ for the (1,4)-projection, and so on. The use of incomplete 1D paired transforms reduces the time necessary to calculate the reconstructed image, when compared with the first method.

In the general $N = 2^r$ case, where $r \geq 2$, the system of equations for the geometrical rays is defined in the following way. First, consider the arithmetical rays $l(t) = l_{p,s}(t)$, which are described by

$$l_{p,s,t}(x,y) = \left\{(x,y); xp + ys = \frac{t}{N} + \frac{p+s}{2N}\right\}.$$

To reduce the number of line integrals, the set of all generators of $J_{N,N}$ can be divided by four subsets. The first and second subsets contain all generators $(1,s)$, where $s = 0\colon N/2$ and $s = N/2+1\colon N-1$, respectively. The third and fourth subsets contain all generators $(p,1)$, where $p = 2p_1$, and $p_1 = 0\colon N/4$ and $p_1 = N/4+1\colon N/2-1$, respectively. For each of these subsets, consider the case when the arithmetical rays are shifted to the right to define the corresponding geometrical rays. Another case involves geometrical rays that are defined from arithmetical rays by shifting to the left. Thus, the geometrical rays are defined from the arithmetical rays by shifting $\tilde{l}(t) = l(t-t_0)$ (or $\tilde{l}(t) = l(t+t_0)$). Here, the shift t_0 is calculated by

$$t_0 = \begin{cases} \dfrac{p+s}{2} - 1, & \text{when} \quad 0 < p,s \leq N/2, \\ \dfrac{p+(N-s)}{2} - 1, & \text{when} \quad p = 1, s > N/2, \\ \dfrac{(N-p)+s}{2} - 1, & \text{when} \quad p > N/2, s = 1. \end{cases} \tag{11.26}$$

11.5.2 Convolution equations

For all generators $(p,s) \in J_{N,N}$, the relationships between the sums $v(t) = v_{p,s}(t)$ of the discrete image $f_{n,m}$ and line integrals $w(t) = w_{p,s}(t)$ of the original unknown image $f(x,y)$ are described in matrix form by the Toeplitz matrices, which are similar to the described $N = 8$ example. Consider separately the convolution equations of line integrals for different sets of (p,s)-projections.

Subset I: $p = 1$ and $s \leq N/2$
When $s \neq 0$, the convolution equation $\mathbf{w} = \mathbf{A}\mathbf{v}$ is described as

$$w(t) = N\frac{\sqrt{1+s^2}}{s}[v(t) + v(t-1) + v(t-2) + \ldots + v(t-s+1)], t = 0:(1+s)(N-1), \tag{11.27}$$

where it is assumed that $v(-1) = v(-2) = \ldots = v(-s+1) = 0$. The inverse transform $\mathbf{b} = \mathbf{A}^{-1}\mathbf{w}$ is calculated by the following recurrent form:

$$\begin{aligned}
b_0 &= w(0),\\
b_1 &= w(1) - b_0,\\
b_2 &= w(2) - (b_1 + b_0),\\
b_3 &= w(3) - (b_2 + b_1 + b_0),\\
&\ldots\\
b_{s-1} &= w(s-1) - (b_{s-2} + b_{s-3} + \ldots + b_1 + b_0),\\
b_t &= w(t) - (b_{t-1} + b_{t-2} + \ldots + b_{t-s+2} + b_{t-s+1}),\\
t &= s, s+1, \ldots, (1+s)(N-1).
\end{aligned} \tag{11.28}$$

The required sums $v(t)$ of the discrete image are calculated by

$$v(t) = b_t \frac{s}{N\sqrt{1+s^2}}, \quad t = 0 : (1+s)(N-1). \tag{11.29}$$

In the $s = 0$ case, there is the following simple solution: $v(t) = w(t)/N$, $t = 0 : (N-1)$.

Subset II: $p = 1$ and $s > N/2$
The $(1, s)$-projection is described as the $(1,\ s-N)$-projection, i.e., s is considered as $s - N$. Therefore, the following convolution equation $\mathbf{w} = \mathbf{A}\mathbf{v}$ is used:

$$\begin{aligned}
w(t) &= N\frac{\sqrt{1+\bar{s}^2}}{\bar{s}}[v(t) + v(t-1) + v(t-2) +, \ldots, + v(t-\bar{s}+1)],\\
t &= (N-1) : -1 : -\bar{s}(N-1),
\end{aligned} \tag{11.30}$$

where $\bar{s} = N - s$, and $v(-k) = 0$, when $k > \bar{s}\ (N-1)$.

M denotes the negative number $-\bar{s}\ (N-1)$. The inverse transform $\mathbf{b} = \mathbf{A}^{-1}\mathbf{w}$ can be calculated by the following recurrent form:

$$\begin{aligned}
b_M &= w(M),\\
b_{M+1} &= w(M+1) - b_M,\\
b_{M+2} &= w(M+2) - (b_{M+1} + b_M),\\
b_{M+3} &= w(M+3) - (b_{M+2} + b_{M+1} + b_M),\\
&\ldots\\
b_{M+\bar{s}-1} &= w(M+\bar{s}-1) - (b_{M+\bar{s}-2} + b_{M+\bar{s}-3} + \ldots + b_{M+1} + b_M),\\
b_t &= w(t) - (b_{t-1} + b_{t-2} + \ldots + b_{t-\bar{s}+2} + b_{t-\bar{s}+1}),\\
t &= M+\bar{s}, M+\bar{s}+1, \ldots, (N-1).
\end{aligned} \tag{11.31}$$

The required sums $v(t)$ of the discrete image are calculated by

$$v(t) = b_t \frac{\bar{s}}{N\sqrt{1+\bar{s}^2}}, t = (N-1) : -1 : -\bar{s}(N-1). \tag{11.32}$$

The remaining two subsets of projections—*Subset III*, when $p \leq N/2$ and $s = 1$, and *Subset IV*, when $p > N/2$ and $s = 1$—are described like *Subset I* and *Subset II*, respectively.

11.5.3 Preliminary results

This section illustrates a few results of reconstructing the image $f(x, y)$ on the lattice of size $N \times N$, $N = 2^r$, from $3N/2$ projections. As an example, Fig. 11.24(a) shows the image, 11.24(b) shows the graph of all line integrals $w(t)$ of the (1,4)-projection, and 11.24(c) shows ray sums $v(t)$ calculated from line integrals.

The splitting-signal $\{f_{1,4,t};\ t = 0 : 255\}$ is shown in Fig. 11.25(a). This signal is in the tensor representation, and it defines the set of paired splitting-signals that are shown in Fig. 11.25(b). The first signal of length 128 is $\{f'_{1,4,t};\ t = 0 : 127\}$, the second signal of length 64 is $\{f'_{2,8,2t};\ t = 0 : 63\}$, and so on. The first four of these splitting-signals are separated in the figure by the vertical dashed lines.

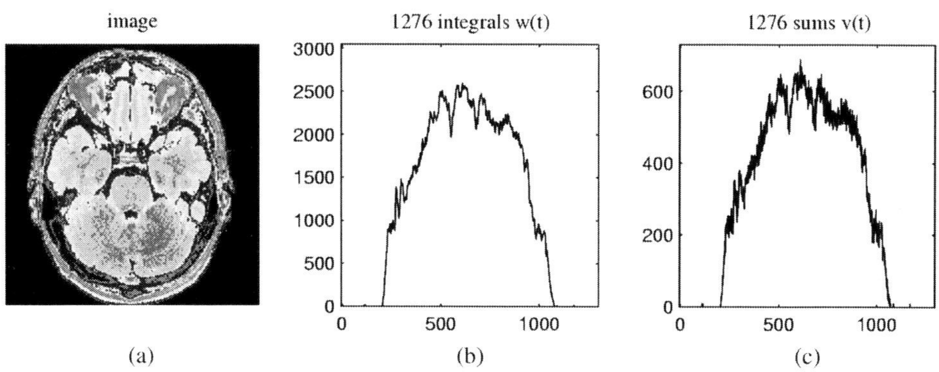

Figure 11.24 (a) Image, (b) line integrals, and (c) ray sums of the (1, 4)-projection.

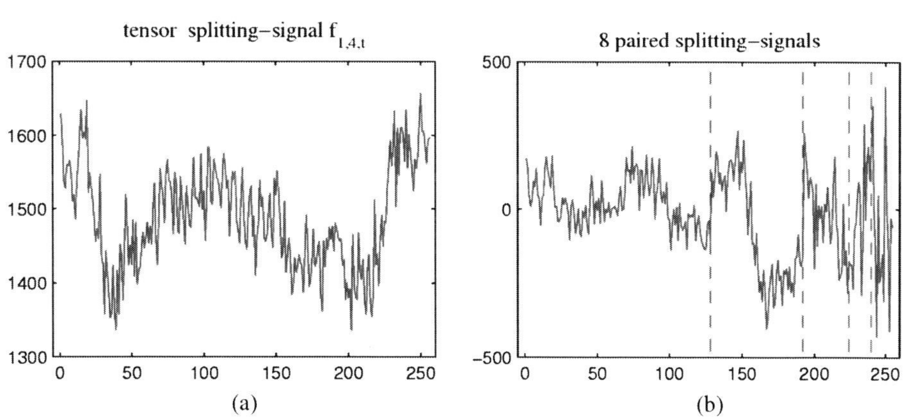

Figure 11.25 Splitting-signals in (a) tensor and (b) paired representations.

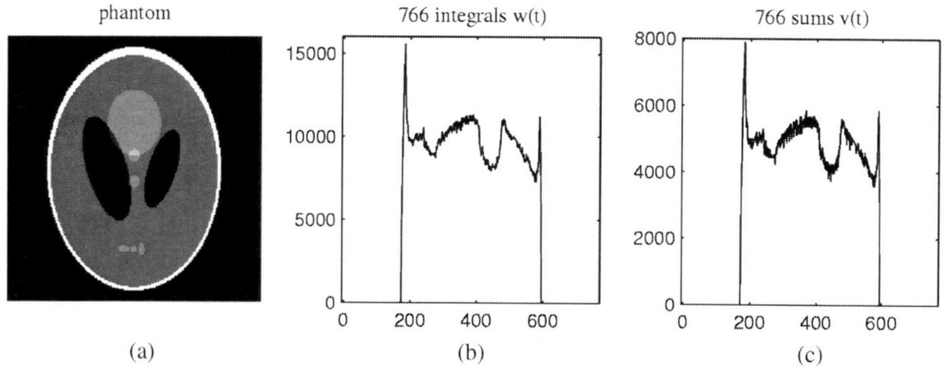

Figure 11.26 (a) Phantom image, (b) line integrals, and (c) ray sums of the (1, 2)-projection.

Figure 11.26(a) shows the MATLAB model of a modified Shepp–Logan phantom image, 11.26(b) shows the graph of all line integrals $w(t)$ of the (1, 2)-projection, and 11.26(c) shows the ray sums $v(t)$ calculated from line integrals.

The tensor splitting-signal $\{f_{1,2,t};\ t = 0 : 255\}$ is shown in Fig. 11.27(a), and 11.27(b) shows the corresponding set of eight paired splitting-signals $\{f'_{1,2,t};\ t = 0 : 127\}$, $\{f'_{2,4,2t};\ t = 0 : 63\}$, $\{f'_{4,8,4t};\ t = 0 : 63\}, \ldots,$ $\{f'_{64,64,0}, f'_{64,64,64}\}$, $\{f'_{128,128,0}\}$, and $\{f'_{0,0,0}\}$. The first four splitting-signals are separated by the vertical lines.

Now consider the results of image reconstruction on examples with random rectangles in the square $[0, 1] \times [0, 1]$, and the objects with sharp edges, in particular. Figure 11.28 shows ten rectangles on the square with different intensities on the left. These rectangles are on the grid 128×128. The image calculated from 192 projections is shown on the right. The reconstruction is

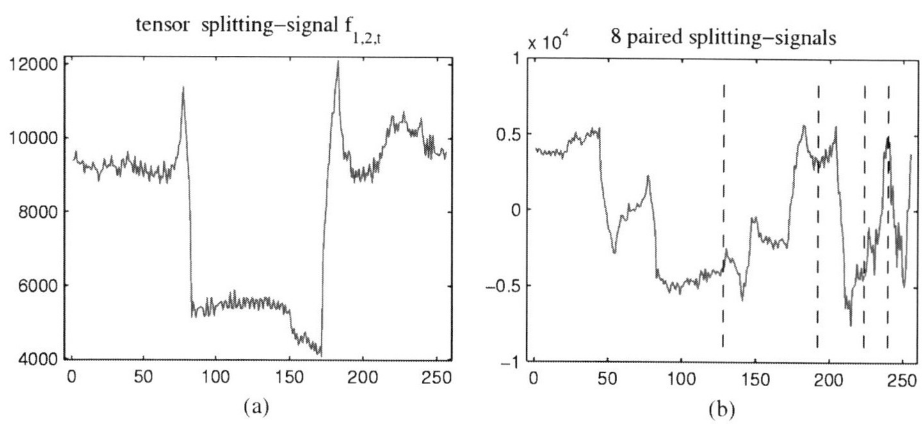

Figure 11.27 (a) Splitting-signal $\{f_{1,2,t}\}$ and (b) eight paired splitting-signals.

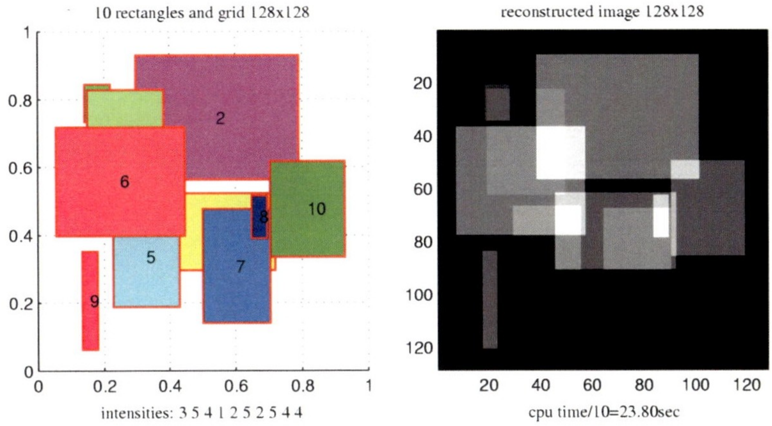

Figure 11.28 Ten random rectangles (left) and the reconstructed image (right).

exact, and the main program required 24 s to calculate all projections, i.e., line integrals $w(t)$, and reconstruct the image. Incomplete 1D DPTs have been used in calculations. The CPU time is calculated as the average time for one rectangle. The process of calculation is organized in such a way that for each (p, s)-projection, the integrals are calculated for each rectangle separately, and then the results are added. The inverse 2D 128×128-point DPT requires 0.5304 s.

Figure 11.29 shows ten rectangles on the square $[0, 1] \times [0, 1]$ on the left; the coordinates of these rectangles are on the grid 256×256. The image calculated from 384 projections is shown on the right. The reconstruction is exact, and the main program required 4.56 min to calculate all projections and reconstruct the image. Incomplete 1D DPT has been used in calculations. The inverse 2D 256×256-point DPT requires 4.4460 s.

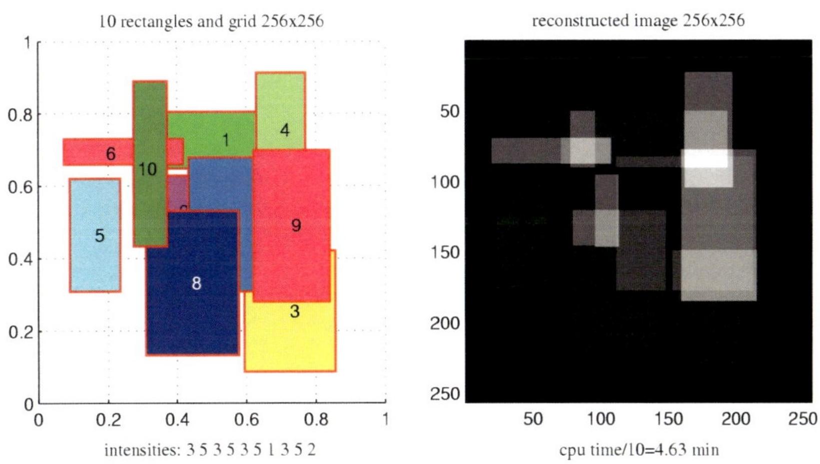

Figure 11.29 Ten random rectangles (left) and the reconstructed image (right).

Table 11.3 Time for scanning and reconstruction.

$N \times N$	in MATLAB	in Ci+
32×32	00.48 s/rec	00.03 s
64×64	02.58 s/rec	00.11 s
128×128	23.80 s/rec	00.79 s
256×256	04.63 min/rec	06.17 s
512×512	–	49.76 s

Time characteristics of calculation, when the proposed method was implemented in MATLAB and Ci+, are given in Table 11.3. The time includes the calculation of all line integrals $w(t)$ and line sums $v(t)$, and the calculation of the 2D direct and inverse paired transforms. Incomplete 1D DPTs are used. The data in these tables is obtained by programs on a personal computer with an Intel® Dual-CPU processor at 3.20-GHz speed. This is the first realization of the method in MATLAB and Ci+, which can be improved in order to achieve a fast reconstruction of images on large-size Cartesian lattices.

To achieve fast reconstruction of the image, each arithmetical ray, as well as each geometrical ray, can be described as the unique set of IEs through which the ray passes. This approach allows for the calculation of the line integrals of the entire modeled image in one pass. As an example, Fig. 11.30 shows the image of 20 random rectangles with the reconstruction. The total time of reconstruction is 4.98 min, which is ten times smaller than the time given in Table 11.3.

Figure 11.31 shows the result of reconstructing an image composed of 13 ellipses and circles. The reconstruction is on the Cartesian lattice of size 256×256.

The 256×256 reconstruction of the modified Shepp–Logan phantom is shown in Fig. 11.32. The reconstruction is exact, and all calculations that

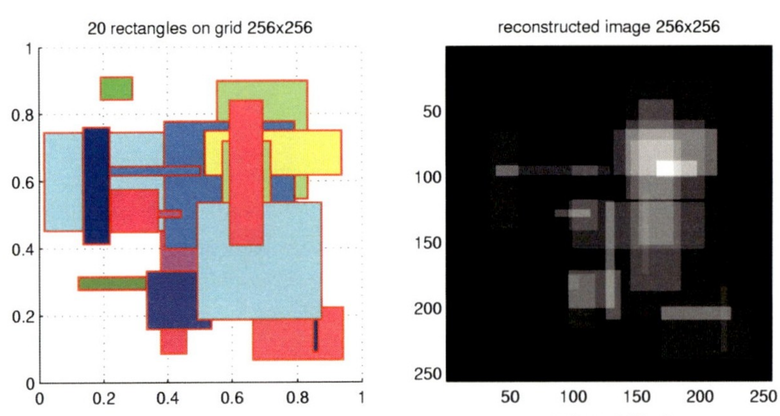

Figure 11.30 20 random rectangles (left) and the reconstructed image (right).

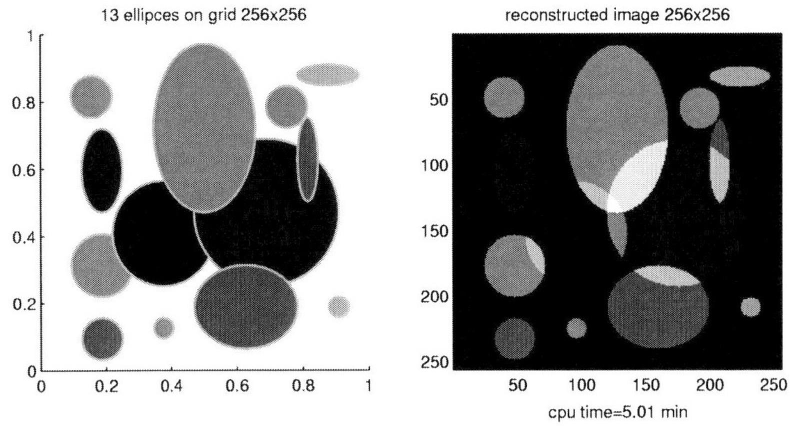

Figure 11.31 Image with 13 ellipses (left) and the reconstructed image (right).

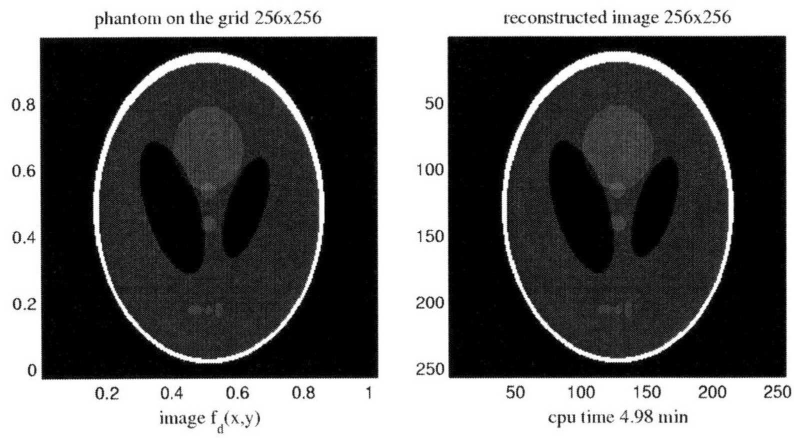

Figure 11.32 Image (left) and its reconstruction (right).

include the line integrals, solutions of the convolution equations, and the inverse 2D paired transform, require about five minutes when running the code in MATLAB.

11.6 Conclusion

The well-known model of image reconstruction has been analyzed when the image is represented by a set of small image elements with constant intensities each, and the rays are considered to have zero width. In the framework of the considered model of the image $f(x, y)$ on the unit square, the solution to the problem of image reconstruction from a finite number of projections is given. The solution is based on the properties of the 2D tensor and paired transforms, each basis function of which is binary and defined by a set of parallel rays on the lattice. The main task in image reconstruction is the transformation of the geometry from the image plane to the Cartesian lattice, on which the reconstructed image is calculated. This problem is solved by introducing the set of geometrical rays for each projection and the triangle Toeplitz-matrix-based system of convolution equations. The cases where image size is $N \times N$, and N is a power of two and a prime, is described. The proposed method can be used in tomographic imaging, and the main idea of transferring geometry can be generalized for the parallel rays with nonzero widths, as well as for the fan beam projections.

References

1. A. C. Kak and M. Slaney, *Principles of Computerized Tomographic Imaging*, IEEE Press, New York (1987).
2. G. T. Herman, *Image Reconstruction from Projections*, Academic Press, New York (1979).
3. D. E. Olins, A. L. Olins, H. A. Levy, R. C. Durfee, S. M. Margle, E. P. Tinnel, and S. D. Dover, "Electron microscope tomography: transcription in three dimensions," *Science* **220**, 498–500 (1983).
4. P. Suetens, *Fundamentals of Medical Imaging*, Cambridge University Press, Cambridge, UK (2002).
5. B. Chen and R. Ning, "Cone-beam volume CT breast imaging: feasibility study," *Med Phys.* **29**, 755–770 (2002).
6. A. Iagaru, R. Masamed, S. Keesara, and P. S. Conti, "Breast MRI and 18F FDG PET/CT in the management of breast cancer," *Annals Nuclear Med.* **21**(1), 3338 (2007).
7. P. Pettigrew and J. L. Berry, "The Role of CT in Breast Imaging," www.eradimaging.com/site/article.cfm?ID=739.
8. S. Glick, "Breast CT," *Annu. Rev. Biomed. Eng.* **9**, 501–526 (2007).
9. C. Lai et al., "Comparison of slot scanning digital mammography system with full-field digital mammography system," *Med. Phys.* **35**, 2339–2346 (2008).
10. T. R. Nelson, L. I. Cervina, and J. M. Boone, "Classification of breast computed tomography data," *Med. Phys.* **35**, 1078–1086 (2008).
11. X. Gong et al., "A comparison simulation study comparing lesion detection accuracy with digital mammography, breast tomosynthesis, and cone-beam CT breast imaging," *Med. Phys.* **33**, 1041–1052 (2006).
12. Z. Chen and R. Ning, "Why should breast tumor detection go three dimensional?" *Phys. Med. Biol.* **48**, 2217–2228 (2003).

13. B. Chen and R. Ning, "Cone-beam volume CT breast imaging: feasibility study," *Med. Phys.* **29**, 755–770 (2002).
14. U. C. Davis Health System, "Breast CT reaches clinical testing: may improve on mammography," www.ucdmc.ucdavis.edu/newsroom/releases/archives/cancer/2005/breast-ct5-2005.html, accessed September 26, 2009.
15. A. Shimauchi et al., "Comparison of MDCT and MRI for evaluating the intraductal component of breast cancer, *Am. J. Roentgenol.* **187**, 322–329 (2006).
16. M. Inoue et al., "Dynamic multidetector CT of breast tumors: diagnostic features and comparison with conventional techniques," *Am. J. Roentgenol.* **181**, 679–686 (2003).
17. K. K. Lindfors et al., "Dedicated breast CT: initial clinical experience," *Radiology* **246**, 725–733 (2008).
18. J. Radon, "On the determination of function from their integrals along certain manifolds," *Ber. Saechs. Akad. Wiss* **69**, 262–277 (1917).
19. S. Helgason, *The Radon Transform, Progress in Mathematics 5*, Birkhauser, Boston, MA (1980).
20. R. Gordon, "A tutorial on ART (Algebraic reconstruction techniques)," *IEEE Trans. Nuclear Science* **21**, 78–93 (1974).
21. R. Gordon, R. Bender, and G. T. Herman, "Algebraic reconstruction techniques (ART) for three-dimensional electron microscopy and X-ray photography," *J. Theor. Biol.* **29**, 471–481 (1970).
22. G. T. Herman and A. Lent, "Iterative reconstruction algorithms," *Comput. Biol. Med.* **6**, 273–294 (1976).
23. Y. Censor, "Finite series-expansion reconstruction methods," *Proc. IEEE* **71**(3), 409–419 (1983).
24. R. H. T. Bates and M. J. McDonnell, *Image Restoration and Reconstruction*, Charedon Press, Oxford, UK (1986).
25. A. H. Andersen and A. C. Kak, "Simultaneous algebraic reconstruction technique (SART): a superior implementation of the ART algorithm," *Ultrasonic Imaging* **6**, 81–94 (1984).
26. The National Academies, *Mathematics and Physics of Emerging Biomedical Imaging*, National Academic Press, Washington, DC (1996).
27. F. Natterer, *Mathematical Methods in Image Reconstruction*, SIAM, Philadelphia, PA (2001).
28. Y. Lifeng and S. Leng, "Image Reconstruction Techniques," www.imagewisely.org/Imaging-Professionals/Medical-Physicists/Articles/Image-Reconstruction-Techniques.aspx (2010).
29. A. M. Grigoryan, "An algorithm of the two-dimensional Fourier transform," *Radioelectronica* **27**(10), 52–57, Izvestiya vuzov SSSR, Kiev,

USSR (1984) [Translation available at fasttransforms.com/Art-USSR-papers/Art-IzvuzovSSSR1984.pdf].

30. A. M. Grigoryan and M. M. Grigoryan, "Two-dimensional Fourier transform in the tensor presentation and new orthogonal functions," *Avtometria* **1**, AS USSR Siberian Section, pp. 2127 (1986).
31. A. M. Grigoryan, "New algorithms for calculating discrete Fourier transforms," Journal Vichislitelnoi Matematiki i Matematicheskoi Fiziki, vol. 26, no 9, pp. 1407-1412, AS USSR, Moscow 1986. (translated in http://fasttransforms.com/Art-USSR-papers/Art-JVMA-TofASUSSR1986.pdf)
32. A. M. Grigoryan, "An algorithm for computing the two-dimensional Fourier transform of equal orders," Proceedings of the IX All-Union Conference on Coding and Transmission of Information, Part 2, pp. 49–52, USSR, Odessa 1988. (translated in http://fasttransforms.com/Art-USSR-papers/Art-AllUnConf1988.pdf)
33. A. M. Grigoryan, "2D and 1D multi-paired transforms: Frequency-time type wavelets," *IEEE Trans. Signal Process.* **49**(2), 344–353 (2001).
34. A. M. Grigoryan, and S. S. Agaian, "Split manageable efficient algorithm for Fourier and Hadamard transforms," *IEEE Trans. Signal Process.* **48**(1), 172–183 (2000).
35. A. M. Grigoryan and S. S. Agaian, "Shifted Fourier transform based tensor algorithms for 2D DCT," *IEEE Trans. Signal Process.* **49**(9), 2113–2126 (2001).
36. S. S. Agaian, "Hadamard matrices and their applications," *Lecture Notes in Mathematics* **1168**, Springer–Verlag, Berlin (1985).
37. S. S. Agaian, "Advances and problems of the fast orthogonal transforms for signal-images processing applications (part 1)," *Pattern Recognition, Classification, Forecasting* **3**, 146–215, The Russian Academy of Sciences, Nauka, Moscow (1990) (in Russian).
38. S. S. Agaian, "Advances and problems of the fast orthogonal transforms for signal-images processing applications (part 2)," *Pattern Recognition, Classification, Forecasting* **4**, 156–246, The Russian Academy of Sciences, Nauka, Moscow (1991) (in Russian).
39. R. Kogan, S. S. Agaian, and K. A. Panetta, "Visualization using rational morphology and zonal magnitude reduction," *Proc. SPIE* **3304**, 153–163 (1998) [doi: 10.1117/12.304595].
40. S. S. Agaian, K. Panetta, and A. M. Grigoryan, "Transform-based image enhancement algorithms," *IEEE Trans. Image Process.* **10**(3), 367–382 (2001).
41. S. S. Agaian, K. A. Panetta, and A. M. Grigoryan, "A new measure of image enhancement," *SPPRA 2001*, Rhodes, Greece (2001).

42. A. M. Grigoryan and S. S. Agaian, "Transform-based image enhancement algorithms with performance measure," *Advances in Imaging and Electron Physics* **130**, Academic Press, New York, 165–242 (2004).
43. F. T. Arslan and A. M. Grigoryan, "Fast splitting -rooting method of image enhancement: Tensor representation," *IEEE Trans. Image Process.* **15**(11), 3375–3384 (2006).
44. A. M. Grigoryan, "Method of paired transforms for reconstruction of images from projections: Discrete model," *IEEE Trans. Image Process.* **12**(9), 985–994 (2003).
45. A. M. Grigoryan and M. M. Grigoryan, *Image Processing: Tensor Transform and Discrete Tomography with MATLAB*, CRC Press, Boca Raton, FL (2012).
46. A. M. Grigoryan and N. Du, "Principle of superposition by direction images," *IEEE Trans. Image Process.* **20**(9), 2531–2541 (2011).
47. A. M. Grigoryan and N. Du, "2D images in frequency-time representation: Direction images and resolution map," *J. Electron. Imaging* **19**(3), 1–14 (2010).
48. A. M. Grigoryan, "Multidimensional Discrete Unitary Transforms," Chapter 19 in *The Transforms and Applications Handbook*, 3rd ed., The Electrical Engineering Handbook Series, A. Poularikas, Ed., CRC Press, Boca Raton, FL (2010).
49. A. M. Grigoryan and S. S. Agaian, *Multidimensional Discrete Unitary Transforms: Representation, Partitioning, and Algorithms*, Marcel Dekker, New York (2003).
50. A. M. Grigoryan and M. M. Grigoryan, *Brief Notes in Advanced DSP: Fourier Analysis with MATLAB*, CRC Press, Boca Raton, FL (2009).

Artyom M. Grigoryan received M.S. degrees in Mathematics from Yerevan State University (YSU), Armenia in 1978; in Imaging Science from the Moscow Institute of Physics and Technology in 1980; and in Electrical Engineering from Texas A&M University in 1999. He received his Ph.D. in Mathematics and Physics from YSU in 1990. From 1990–1996, he was a senior researcher with the Department of Signal and Image Processing at the Institute for Informatics and Automation Problems (IAAP) at the National Academy of Sciences of the Republic of Armenia (NAS RA). From 1996–2000, he was a Research Engineer with the Department of Electrical Engineering at Texas A&M. In December 2000, he joined the Department of Electrical Engineering at the University of Texas, San Antonio, where he is currently an Associate Professor. He holds two patents for developing an algorithm of automated 3D fluorescent *in situ* hybridization spot counting, and one patent for fast calculation of cyclic convolution. He is the author of three books, three book chapters, two patents, and many journal papers and specializes in the theory and application of fast one- and multidimensional Fourier transforms, elliptic Fourier transforms, tensor and paired transforms, unitary heap transforms, design of robust linear and nonlinear filters, image enhancement, encoding, computerized 2D and 3D tomography, processing biomedical images, and image cryptography.

Index